GRADUATE TEXTS IN COMPUTER SCIENCE

Editors
David Gries
Fred B. Schneider

Springer-Science+Business Media, LLC

GRADUATE TEXTS IN COMPUTER SCIENCE

Alagar and Periyasamy, Specification of Software Systems

Apt and Olderog, Verification of Sequential and Concurrent Programs, Second Edition

Back and von Wright, Refinement Calculus

Fitting, First-Order Logic and Automated Theorem Proving, Second Edition

Immerman, Descriptive Complexity

Li and Vitányi, An Introduction to Kolmogorov Complexity and Its Applications, Second Edition

Munakata, Fundamentals of the New Artificial Intelligence

Nerode and Shore, Logic for Applications, Second Edition

Schneider, On Concurrent Programming

Smith, A Recursive Introduction to the Theory of Computation

Socher-Ambrosius and Johann, Deduction Systems

Neil Immerman

DESCRIPTIVE COMPLEXITY

With 41 Illustrations

Springer

Neil Immerman
Department of Computer Science
University of Massachusetts
Amherst, MA 01003-4610, USA

Series Editors
David Gries
Fred B. Schneider

Department of Computer Science
Cornell University
Upson Hall
Ithaca, NY 14853-7501, USA

Library of Congress Cataloging-in-Publication Data
Immerman, Neil, 1953–
    Descriptive complexity / Neil Immerman.
      p.   cm. — (Graduate texts in computer science)
    Includes bibliographical references and index.
    ISBN 978-1-4612-6809-3   ISBN 978-1-4612-0539-5 (eBook)
    DOI 10.1007/978-1-4612-0539-5
    1. Computational complexity. 2. Logic, Symbolic and mathematical.
  I. Title. II. Series: Graduate texts in computer science (Springer-
Verlag New York Inc.)
QA267.7.I46   1998
511.3—dc21   98-33563

Printed on acid-free paper.

© 1999 Springer Science+Business Media New York
Originally published by Springer-Verlag New York Berlin Heidelberg in 1999
Softcover reprint of the hardcover 1st edition 1999
All rights reserved. This work may not be translated or copied in whole or in part without the written permission of the publisher (Springer Science+Business Media, LLC), except for brief excerpts in connection with reviews or scholarly analysis. Use in connection with any form of information storage and retrieval, electronic adaptation, computer software, or by similar or dissimilar methodology now known or hereafter developed is forbidden.

The use of general descriptive names, trade names, trademarks, etc., in this publication, even if the former are not especially identified, is not to be taken as a sign that such names, as understood by the Trade Marks and Merchandise Marks Act, may accordingly be used freely by anyone.

Production managed by Allan Abrams; manufacturing supervised by Jacqui Ashri.
Photocomposed copy prepared from the author's LaTeX files.

9 8 7 6 5 4 3 2 1

ISBN 978-1-4612-6809-3   SPIN 10688630

This book is dedicated to Daniel and Ellie.

Preface

This book should be of interest to anyone who would like to understand computation from the point of view of logic. The book is designed for graduate students or advanced undergraduates in computer science or mathematics and is suitable as a textbook or for self study in the area of descriptive complexity. It is of particular interest to students of computational complexity, database theory, and computer aided verification. Numerous examples and exercises are included in the text, as well as a section at the end of each chapter with references and suggestions for further reading.

The book provides plenty of material for a one semester course. The core of the book is contained in Chapters 1 through 7, although even here some sections can be omitted according to the taste and interests of the instructor. The remaining chapters are more independent of each other. I would strongly recommend including at least parts of Chapters 9, 10, and 12. Chapters 8 and 13 on lower bounds include some of the nicest combinatorial arguments. Chapter 11 includes a wealth of information on uniformity; to me, the low-level nature of translations between problems that suffice to maintain completeness is amazing and provides powerful descriptive tools for understanding complexity. I assume that most readers will want to study the applications of descriptive complexity that are introduced in Chapter 14.

Map of the Book

Chapters 1 and 2 provide introductions to logic and complexity theory, respectively. These introductions are fast-moving and specialized. (Alternative sources are suggested at the end of these chapters for students who would prefer more background.) This background material is presented with an eye toward the descriptive point of view. In particular, Chapter 1 introduces the notion of queries. In Chapter 2, all complexity classes are defined as sets of boolean queries. (A boolean query is a query whose answer is a single bit: yes or no. Since traditional complexity classes are defined as sets of yes/no questions, they are exactly sets of boolean queries.)

Chapter 3 begins the study of the relationship between descriptive and computational complexity. All first-order queries are shown to be computable in the low complexity class deterministic logspace (L). Next, notion *first-order reduction* — a first-order expressible translation from one problem to another — is introduced. Problems complete via first-order reductions for the complexity classes L, nondeterministic logspace (NL), and P are presented.

Chapter 4 introduces the least-fixed-point operator, which formalizes the power of making inductive definitions. P is proved equal to the set of boolean queries expressible in first-order logic plus the power to define new relations by induction. It is striking that such a significant descriptive class is equal to P. We thus have a natural, machine-independent view of feasible computation. This means we can understand the P versus NP question entirely from a logical point of view: P is equal to NP iff every second-order expressible query is already expressible in first-order logic plus inductive definitions (Corollary 7.23).

Chapter 5 introduces the notion of parallel computation and ties it to descriptive complexity. In parallel computation, we can take advantage of many different processors or computers working simultaneously. The notion of quantification is inherently parallel. I show that the parallel time needed to compute a query corresponds exactly to its quantifier depth. The number of distinct variables occurring in a first-order inductive query corresponds closely with the amount of hardware — processors and memory — needed to compute this query. The most important tradeoff in complexity theory — between parallel time and hardware — is thus identical to the tradeoff between inductive depth and number of variables.

Chapter 6 introduces a combinatorial game that serves as an important tool for ascertaining what can and cannot be expressed in logical languages. Ehrenfeucht-Fraïssé games offer a semantics for first-order logic that is equivalent to, but more directly applicable than, the standard definitions. These games provide powerful tools for descriptive complexity. Using them, we can often decide whether a given query is or is not expressible in a given language.

Chapter 7 introduces second-order logic. This is much more expressive than first-order logic because we may quantify over an exponentially larger space of objects. I prove Fagin's theorem as well as Stockmeyer's characterization of the polynomial-time hierarchy as the set of second-order describable boolean queries. It follows from previous results that the polynomial-time hierarchy is the set of boolean queries computable in constant time, but using exponentially much hardware (Corollary 7.28). This insight exposes the strange character of the polynomial-time hierarchy and of the class NP.

Chapter 8 uses Ehrenfeucht-Fraïssé games to prove that certain queries are not expressible in some restrictions of second-order logic. Since second-order logic is so expressive, it is surprising that we can prove results about non-expressibility.

However, the restrictions needed on the second-order languages — in particular, that they quantify only monadic relations — are crucial.

Chapter 9 studies the transitive-closure operator, a restriction of the least-fixed-point operator. I show that transitive closure characterizes the power of the class NL. When infinite structures are allowed, both the least-fixed-point and transitive-closure operators are not closed under negation. In this case there is a strict expressive hierarchy as we alternate applications of these operators with negation. However, for finite structures I show that these operators are closed under negation. A corollary is that nondeterministic space classes are closed under complementation. This was a very unexpected result when it was proved. It constitutes a significant contribution of descriptive complexity to computer science.

Chapter 10 studies the complexity class polynomial space, PSPACE, which is the set of all boolean queries that can be computed using a polynomial amount of hardware, but with no restriction on time. Thus PSPACE is beyond the realm of what is feasibly computable. It is obvious that NP is contained in PSPACE, but it is not known whether PSPACE is larger than NP. Indeed, it is not even known that PSPACE is larger than P. PSPACE is a very robust complexity class. It has several interesting descriptive characterizations, which expose more information about the tradeoff between inductive depth and number of variables.

Chapter 11 studies precomputation — the work that may go into designing the program, formula, or circuit before any input is seen. Precomputation — even less well understood than time and hardware — has an especially crisp formulation in descriptive complexity.

In order for a structure such as a graph to be input to a real or idealized machine, it must be encoded as a character string. Such an encoding imposes an ordering on the universe of the structure, e.g., on the vertices of the graph. All first-order, descriptive characterizations of complexity classes assume that a total ordering relation on the universe is available in the languages. Without such an ordering, simple lower bounds from Chapter 6 show that certain trivial properties — such as computing the PARITY of the cardinality of the universe — are not expressible. However, the ordering relation allows us to distinguish isomorphic structures which all plausible queries should treat the same. In addition, an ordering relation spoils the power of Ehrenfeucht-Fraïssé games for most languages. The mathematically rich search for a suitable alternative to ordering is described in Chapter 12.

Chapter 13 describes some interesting combinatorial arguments that provide lower bounds on descriptive complexity. The first is the optimal lower bound due to Håstad on the quantifier depth needed to express PARITY. One of many corollaries is that the set of first-order boolean queries is a strict subset of L. The second two lower bounds are weaker: they use Ehrenfeucht-Fraïssé games without ordering and thus, while quite interesting, do not separate complexity classes.

Chapter 14 describes applications of descriptive complexity to databases and computer-aided verification. Relational databases are exactly finite logical structures, and commercial query languages such as SQL are simple extensions of first-order logic. The complexity of query evaluation, the expressive power of query languages, and the optimization of queries are all important practical issues here, and the tools that have been developed previously can be brought to bear on these issues.

Model checking is a burgeoning subfield of computer-aided verification. The idea is that the design of a circuit, protocol, or program can be automatically translated into a transition system, i.e., a graph whose vertices represent global states and whose edges represent possible atomic transitions. Model checking means deciding whether such a design satisfies a simple correctness condition such as, "Doors are not opened between stations", or, "Division is always performed correctly". In descriptive complexity, we can see on the face of such a query what the complexity of checking it will be.

Finally, Chapter 15 sketchs some directions for future research in and applications of descriptive complexity.

Acknowledgments

I have been intending to write this book for more years than I would like to admit. In the mean time, many researchers have changed and extended the field so quickly that it is not possible for me to really keep up. I have tried to give pointers to some of the many topics not covered.

I am grateful to everyone who has found errors or made suggestions or encouraged me to write this book. All the errors remaining are mine alone. There will be a page on the world wide web with corrections, recent developments, etc., concerning this book and descriptive complexity in general. Just search for "Neil Immerman" on the web and you will find it. All comments, corrections, etc., will be greatly appreciated.

Some of the people who have already provided help and helpful comments are: Natasha Alechina, Jose Balcázar, Dave Mix Barrington, Jonathan Buss, Russ Ellsworth, Miklos Erdelyi-Szabo, Ron Fagin, Erich Grädel, Jens Gramm, Martin Grohe, Brian Hanechak, Lauri Hella, Janos Makowsky, Yiannis Moschovakis, John Ridgway, Jose Antonio Medina, Gleb Naumovich, Sushant Patnaik, Nate Segerlind, Richard Shore, Wolfgang Thomas, and especially Kousha Etessami.

I am grateful to David Gries for taking his role as an editor of this series so seriously that he read this book in detail, making numerous helpful comments and corrections.

Thanks to the following institutions for financial support during the long process of writing this book: NSF Grant CCR-9505446, Cornell University Computer Science Department, and the DIMACS special year in logic and algorithms.

I want to acknowledge my debt to the many inspiring teachers that I have had over the years. I have a vivid memory of Larry Carter on crutches because he had broken a blood vessel in his leg during the previous afternoon's soccer game hopping back and forth in front of the room as he built a Turing machine to multiply two numbers. This was at an NSF sponsored summer program at the University of New Hampshire in 1969. David Kelly was the codirector of that program and has been making magic ever since, creating summer programs where mathematics as a creative and cooperative endeavor is taught and shared. These programs more than anything else taught me the value and pleasure of teaching and research. Larry later roomed with Ron Fagin as a graduate student at Berkeley, and it was because of this connection that I learned of Ron's research connecting logic and complexity.

A few other of my teachers that I would like to thank by name are Shizuo Kakutani, Angus Macintyre, John Hopcroft, and Juris Hartmanis.

I thank my wife, Susan Landau, who, in part to let me pursue my own obscure interests, has given up more than anyone should have to. I am delighted that you finished your book [DL98] first and to such acclaim. Thanks for your love, for your unswerving integrity, and for your amazing ability to keep moving forward.

Amherst, MA *Neil Immerman*

Contents

Preface   vii

Introduction   1

1 Background in Logic   5
    1.1 Introduction and Preliminary Definitions 5
    1.2 Ordering and Arithmetic 12
       1.2.1 FO(BIT) = FO(PLUS, TIMES) 14
    1.3 Isomorphism 16
    1.4 First-Order Queries 17

2 Background in Complexity   23
    2.1 Introduction 23
    2.2 Preliminary Definitions 24
    2.3 Reductions and Complete Problems 27
    2.4 Alternation 34
    2.5 Simultaneous Resource Classes 40
    2.6 Summary 41

3 First-Order Reductions   45
    3.1 $FO \subseteq L$ 45
    3.2 Dual of a First-Order Query 46
    3.3 Complete problems for L and NL 50
    3.4 Complete Problems for P 53

xiv   Contents

4   Inductive Definitions **57**
   4.1   Least Fixed Point 57
   4.2   The Depth of Inductive Definitions 61
   4.3   Iterating First-Order Formulas 63

5   Parallelism **67**
   5.1   Concurrent Random Access Machines 68
   5.2   Inductive Depth Equals Parallel Time 70
   5.3   Number of Variables Versus Number of Processors 74
   5.4   Circuit Complexity 77
   5.5   Alternating Complexity 85
        5.5.1   Alternation as Parallelism 87

6   Ehrenfeucht-Fraïssé Games **91**
   6.1   Definition of the Games 91
   6.2   Methodology for First-Order Expressibility 99
   6.3   First-Order Properties Are Local 102
   6.4   Bounded Variable Languages 104
   6.5   Zero-One Laws 107
   6.6   Ehrenfeucht-Fraïssé Games with Ordering 109

7   Second-Order Logic and Fagin's Theorem **113**
   7.1   Second-Order Logic 113
   7.2   Proof of Fagin's Theorem 115
   7.3   NP-Complete Problems 119
   7.4   The Polynomial-Time Hierarchy 121

8   Second-Order Lower Bounds **125**
   8.1   Second-Order Games 125
   8.2   SO∃(monadic) Lower Bound on Reachability 129
   8.3   Lower Bounds Including Ordering 133

9   Complementation and Transitive Closure **139**
   9.1   Normal Form Theorem for FO(LFP) 139
   9.2   Transitive Closure Operators 143
   9.3   Normal Form for FO(TC) 144
   9.4   Logspace is Primitive Recursive 148
   9.5 $\text{NSPACE}[s(n)] = \text{co-NSPACE}[s(n)]$ 149
   9.6   Restrictions of SO 151

10 Polynomial Space **157**
   10.1 Complete Problems for PSPACE 157
   10.2 Partial Fixed Points 160
   10.3 $\text{DSPACE}[n^k] = \text{VAR}[k+1]$ 162
   10.4 Using Second-Order Logic to Capture PSPACE 165

11 Uniformity and Precomputation — 169
- 11.1 An Unbounded Number of Variables — 170
  - 11.1.1 Tradeoffs Between Variables and Quantifier Depth — 171
- 11.2 First-Order Projections — 171
- 11.3 Help Bits — 176
- 11.4 Generalized Quantifiers — 177

12 The Role of Ordering — 181
- 12.1 Using Logic to Characterize Graphs — 182
- 12.2 Characterizing Graphs Using $\mathcal{L}^k$ — 183
- 12.3 Adding Counting to First-Order Logic — 185
- 12.4 Pebble Games for $\mathcal{C}^k$ — 187
- 12.5 Vertex Refinement Corresponds to $\mathcal{C}^2$ — 189
- 12.6 Abiteboul-Vianu and Otto Theorems — 193
- 12.7 Toward a Language for Order-Independent P — 199

13 Lower Bounds — 203
- 13.1 Håstad's Switching Lemma — 203
- 13.2 A Lower Bound for $REACH_a$ — 208
- 13.3 Lower Bound for Fixed Point and Counting — 213

14 Applications — 221
- 14.1 Databases — 221
  - 14.1.1 SQL — 222
  - 14.1.2 Datalog — 224
- 14.2 Dynamic Complexity — 226
  - 14.2.1 Dynamic Complexity Classes — 227
- 14.3 Model Checking — 234
  - 14.3.1 Temporal Logic — 235
- 14.4 Summary — 239

15 Conclusions and Future Directions — 241
- 15.1 Languages That Capture Complexity Classes — 241
  - 15.1.1 Complexity on the Face of a Query — 243
  - 15.1.2 Stepwise Refinement — 244
- 15.2 Why Is *Finite* Model Theory Appropriate? — 244
- 15.3 Deep Mathematical Problems: P versus NP — 245
- 15.4 Toward Proving Lower Bounds — 246
  - 15.4.1 Role of Ordering — 246
  - 15.4.2 Approximation and Approximability — 247
- 15.5 Applications of Descriptive Complexity — 248
  - 15.5.1 Dynamic Complexity — 248
  - 15.5.2 Model Checking — 248
  - 15.5.3 Abstract State Machines — 248
- 15.6 Software Crisis and Opportunity — 249

    15.6.1 How can Finite Model Theory Help? 250

References   **251**

Index   **263**

Introduction

In the beginning, there were two measures of computational complexity: time and space. From an engineering standpoint, these were very natural measures, quantifying the amount of physical resources needed to perform a computation. From a mathematical viewpoint, time and space were somewhat less satisfying, since neither appeared to be tied to the inherent mathematical complexity of the computational problem.

In 1974, Ron Fagin changed this. He showed that the complexity class NP — those problems computable in nondeterministic polynomial time — is exactly the set of problems describable in second-order existential logic. This was a remarkable insight, for it demonstrated that the computational complexity of a problem can be understood as the richness of a language needed to specify the problem. Time and space are not model-dependent engineering concepts, they are more fundamental.

Although few programmers consider their work in this way, a computer program is a completely precise description of a mapping from inputs to outputs. In this book, will follow database terminology and call such a map a *query* from input structures to output structures. Typically a program describes a precise sequence of steps that compute a given query. However, we may choose to describe the query in some other precise way. For example, we may describe queries in variants of first- and second-order mathematical logic.

Fagin's Theorem gave the first such connection. Using first-order languages, this approach, commonly called descriptive complexity, demonstrated that virtually all measures of complexity can be mirrored in logic. Furthermore, as we will see, the most important classes have especially elegant and clean descriptive characterizations.

Descriptive complexity provided the insight behind a proof of the Immerman-Szelepcsényi Theorem, which states that nondeterministic space classes are closed under complementation. This settled a question that had been open for twenty-five years; indeed, almost everyone had conjectured the negation of this theorem.

Descriptive complexity has long had applications to database theory. A relational database is a finite logical structure, and commonly used query languages are small extensions of first-order logic. Thus, descriptive complexity provides a natural foundation for database theory, and many questions concerning the expressibility of query languages and the efficiency of their evaluation have been settled using the methods of descriptive complexity. Another prime application area of descriptive complexity is to the problems of Computer Aided Verification.

Since the inception of complexity theory, a fundamental question that has bedeviled theorists is the P versus NP question. Despite almost three decades of work, the problem of proving P different from NP remains. As we will see, P versus NP is just a famous and dramatic example of the many open problems that remain. Our inability to ascertain relationships between complexity classes is pervasive. We can prove that more of a given resource, e.g., time, space, nondeterministic time, etc., allows us to compute strictly more queries. However, the relationship between different resources remains virtually unknown.

We believe that descriptive complexity will be useful in these and many related problems of computational complexity. Descriptive complexity is a rich edifice from which to attack the tantalizing problems of complexity. It gives a mathematical structure with which to view and set to work on what had previously been engineering questions. It establishes a strong connection between mathematics and computer science, thus enabling researchers of both backgrounds to use their various skills to set upon the open questions. It has already led to significant successes.

The Case for Finite Models

A fundamental philosophical decision taken by the practitioners of descriptive complexity is that computation is inherently finite. The relevant objects — inputs, databases, programs, specifications — are all finite objects that can be conveniently modeled as finite logical structures. Most mathematical theories study infinite objects. These are considered more relevant, general, and important to the typical mathematician. Furthermore, infinite objects are often simpler and better behaved than their finite cousins. A typical example is the set of natural numbers, $\mathbf{N} = \{0, 1, 2, \ldots\}$. Clearly this has a simpler and more elegant theory than the set of natural numbers representable in 64-bit computer words. However, there is a significant danger in taking the infinite approach. Namely, the models are often wrong! Properties that we can prove about $\mathbf{N}$ are often false or irrelevant if we try to apply them to the objects that computers have and hold. We find that the subject of finite models is quite different in many respects. Different theorems hold and different techniques apply.

Living in the world of finite structures may seem odd at first. Descriptive complexity requires a new way of thinking for those readers who have been brought up on infinite fare. Finite model theory is different and more combinatorial than general model theory. In Descriptive complexity, we use finite model theory to understand computation. We expect that the reader, after some initial effort and doubt, will agree that the theory of computation that we develop has significant advantages. We believe that it is more accurate and more relevant in the study of computation.

I hope the reader has as much pleasure in discovering and using the tools of Descriptive complexity as I have had. I look forward to new contributions in the modeling and understanding of computation to be made by some of the readers of this book.

1
Background in Logic

Mathematics enables us to model many things abstractly. Group theory, for example, abstracts features of such diverse activities as English changeringing and quantum mechanics. Mathematical logic carries the abstraction one level higher: it is a mathematical model of mathematics. This book shows that the computational complexity of all problems in computer science can be understood via the complexity of their logical descriptions. We begin with a high-level introduction to logic. Although much of the material is well-known, we urge readers to at least skim this background chapter as the concentration on finite and ordered structures, i.e., relational databases, is not standard in most treatments of logic.

1.1 Introduction and Preliminary Definitions

All logic books begin with definitions. We have to introduce the language before we start to speak. Thus, a *vocabulary*

$$\tau = \langle R_1^{a_1}, \ldots, R_r^{a_r}, c_1, \ldots, c_s, f_1^{r_1}, \ldots, f_t^{r_t} \rangle$$

is a tuple of relation symbols, constant symbols, and function symbols. R_i is a relation symbol of arity a_i and f_j is a function symbol of arity r_j. Two important examples are $\tau_g = \langle E^2, s, t \rangle$, the vocabulary of graphs with specified source and terminal nodes, and $\tau_s = \langle \leq^2, S^1 \rangle$, the vocabulary of binary strings.

A *structure* with vocabulary τ is a tuple,

$$\mathcal{A} = \langle |\mathcal{A}|, R_1^{\mathcal{A}}, \ldots, R_r^{\mathcal{A}}, c_1^{\mathcal{A}}, \ldots, c_s^{\mathcal{A}}, f_1^{\mathcal{A}}, \ldots, f_t^{\mathcal{A}} \rangle$$

whose universe is the nonempty set $|\mathcal{A}|$. For each relation symbol R_i of arity a_i in τ, $\mathcal{A}$ has a relation $R_i^\mathcal{A}$ of arity a_i defined on $|\mathcal{A}|$, i.e., $R_i^\mathcal{A} \subseteq |\mathcal{A}|^{a_i}$. For each constant symbol $c_j \in \tau$, $\mathcal{A}$ has a specified element of its universe $c_j^\mathcal{A} \in |\mathcal{A}|$. For each function symbol $f_i \in \tau$, $f_i^\mathcal{A}$ is a total function from $|\mathcal{A}|^{r_i}$ to $|\mathcal{A}|$. A vocabulary without function symbols is called a *relational vocabulary*. In this book, unless stated otherwise, all vocabularies are relational. The notation $\|\mathcal{A}\|$ denotes the cardinality of the universe of $\mathcal{A}$.

In the history of mathematical logic most interest has concentrated on infinite structures. Indeed, many mathematicians consider the study of finite structures trivial. Yet, the objects computers have and hold are always finite. To study computation we need a theory of finite structures.

Logic restricted to finite structures is rather different from the theory of infinite structures. We mention infinite structures from time to time, most often when we comment on whether a given theorem also holds in the infinite case. However, we concentrate on finite structures. We define STRUC[τ] to be the set of finite structures of vocabulary τ.

As an example, the graph $G = \langle V^G, E^G, 1, 3 \rangle$ defined by,

$$V^G = \{0, 1, 2, 3, 4\}, \qquad E^G = \{(1,2), (3,0), (3,1), (3,2), (3,4), (4,0)\}$$

is a structure of vocabulary τ_g consisting of a directed graph with two specified vertices s and t. G has five vertices and six edges. (See Figure 1.1, which shows G as well as another graph H which is isomorphic but not equal to G.)

For another example, consider the binary string $w = $ "01101". We can code w as the structure $\mathcal{A}_w = \langle \{0, 1, \ldots, 4\}, \leq, \{1, 2, 4\} \rangle$ of vocabulary τ_s. Here $\leq$ represents the usual ordering on $0, 1, \ldots, 4$. Relation $S^w = \{1, 2, 4\}$ represents the positions where w is one. (Relation symbols of arity one, such as S^w, are sometimes called monadic.)

A relational database is exactly a finite relational structure. The following begins a running example of a genealogical database.

Example 1.2 Consider a genealogical database $\mathcal{B}_0 = \langle U_0, F_0, P_0, S_0 \rangle$; where U_0 is a finite set of people,

$$U_0 = \{\text{Abraham, Isaac, Rebekah, Sarah, }\ldots\}$$

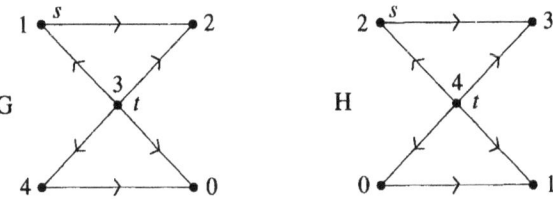

Figure 1.1: Graphs G and H

F_0 is a monadic relation that is true of the female elements of U_0,

$$F_0 = \{\text{Sarah, Rebekah, }\ldots\}$$

P_0 and S_0 are the binary relations for parent and spouse, respectively, e.g.,

$$P_0 = \{\langle\text{Abraham,Isaac}\rangle, \langle\text{Sarah,Isaac}\rangle, \ldots\}$$
$$S_0 = \{\langle\text{Abraham,Sarah}\rangle, \langle\text{Isaac,Rebekah}\rangle, \ldots\}$$

Thus, $\mathcal{B}_0$ is a structure of vocabulary $\langle F^1, P^2, S^2 \rangle$. □

For any vocabulary τ, define the *first-order language* $\mathcal{L}(\tau)$ to be the set of formulas built up from the relation and constant symbols of τ; the logical relation symbol =; the boolean connectives $\wedge, \neg$; variables: VAR = $\{x, y, z, \ldots\}$; and quantifier $\exists$.

We say that an occurrence of a variable v in φ is *bound* if it lies within the scope of a quantifier ($\exists v$) or ($\forall v$), otherwise it is *free*. Variable v is *free* in φ iff it has a free occurrence in φ. For example, the free variables in the following formula are x and y. We use the symbol "$\equiv$" to define or denote equivalence of formulas.

$$\alpha \quad \equiv \quad [(\exists y)(y + 1 = x)] \wedge x < y$$

In a similar way we sometimes "$\Leftrightarrow$" to indicate that two previously defined formulas or conditions are equivalent.

Bound variables are "dummy" variables and may be renamed to avoid confusion. For example, α is equivalent to the following α' which also has free variables x and y,

$$\alpha' \quad \equiv \quad [(\exists z)(z + 1 = x)] \wedge x < y$$

We write $\mathcal{A} \models \varphi$ to mean that $\mathcal{A}$ satisfies φ, i.e., that φ is true in $\mathcal{A}$. Since φ may contain some free variables, we will let an *interpretation* into $\mathcal{A}$ be a map $i : V \to |\mathcal{A}|$ where V is some finite subset of VAR. For convenience, for every constant symbol $c \in \tau$ and any interpretation i for $\mathcal{A}$, we let $i(c) = c^{\mathcal{A}}$. If τ has function symbols, then the definition of i extends to all terms via the recurrence,

$$i(f_j(t_1, \ldots, t_{r_j})) = f_j^{\mathcal{A}}(i(t_1), \ldots, i(t_{r_j})) .$$

We can be completely precise about the semantics of mathematical logic. In particular, we can definitively define what it means for a sentence φ to be true in a structure $\mathcal{A}$.

Definition 1.3 (**Definition of Truth**) Let $\mathcal{A} \in \text{STRUC}[\tau]$ be a structure, and let i be an interpretation into $\mathcal{A}$ whose domain includes all the relevant free variables. We inductively define whether a formula $\varphi \in \mathcal{L}(\tau)$ is true in $(\mathcal{A}, i)$:

$$(\mathcal{A}, i) \models t_1 = t_2 \quad \Leftrightarrow \quad i(t_1) = i(t_2)$$
$$(\mathcal{A}, i) \models R_j(t_1, \ldots, t_{a_j}) \quad \Leftrightarrow \quad \langle i(t_1), \ldots, i(t_{a_j}) \rangle \in R_j^{\mathcal{A}}$$
$$(\mathcal{A}, i) \models \neg\varphi \quad \Leftrightarrow \quad \text{it is not the case that } (\mathcal{A}, i) \models \varphi$$
$$(\mathcal{A}, i) \models \varphi \wedge \psi \quad \Leftrightarrow \quad (\mathcal{A}, i) \models \varphi \text{ and } (\mathcal{A}, i) \models \psi$$

$(\mathcal{A}, i) \models (\exists x)\varphi \iff$ (there exists $a \in |\mathcal{A}|)(\mathcal{A}, i, a/x) \models \varphi$

$$\text{where}(i, a/x)(y) = \begin{cases} i(y) & \text{if } y \neq x \\ a & \text{if } y = x \end{cases}$$

Write $\mathcal{A} \models \varphi$ to mean that $(\mathcal{A}, \emptyset) \models \varphi$. □

Definition 1.3 is our first example of an inductive definition, a device that is often used by logicians. It deserves a few comments. Note that the equality symbol (=) is not treated as an ordinary binary relation symbol — the definition insists that this symbol be interpreted as equality. Many students, on first seeing this definition, feel that it is circular. It is not. We are defining the meaning of the symbol "=" in terms of the intuitively well-understood standard equality. In the same way, we define the meaning of "¬", "∧", and "∃" in terms of their intuitive counterparts.

We define the "for all" quantifier as the dual of ∃ and the boolean "or" as the dual of ∧,

$$(\forall x)\varphi \equiv \neg(\exists x)\neg\varphi; \qquad \alpha \vee \beta \equiv \neg(\neg\alpha \wedge \neg\beta)$$

It is convenient to introduce other abbreviations into our formulas. For example, "$y \neq z$" is an abbreviation for "$\neg y = z$". Similarly "$\alpha \to \beta$" is an abbreviation for "$\neg\alpha \vee \beta$", and "$\alpha \leftrightarrow \beta$" is an abbreviation for "$\alpha \to \beta \wedge \beta \to \alpha$". In some sense, the symbols we introduce formally into our language are part of our low-level "machine language", and abbreviations are analogous to what computer scientists call macros. Abbreviations are directly translatable into the real language, and they make formulas more readable. Without abbreviations and the breaking of formulas into modular descriptions, it would be impossible to communicate complicated ideas in first-order logic.

We use spacing and parentheses to make the order of operations clear. Our convention for operator precedence is that "¬" has highest precedence, then "∧" and "∨", and finally, "→" and "↔", and operators of equal precedence are evaluated left to right. For example, the following two formulas are equivalent,

$$\neg R(a) \to R(b) \wedge R(c) \leftrightarrow R(d)$$

$$((((\neg R(a)) \to (R(b)) \wedge R(c))) \leftrightarrow R(d))$$

A *sentence* is a formula with no free variables. Every sentence $\varphi \in \mathcal{L}(\tau)$ is either true or false in any structure $\mathcal{A} \in \text{STRUC}[\tau]$.

Example 1.4 We give a few examples of first-order formulas in the language of graphs:

$$\varphi_{undir} \equiv (\forall x)(\forall y)(\neg E(x, x) \wedge E(x, y) \to E(y, x))$$

Formula φ_{undir} says that the graph in question is undirected and has no loops.

$$\varphi_{out2} \equiv (\forall x)(\exists yz)(y \neq z \wedge E(x, y) \wedge E(x, z) \wedge$$
$$(\forall w)(E(x, w) \to (w = y \vee w = z)))$$

$$\varphi_{deg2} \equiv \varphi_{undir} \wedge \varphi_{out2}$$

Formula φ_{out2} says that every vertex has exactly two edges leaving it. Thus, φ_{deg2} says that the graph in question is undirected, has no loops, and is regular of degree two, i.e., every vertex has exactly two neighbors.

$$\varphi_{dist1} \equiv x = y \vee E(x, y)$$
$$\varphi_{dist2} \equiv (\exists z)(\varphi_{dist1}(x, z) \wedge \varphi_{dist1}(z, y))$$
$$\varphi_{dist4} \equiv (\exists z)(\varphi_{dist2}(x, z) \wedge \varphi_{dist2}(z, y))$$
$$\varphi_{dist8} \equiv (\exists z)(\varphi_{dist4}(x, z) \wedge \varphi_{dist4}(z, y))$$

Formulas $\varphi_{dist1}, \varphi_{dist2}$, and so on say that there is a path from x to y of length at most 1, 2, 4, and 8, respectively. Note that these formulas have free variables x and y.

Formulas express properties about their free variables. For example, for a pair of vertices a, b from the universe of a graph G, the meaning of

$$(G, a/x, b/y) \models \varphi_{dist8}$$

is that the distance from a to b in G is at most 8.

Sometimes we will make the free variables in a formula explicit, e.g., writing $\varphi_{dist8}(x, y)$ instead of just φ_{dist8}. This offers the advantage of making substitutions more readable: we can write $\varphi_{dist8}(a, b)$ instead of $\varphi_{dist8}(a/x, b/y)$. □

Exercise 1.5 For $n \in \mathbb{N}$, consider the logical structures

$$\mathcal{A}_n = \langle \{0, 1, \ldots, n-1\}, \text{PLUS}^{\mathcal{A}_n}, \text{TIMES}^{\mathcal{A}_n}, 0, 1, n-1 \rangle$$

of vocabulary $\tau_a = \langle \text{PLUS}^3, \text{TIMES}^3, 0, 1, max \rangle$, where PLUS and TIMES are the arithmetic relations, i.e., for $i, j, k < n$,

$$\mathcal{A}_n \models \text{PLUS}(i, j, k) \Leftrightarrow i + j = k$$
$$\mathcal{A}_n \models \text{TIMES}(i, j, k) \Leftrightarrow i \cdot j = k$$

Write formulas in $\mathcal{L}(\tau)$ that represent the following arithmetic relations,

1. DIVIDES(x, y), meaning that y is a multiple of x.
2. PRIME(x), meaning that x is a prime number.
3. $p_2(x)$, meaning that x is a power of 2.

[Hint for (3): x is a power of 2 iff 2 is the only prime divisor of x.] □

Example 1.6 Here are a few formulas in the language of strings. The first describes the set of strings that have no consecutive "1"s. It uses the abbreviation "$x < y$", meaning "$x \leq y \wedge x \neq y$".

$$\varphi_{no11} \equiv (\forall x)(\forall y)(\exists z)((S(x) \wedge S(y) \wedge x < y) \rightarrow (x < z < y \wedge \neg S(z)))$$

Formula φ_{five1} below says that the given string contains at least five "1"s. To do so, it uses the abbreviation "distinct":

$$\text{distinct}(x_1, \ldots, x_k) \equiv (x_1 \neq x_2 \wedge \wedge \cdots \wedge x_1 \neq x_k \wedge \cdots \wedge x_{k-1} \neq x_k)$$

$\varphi_{five1} \equiv (\exists uvwxy)(\text{distinct}(u, v, w, x, y) \wedge S(u) \wedge S(v) \wedge S(w) \wedge S(x) \wedge S(y))$

Note that φ_{five1} uses five variables to say that there are five "1"s. Using the ordering relation, we can reduce the number of variables. The following formula is equivalent to φ_{five1} but uses only two variables:

$$(\exists x)\Big(S(x) \wedge (\exists y)\Big(x < y \wedge S(y) \wedge (\exists x)\big(y < x \wedge S(x) \wedge \\ (\exists y)(x < y \wedge S(y) \wedge (\exists x)y < x \wedge S(x))\big)\Big)\Big)$$

Read the above sentence carefully. A good way to think of it is that we have two fingers and are trying to count the number of "1"s in a string. We put finger x down on the first "1". Then we put finger y down on the next "1" to the right. Now we don't need x anymore so we can move it to the next "1" to the right of y, and so on.

We will see later that the number of variables is an important descriptive resource. Note that the standard semantics of first-order logic (Definition 1.3) allows us to requantify variables. Each quantifier $(\exists x)$ or $(\forall x)$ bounds only the free occurrences of x within its scope. We will see in Theorem 6.32 that every first-order sentence in $\mathcal{L}(\tau_s)$ — i.e., every sentence about strings — is equivalent to a sentence with only three distinct variables. □

Exercise 1.7 Prove that if interpretations i and i' agree on all the free variables in φ then

$$(\mathcal{A}, i) \models \varphi \quad \Leftrightarrow \quad (\mathcal{A}, i') \models \varphi$$

[Hint: by induction on φ using Definition 1.3.] □

Exercise 1.8 Let $(\exists! x)\alpha(x)$ mean that there exists a unique x such that α. Show how to write $(\exists! x)\alpha(x)$ using the usual quantifiers $\forall, \exists$. □

As another example, let $\tau_{ab} = \langle \leq^2, A^1, B^1 \rangle$ consist of an ordering relation and two monadic relation symbols A and B, each serving the same role as the symbol S in τ_s. Let $\mathcal{A} \in \text{STRUC}[\tau_{ab}]$, and let $n = \|\mathcal{A}\|$. Then $\mathcal{A}$ is a pair of binary strings A, B, each of length n. These binary strings represent natural numbers, where we think of the bit zero as most significant and bit $n - 1$ as least significant. Here $A(i)$ is true iff bit i of A is "1".

The following sentence expresses the ordering relation on such natural numbers represented in binary.

$$\text{LESS}(A, B) \equiv (\exists x)(B(x) \wedge \neg A(x) \wedge (\forall y. y < x)(A(y) \to B(y)))$$

The above sentence uses a very useful abbreviation, that of restricted quantifiers,

$$(\forall x.\alpha)\varphi \equiv (\forall x)(\alpha \to \varphi); \quad (\exists x.\alpha)\varphi \equiv (\exists x)(\alpha \wedge \varphi)$$

In the next proposition we show that addition is first-order expressible. Addition of natural numbers represented in binary is one of the most basic computations.

We will see in Theorem 5.2 that the first-order queries characterize the problems computable in constant parallel time. Thus the following may be thought of as an addition algorithm that runs in constant parallel time.

Proposition 1.9. *Addition of natural numbers, represented in binary, is first-order expressible.*

Proof We use the well-known "carry-look-ahead" algorithm. In order to express addition, we first express the carry bit,

$$\varphi_{carry}(x) \equiv (\exists y.x < y)[A(y) \wedge B(y) \wedge (\forall z.x < z < y)A(z) \vee B(z)]$$

The formula $\varphi_{carry}(x)$ holds if there is a position y to the right of x where $A(y)$ and $B(y)$ are both one (i.e. the carry is generated) and for all intervening positions z, at least one of $A(z)$ and $B(z)$ holds (that is, the carry is propagated). Let $\oplus$ be an abbreviation for the commutative and associative "exclusive or" operation.

We can express φ_{add} as follows,

$$\alpha \oplus \beta \equiv \alpha \leftrightarrow \neg \beta$$
$$\varphi_{add}(x) \equiv A(x) \oplus B(x) \oplus \varphi_{carry}(x)$$

Note that the formula $\varphi_{add}(x)$ has the free variable x. Thus, φ_{add} is a description of n bits: one for each possible value of x. □

An important relation between two structures of the same type is that one may be a substructure of the other. $\mathcal{A}$ is a substructure of $\mathcal{B}$ if the universe of $\mathcal{A}$ is a subset of the universe of $\mathcal{B}$ and the relations and constants on $\mathcal{A}$ are inherited from $\mathcal{B}$.

Definition 1.10 (Substructure) Let $\mathcal{A}$ and $\mathcal{B}$ be structures of the same vocabulary $\tau = \langle R_1^{a_1}, \ldots, R_r^{a_r}, c_1, \ldots, c_s \rangle$. We say that $\mathcal{A}$ is a *substructure* of $\mathcal{B}$, written $\mathcal{A} \leq \mathcal{B}$, iff the following conditions hold,

1. $|\mathcal{A}| \subseteq |\mathcal{B}|$
2. For $i = 1, 2, \ldots, r$, $R_i^{\mathcal{A}} = R_i^{\mathcal{B}} \cap |\mathcal{A}|^{a_i}$
3. For $j = 1, 2, \ldots, s$, $c_j^{\mathcal{A}} = c_j^{\mathcal{B}}$. □

See Figure 1.11 where A and B are substructures of G. Note that C is not a substructure of G for two reasons: it doesn't contain the constant t and the induced edge from vertex 1 to vertex 2 is missing.

Exercise 1.12 Let $\mathcal{A} \in \text{STRUC}[\tau]$ be a structure and let $\alpha(x)$ be a formula such that $\mathcal{A} \models (\exists x)\alpha(x)$. Assume also that for every constant symbol c in τ, $\mathcal{A} \models \alpha(c)$. Let $\mathcal{B}$ be the substructure of $\mathcal{A}$ with universe

$$|\mathcal{B}| = \{a \in |\mathcal{A}| \mid \mathcal{A} \models \alpha(a)\}$$

Let φ be a sentence in $\mathcal{L}(\tau)$. Define the *restriction* of φ to α to be the sentence φ^α, the result of changing every quantifier $(\forall y)$ or $(\exists y)$ in φ to the restricted quantifier $(\forall y.\alpha(y))$ or $(\exists y.\alpha(y))$ respectively. Prove the following,

$$\mathcal{A} \models \varphi^\alpha \quad \Leftrightarrow \quad \mathcal{B} \models \varphi \qquad \qquad □$$

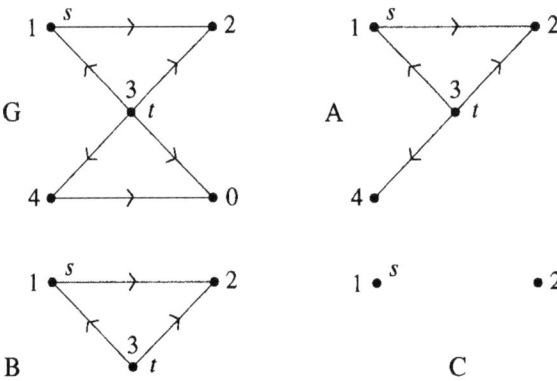

Figure 1.11: $\mathcal{A}$ and $\mathcal{B}$ but not $\mathcal{C}$ are substructures of $\mathcal{G}$.

We say that φ is *universal* iff it can be written in prenex form — i.e. with all quantifiers at the beginning — using only universal quantifiers. Similarly, we say that φ is *existential* iff it can be written in prenex form with only existential quantifiers.

The following "preservation theorems" provide a good way of proving that a formula is existential or universal.

Exercise 1.13 Prove the following preservation theorems. Let $\mathcal{A} \leq \mathcal{B}$ be structures and φ a first-order sentence.

1. Suppose φ is existential. If $\mathcal{A} \models \varphi$ then $\mathcal{B} \models \varphi$.
2. Suppose φ is universal. If $\mathcal{B} \models \varphi$ then $\mathcal{A} \models \varphi$.

[Hint: by induction on φ using Definition 1.3.] □

1.2 Ordering and Arithmetic

A logical structure such as a graph does not need to have an ordering on its vertices. However, if we use a computer to store or manipulate this graph, it must be encoded in some way that imposes an ordering on the vertices. In order to discuss computation in general, it is necessary to assume that the universes of our structures are ordered. This section introduces the issue of ordering and explains what we will assume about the ordering of structures in the remainder of this book.

When we code an input to a computer, we do so as a string of characters. There is always an ordering here: the first character, the second character, and so on. Indeed the concept of ordering is deeply embedded in the concepts of string and of computation.

For this reason, the binary relation symbol "$\leq$" will play a special role in descriptive complexity. When "$\leq$" is an element of τ, and $\mathcal{A} \in \text{STRUC}[\tau]$, then $\mathcal{A}$ must interpret $\leq$ as a total ordering on its universe. In this case, we also place constant symbols 0, 1, *max* in τ and insist that these be interpreted as the minimum, second, and maximum elements under the $\leq$ ordering. In order for formulas in first-order logic to express general computation, they need access to a total ordering of the universe. The requirement of a total ordering in descriptive complexity is analogous to the assumption of the Axiom of Choice in set theory.

Let $\mathcal{A} \in \text{STRUC}[\tau]$ be an ordered structure. Let $n = \|\mathcal{A}\|$. Let the elements of $|\mathcal{A}|$ in increasing order be $a_0, a_1, \ldots, a_{n-1}$. Then there is a 1:1 correspondence $i \mapsto a_i, i = 0, 1, \ldots n - 1$. We usually identify the elements of the universe with the set of natural numbers less than n. In a computer these would be represented as $\lceil \log n \rceil$-bit words, and the operations plus, times, and even picking out bit j of such a word would all be wired in. The following numeric relations are useful:

1. PLUS(i, j, k) meaning $i + j = k$
2. TIMES(i, j, k) meaning $i \times j = k$
3. BIT(i, j) meaning bit j in the binary representation of i is 1

In the definition of BIT we will take bit 0 to be the low order bit, so BIT$(i, 0)$ holds iff i is odd. We will see in Chapter 11 that adding BIT (or equivalently PLUS and $\times$) to our vocabularies makes the set of first-order definable boolean queries a more robust complexity class.

When working with very weak reductions or proving normal form theorems, we will sometimes use the successor relation SUC in lieu of or in addition to $\leq$. Of course, SUC is first-order definable from $\leq$.

$$\text{SUC}(x, y) \equiv (x < y) \wedge (\forall z)(\neg(x < z \wedge z < y))$$

The symbols $\leq$, PLUS, TIMES, BIT, SUC, 0, 1, *max* are called *numeric* relation and constant symbols. They depend only on the size of the universe. We call the remainder of τ the *input* relation and constant symbols. We will see in Chapter 5 that the choice of numeric relations for weak languages such as FO corresponds to the definition of uniformity for complexity classes defined by uniform sequences of circuits. The numeric relations and constants are not explicitly given in the input since they are easily computable as functions of the size of the input. Whenever any of the numeric relation or constant symbols occur, they are required to have their standard meanings.

Proviso 1.14. (Ordering Proviso) *From now on, unless stated otherwise, we assume that the numeric relations and constants: $\leq$, PLUS, TIMES, BIT, SUC, 0, 1, max are present in all vocabularies. When we define vocabularies, we do not explicitly mention or show these symbols unless they are not present. In Chapter 6 we prove lower bounds on what can be expressed in some first-order language. We use the notation $\mathcal{L}(wo\leq)$ to indicate language $\mathcal{L}$ without any of the numeric relations. We will write $\mathcal{L}(wo\,BIT)$ to indicate language $\mathcal{L}$, including ordering, but not arithmetic, i.e., only the numeric relations $\leq$ and SUC and the constants 0, 1, max are included.*

The following proviso is useful. It eliminates the trivial and sometimes annoying case of the structure with only one element which would thus satisfy the equation

14   1. Background in Logic

0 = 1. We assume this proviso unless otherwise noted. (The only time we do not assume the existence of boolean constants is in Section 6.5.)

Proviso 1.15. (**Boolean Constants**) *From now on, we assume that all structures have at least two elements. In particular, we will assume that we have two unequal constants denoted by* 0 *and* 1.

Next, we define what it means to have a boolean variable in a first-order formula. Boolean variables allow a more robust measure of the number of first-order variables needed to express a query. When we measure the number of first-order variables needed, we discount the (bounded) number of boolean variables.

Definition 1.16 A *boolean variable* in a first-order formula is a variable that is restricted to being either 0 or 1. Here 0 is identified with **false** and 1 is identified with **true**. We typically use the letters b, c, d, e for boolean variables. We use the following abbreviations:

$$\text{bool}(b) \equiv b \leq 1$$
$$(\exists b) \equiv (\exists b.\text{bool}(b))$$
$$(\forall b) \equiv (\forall b.\text{bool}(b))$$

□

1.2.1 FO(BIT) = FO(PLUS, TIMES)

In the remainder of this section we prove that adding BIT to first-order logic is equivalent to adding PLUS and TIMES. In order to prove this, we also need to prove the Bit Sum Lemma which is interesting in its own right. The proofs in this subsection are very technical and may safely be skipped at first reading.

Theorem 1.17. *Let τ be a vocabulary that includes ordering. Then*

1. *If* BIT $\in \tau$ *then* PLUS *and* TIMES *are first-order definable.*
2. *If* PLUS, TIMES $\in \tau$ *then* BIT *is first-order definable.*

Proof To prove (1), we have essentially seen in Proposition 1.9 that PLUS is expressible using BIT. To prove that TIMES is expressible we first need the following:

Lemma 1.18. (**Bit Sum Lemma**) *Let* BSUM(x, y) *be true iff y is equal to the number of ones in the binary representation of x.* BSUM *is first-order expressible using ordering and* BIT.

Proof The bit-sum problem is to add a column of $\log n$ 0's and 1's. The idea is to keep a running sum. Since the sum of $\log n$ 1's requires at most $\log \log n$ bits to record, we maintain running sums of $\log \log n$ bits each. With one existentially quantified variable, we can guess $\log n / \log \log n$ of these. Thus, to express BSUM(x, y) we existentially quantify s — the $\log \log n \cdot (\log n / \log \log n)$ bits of running sums. In the following example, $n = 2^{16}$, so x and y each have 16 bits. To assert BSUM(0110110110101101, 1010) we would guess $s = 0010010101111010$ as our partial sum bit string.

```
0
1
1
0   0010
1
1
0
1   0101
1
0
1
0   0111
1
1
0
1   1010                        BSUM(0110110110101101, 1010)
```

Next we say that for all i where $i \leq \log n / \log \log n$, running sum i, plus the number of 1's in segment $(i + 1)$ is equal to the running sum $(i + 1)$.

Thus, it suffices to express the bit sum of a segment of length $\log \log n$. This we can do by keeping a running sum at every position because this requires only $\log \log \log n \cdot \log \log n$, which is less than $\log n$ for sufficiently large n. □

We next show that TIMES is first-order expressible using BIT. TIMES is equivalent to the addition of $\log n$ $\log n$-bit numbers,

$$A = A_1 + A_2 + \cdots + A_{\log n}$$

The first trick we employ is to split each A_i into a sum of two numbers, $A_i = B_i + C_i$, so that B_i and C_i have blocks of $\log \log n$ bits separated by $\log \log n$ 0's. We compute the sum of the B_i's and of the C_i's. In this way, we insure that no carries extend more than $\log \log n$ bits. Finally, we add the two sums with a single use of PLUS. In the following, let $\ell = \lceil \log \log n \rceil$.

B_i	$=$	$a_{i,1}$	$\cdots$	$a_{i,\ell}$	0	$\cdots$	0	$\cdots$	$a_{i,\log n+1-\ell}$	$\cdots$	$a_{i,\log n}$
$+ C_i$	$=$	0	$\cdots$	0	$a_{i,\ell+1}$	$\cdots$	$a_{i,2\ell}$	$\cdots$	0	$\cdots$	0
A_i	$=$	$a_{i,1}$	$\cdots$	$a_{i,\ell}$	$a_{i,\ell+1}$	$\cdots$	$a_{i,2\ell}$	$\cdots$	$a_{i,\log n+1-\ell}$	$\cdots$	$a_{i,\log n}$

1. Background in Logic

Position	16	15	14	13	12	11	10	9	8	7	6	5	4	3	2	1	0
Y	0	1	0	0	0	0	0	0	0	1	0	0	0	1	0	1	0
I	0	1	1	1	1	0	0	0	0	1	1	1	0	1	1	0	0

Table 1.19: Encoding of an arithmetic fact: $2^{15} = 32,768$.

In this way, we have reduced the problem of adding $\log n$ $\log n$-bit numbers to that of adding $\log n$ $\log \log n$-bit numbers. We can simultaneously guess the sums of each of the $\log \log n$ columns in a single variable, c. Using BSUM and a universal quantifier we can verify that each section of c is correct. Finally, we can add the $\log \log n$ numbers in c maintaining all the running sums as in the last paragraph of the proof of the Bit Sum Lemma.

2. In this direction, we want to show that BIT is first-order expressible using PLUS and TIMES. We do this with a series of definitions. First, let $p_2(y)$ mean that y is a power of 2. (See Exercise 1.5).

Next, define $\text{BIT}'(x, y)$ to mean that for some i, $y = 2^i$ and $\text{BIT}(x, i)$,

$$\text{BIT}'(x, y) \equiv p_2(y) \wedge (\exists uv)(x = 2uy + y + v \wedge v < y)$$

Using BIT' we can copy a sequence of bits. For example, the following formula says that if $y = 2^i$ and $z = 2^j$, then bits $i + j..i$ of x are the same as bits $j..0$ of c:

$$\text{COPY}(x, y, z, c) \equiv (\forall u.p_2(u) \wedge u \leq z)(\text{BIT}'(x, yu) \leftrightarrow \text{BIT}'(c, u))$$

Finally, to express BIT, we would like to express the relation $2^i = y$. We express this using the following recurrence,

$$2^i = y \leftrightarrow (\exists j)(\exists z.2^j = z)(i = 2j + 1 \wedge y = 2z^2 \vee i = 2j \wedge y = z^2). \quad (1.20)$$

We can guess two variables, Y, I, that simultaneously include all but a bounded number of the $\log i$ computations indicated by Equation (1.20), namely all those such that $i > 2 \log i$. This is done as follows: Place a "1" in positions i, j, etc., of Y. Place the binary encoding of i starting at position i of I, the binary encoding of j starting at position j of I and so on. Using a universal quantifier we say that the variables Y and I encode all the relevant and sufficiently large computations of Equation (1.20). Table 1.19 shows the encodings Y and I for the proposition that $2^{15} = 32,768$. Note that I records the exponent 15, which is 1111 in binary, starting at position 15; 7 which is 111 in binary, starting at position 7; and 3 which is 11 in binary, starting at position 3. We leave the details of actually writing the relevant first-order formula as an exercise. □

1.3 Isomorphism

When we impose an ordering on the universe of a structure, we have essentially labeled its elements 0, 1, and so on. It becomes interesting and important to know when we have used this ordering in an essential way. For this we need the concept of isomorphism. Two structures are isomorphic iff they are identical except perhaps for the names of the elements of their universes:

Definition 1.21 (**Isomorphism of Unordered Structures**) Let $\mathcal{A}$ and $\mathcal{B}$ be structures of vocabulary $\tau = \langle R_1^{a_1}, \ldots, R_r^{a_r}, c_1, \ldots, c_s \rangle$. We say that $\mathcal{A}$ is *isomorphic* to $\mathcal{B}$, written, $\mathcal{A} \cong \mathcal{B}$, iff there is a map $f : |\mathcal{A}| \to |\mathcal{B}|$ with the following properties:

1. f is 1:1 and onto.
2. For every input relation symbol R_i and for every a_i-tuple of elements of $|\mathcal{A}|$, $e_1, \ldots, e_{a_i}$,

$$\langle e_1, \ldots, e_{a_i} \rangle \in R_i^{\mathcal{A}} \quad \Leftrightarrow \quad \langle f(e_1), \ldots, f(e_{a_i}) \rangle \in R_i^{\mathcal{B}}$$

3. For every input constant symbol c_i, $f(c_i^{\mathcal{A}}) = c_i^{\mathcal{B}}$

The map f is called an *isomorphism*. □

As an example, see graphs G and H in Figure 1.1 which are isomorphic using the map that adds one mod five to the numbers of the vertices of G.

Note that we have defined isomorphisms so that they need only preserve the input symbols, not the ordering and other numeric relations. If we included the ordering relation then we would have $A \cong B$ iff $A = B$. To be completely precise, we should call the mapping f defined above an "isomorphism of unordered structures" and say that $\mathcal{A}$ and $\mathcal{B}$ are "isomorphic as unordered structures". (Note also that, since "unordered string" does not make sense, neither does the concept of isomorphism for strings. By a strict interpretation of Definition 1.21, two strings would be isomorphic as unordered structures iff they had the same number of each symbol.)

The following proposition is basic.

Proposition 1.22. *Suppose $\mathcal{A}$ and $\mathcal{B}$ are isomorphic. Then for all sentences $\varphi \in \mathcal{L}(\tau - \{\leq\})$, $\mathcal{A}$ and $\mathcal{B}$ agree on φ.*

Exercise 1.23 Prove Proposition 1.22. [Hint: do this by induction using Definition 1.3.] □

1.4 First-Order Queries

As mentioned in the introduction, we use the concept of query as the fundamental paradigm of computation:

Definition 1.24 A *query* is any mapping $I : \text{STRUC}[\sigma] \to \text{STRUC}[\tau]$ from structures of one vocabulary to structures of another vocabulary, that is polynomially bounded. That is, there is a polynomial p such that for all $\mathcal{A} \in \text{STRUC}[\sigma]$, $\|I(\mathcal{A})\| \leq p(\|\mathcal{A}\|)$. A *boolean query* is a map $I_b : \text{STRUC}[\sigma] \to \{0, 1\}$. A boolean query may also be thought of as a subset of $\text{STRUC}[\sigma]$ — the set of structures $\mathcal{A}$ for which $I(\mathcal{A}) = 1$.

An important subclass of queries are the order-independent queries. (In database theory the term "generic" is often used instead of "order-independent".) Let I be a query defined on STRUC$[\sigma]$. Then I is *order-independent* iff for all isomorphic structures $\mathcal{A}, \mathcal{B} \in$ STRUC$[\sigma]$, $I(\mathcal{A}) \cong I(\mathcal{B})$. For boolean queries, this last condition translates to $I(\mathcal{A}) = I(\mathcal{B})$. □

From our point of view, the simplest kind of query is a first-order query. As an example, any first-order sentence $\varphi \in \mathcal{L}(\tau)$ defines a boolean query I_φ on STRUC$[\tau]$ where $I_\varphi(\mathcal{A}) = 1$ iff $\mathcal{A} \models \varphi$.

For example, let DIAM[8] be the query on graphs that is true of a graph iff its diameter is at most eight. This is a first-order query given by the formula,

$$\text{DIAM[8]} \equiv (\forall xy)\varphi_{dist8}$$

where φ_{dist8}, meaning that there is a path from x to y of length at most eight, was written in Example 1.4.

As another example, consider the query I_{add}, which, given a pair of natural numbers represented in binary, returns their sum. This query is defined by the first-order formula φ_{add} from Proposition 1.9. More explicitly, let $\mathcal{A} = \langle |\mathcal{A}| \leq, A, B \rangle$ be any structure in STRUC$[\tau]_{ab}$. $\mathcal{A}$ is a pair of natural numbers each of $n = \|\mathcal{A}\|$ bits. Their sum is given by $I_{add}(\mathcal{A}) = \langle |\mathcal{A}|, S \rangle$ where,

$$S = \{a \in |\mathcal{A}| \mid (\mathcal{A}, a/x) \models \varphi_{add}\} \tag{1.25}$$

The first-order query I_{add} : STRUC$[\tau_{ab}] \to$ STRUC$[\tau_s]$ maps structure $\mathcal{A}$ to another structure with the same universe, i.e., $|\mathcal{A}| = |I_{add}(\mathcal{A})|$. The following is a general definition of a k-ary first-order query. Such a query maps any structure $\mathcal{A}$ to a structure whose universe is a first-order definable subset of all k-tuples from $|\mathcal{A}|$. Each relation R_i over $I(\mathcal{A})$ is a first-order definable subset of $|I(\mathcal{A})|^{a_i}$. The constants of $I(\mathcal{A})$ are first-order definable elements of $|\mathcal{A}|^k$.

Definition 1.26 (First-Order Queries) Let σ and τ be any two vocabularies where $\tau = \langle R_1^{a_1}, \ldots, R_r^{a_r}, c_1, \ldots, c_s \rangle$, and let k be a fixed natural number. We want to define the notion of a first-order query,

$$I : \text{STRUC}[\sigma] \to \text{STRUC}[\tau] .$$

I is given by an $r + s + 1$-tuple of formulas, $\varphi_0, \varphi_1, \ldots, \varphi_r, \psi_1, \ldots, \psi_s$, from $\mathcal{L}(\sigma)$. For each structure $\mathcal{A} \in$ STRUC$[\sigma]$, these formulas describe a structure $I(\mathcal{A}) \in$ STRUC$[\tau]$,

$$I(\mathcal{A}) = \langle |I(\mathcal{A})|, R_1^{I(\mathcal{A})}, \ldots, R_r^{I(\mathcal{A})}, c_1^{I(\mathcal{A})}, \ldots, c_s^{I(\mathcal{A})} \rangle .$$

The universe of $I(\mathcal{A})$ is a first-order definable subset[1] of $|\mathcal{A}|^k$,

$$|I(\mathcal{A})| = \{\langle b^1, \ldots, b^k \rangle \mid \mathcal{A} \models \varphi_0(b^1, \ldots, b^k)\}$$

Each relation $R_i^{I(\mathcal{A})}$ is a first-order definable subset of $|I(\mathcal{A})|^{a_i}$,

---

[1] Usually we will take $\varphi_0 \equiv$ **true**, thus letting $|I(\mathcal{A})| = |\mathcal{A}|^k$, cf. Remark 1.32.

1.4 First-Order Queries   19

$$R_i^{I(\mathcal{A})} = \{(\langle b_1^1, \ldots, b_1^k\rangle, \ldots, \langle b_{a_i}^1, \ldots, b_{a_i}^k\rangle) \in |I(\mathcal{A})|^{a_i} \mid \mathcal{A} \models \varphi_i(b_1^1, \ldots, b_{a_i}^k)\}.$$

Each constant symbol $c_j^{I(\mathcal{A})}$ is a first-order definable element of $|I(\mathcal{A})|$,

$$c_j^{I(\mathcal{A})} = \text{the unique } \langle b^1, \ldots, b^k\rangle \in |I(\mathcal{A})| \text{ such that } \mathcal{A} \models \psi_j(b^1, \ldots, b^k).$$

When we need to be formal, we let $a = \max\{a_i \mid 1 \leq i \leq r\}$ and let the free variables of φ_i be $x_1^1, \ldots x_1^k, \ldots, x_{a_i}^1, \ldots, x_{a_i}^k$. The free variables of φ_0 and the ψ_j's are $x_1^1, \ldots, x_1^k$.

If the formulas ψ_j have the property that for all $\mathcal{A} \in \text{STRUC}[\sigma]$,

$$\left|\{\langle b^1, \ldots, b^k\rangle \in |\mathcal{A}|^k \mid (\mathcal{A}, b^1/x_1^1, \cdots, b^k/x_1^k) \models \varphi_0 \wedge \psi_j\}\right| = 1$$

then we write $I = \lambda_{x_1^1 \ldots x_a^k}\langle\varphi_0, \ldots, \psi_s\rangle$ and say that I is a k-ary first-order query from STRUC[σ] to STRUC[τ].

It is often possible to name constant $c_j^{I(\mathcal{A})}$ explicitly as a k-tuple of constants, $\langle t^1, \ldots, t^k\rangle$. In this case, we may simply write this tuple in place of its corresponding defining formula,

$$\psi_j \equiv x_1^1 = t^1 \wedge \cdots \wedge x_1^k = t^k.$$

As another example, in a 3-ary query I, the numerical constants 0, 1, and *max* will be mapped to the following:

$$0^{I(\mathcal{A})} = \langle 0, 0, 0\rangle; \quad 1^{I(\mathcal{A})} = \langle 0, 0, 1\rangle; \quad max^{I(\mathcal{A})} = \langle max, max, max\rangle$$

A *first-order query* is either boolean, and thus defined by a first-order sentence, or is a k-ary first-order query, for some k.

Let FO be the set of first-order boolean queries. Let $Q(\text{FO})$ be the set of all first-order queries. □

Example 1.27 Consider the genealogical database from Example 1.2. The following pair of formulas define a unary query, $I_{sa} = \lambda_{xy}\langle\text{true}, \varphi_{sibling}, \varphi_{aunt}\rangle$, from genealogical databases to structures of vocabulary $\langle\text{SIBLING}^2, \text{AUNT}^2\rangle$:

$$\varphi_{sibling}(x, y) \equiv (\exists fm)(x \neq y \wedge f \neq m \wedge P(f, x) \wedge P(f, y) \wedge P(m, x) \wedge P(m, y))$$

$$\varphi_{aunt}(x, y) \equiv (\exists ps(P(p, y) \wedge \varphi_{sibling}(p, s)$$
$$\wedge (s = x \vee S(x, s))) \wedge F(x)$$

Codd defined a database query language as "complete" if it could express all first-order queries. As we will see, many queries of interest are not first-order. One such example is the ancestor query on genealogical databases (Exercise 6.47). □

As another example, the first-order query I_{add} (Equation (1.25)) is a unary query, i.e., $k = 1$, given by $I_{add} = \lambda_{xy}\langle\text{true}, \varphi_{add}\rangle$. In this case, $\varphi_0 = \text{true}$ means that the universe of $I_{add}(\mathcal{A})$ is equal to the universe of $\mathcal{A}$.

We will see later that $Q(\text{FO})$ is a very robust class of queries. For now, the reader should check the following proposition, which says that first-order queries are closed under composition.

Proposition 1.28. *Let* $I_1 : \text{STRUC}[\sigma] \to \text{STRUC}[\tau]$ *be a k-ary first-order query and let* $I_2 : \text{STRUC}[\tau] \to \text{STRUC}[\upsilon]$ *be an m-ary first-order query. Then* $I_2 \circ I_1 : \text{STRUC}[\sigma] \to \text{STRUC}[\upsilon]$ *is an mk-ary first-order query.*

Exercise 1.29 Consider the following binary first-order query from graphs to graphs: $I = \lambda_{x,y,x',y'}\langle \mathbf{true}, \alpha, \langle 0, 0 \rangle, \langle max, max \rangle \rangle$, where

$$\alpha(x, y, x', y') \equiv (x = x' \wedge E(y, y')) \vee (\text{SUC}(x, y) \wedge x' = y' = y)$$

Recall that part of the meaning of this query is that given a structure $\mathcal{A} \in \text{STRUC}[\tau_g]$, with $n = \|\mathcal{A}\|$,

$$|I(\mathcal{A})| = \{\langle i, j \rangle \mid i, j \in |\mathcal{A}|\}; \quad s^{I(\mathcal{A})} = \langle 0, 0 \rangle; \quad t^{I(\mathcal{A})} = \langle n - 1, n - 1 \rangle \,.$$

1. Show that I has the following interesting property: For all undirected graphs G,

   $(G$ is connected$) \iff (t$ is reachable from s in $I(G))$

2. Recall that a graph is *strongly connected* iff for every pair of vertices g and h, there is a path in G from g to h. Modify I to be a 3-ary query I' such that for all directed graphs G,

   $(G$ is strongly connected$) \iff (t$ is reachable from s in $I'(G))\,.$

[Hint: I almost works, but we need to also make sure that there is a path in G from *max* to 0.] □

Exercise 1.30 Show that the set of first-order queries is closed under composition, i.e., prove Proposition 1.28. □

Exercise 1.31 The first-order query I_{add} defined in Equation (1.25) has the defect that it ignores the possibility that the sum of two n-bit numbers might be $n + 1$ bits. Show how to define a more robust first-order query that returns the always correct $n + 1$-bit sum. Going further, show how to define the first-order query that always returns the correct sum and has no superfluous leading 0's. □

Remark 1.32. If I is a first-order query on ordered structures, then it must include first-order definitions of the numeric relations and constants. Unless we state otherwise, the ordering on $I(\mathcal{A})$ will be the lexicographic ordering of k-tuples $\leq^k$ inherited from $\mathcal{A}$: $\leq^1 = \leq$ and inductively,

$$\langle x_1, \ldots, x_k \rangle \leq^k \langle y_1, \ldots, y_k \rangle \equiv x_1 < y_1 \vee (x_1 = y_1 \wedge$$
$$\langle x_2, \ldots, x_k \rangle \leq^{k-1} \langle y_2, \ldots, y_k \rangle)$$

In the following exercise, you are asked to write the definition of the remaining numeric relations and constants, assuming that $\varphi_0 \equiv \mathbf{true}$. For the first-order queries in this book, we usually limit ourselves to the case that $\varphi_0 \equiv \mathbf{true}$. If not, we must express the new numeric relations explicitly.

Exercise 1.33 Let I be a first-order query on ordered structures. The successor and bit relations must be defined.

1. Give the formulas defining 0, 1, and *max* the minimum, second, and maximum elements of the new universe under the lexicographical ordering. Note that if $\varphi_0 \equiv$ **true**, then the resulting constants are just k-tuples of constants:

   $$0^{I(\mathcal{A})} = \langle 0, \ldots, 0 \rangle; \quad 1^{I(\mathcal{A})} = \langle 0, \ldots, 0, 1 \rangle; \quad max^{I(\mathcal{A})} = \langle max, \ldots, max \rangle$$

   However, in the more general case you must use quantifiers to say that the given element is the minimum, second, maximum in the lexicographical ordering.
2. Assuming that $\varphi_0 \equiv$ **true**, write a quantifier-free formula defining the new SUC relation.
3. Assuming that $\varphi_0 \equiv$ **true**, write the formula defining the new BIT relation. [Hint: by Theorem 1.17 you may define addition and multiplication on k-tuples.]

□

Without the assumption that $\varphi_0 \equiv$ **true**, BIT need not be first-order definable in the image structures. For example, if $\sigma = \tau_s$ and $\varphi_0(x) \equiv S(x)$, then the parity of the universe of $I(\mathcal{A})$ is not first-order expressible in $\mathcal{A}$ (Theorem 13.1). If BIT were definable in $I(\mathcal{A})$ then so would the parity of its universe. For this reason, when we define first-order reductions, we restrict our attention to very simple formulas, φ_0, that define the universe of the image structure.

Historical Notes and Suggestions for Further Reading

There are many excellent introductions to logic. We especially recommend [End72] and [EFT94]. The recent books on finite model theory [EF95], [LR96] and [Ott97] complement this book. For history of logic, it is wonderful to go back to some of the original sources, carefully translated and annotated in [vH].

The definition of semantics for first-order logic (Definition 1.3) is usually attributed to Tarski [Tar36].

Lemma 1.18 was originally proved by Barrington, Immerman, and Straubing in [BIS88]. We got part 2 of Theorem 1.17 from Lindell [L], who says that the result comes from page 299 of Hajek and Pudlak [HP93]. However, on page 406 of [HP93] the result is attributed to Bennett [Ben62]. See also [DLW96], where Dawar, Lindell, and Weinstein prove that ordering is definable when the only given numeric predicate is BIT. It follows that Theorem 1.17 remains valid when ordering is not given because ordering is easily definable from PLUS.

Exercise 1.12 is from [vD94].

One very important topic in a standard course on first-order logic that is omitted here is the study of proofs. In particular, the following two theorems are basic and appear in every standard logic book. They were originally proved by Gödel in his Ph.D. thesis [Göd30].

Theorem 1.34. (**Completeness Theorem for First-Order Logic**) *There is a complete recursive axiomitization for the set of formulas valid in all — finite and infinite — structures.*

Theorem 1.35. (**Compactness Theorem for First-Order Logic**) *Let Γ be a set of first-order formulas with the property that every finite subset of Γ has a (perhaps infinite) model. Then Γ has a (perhaps infinite) model.*

These theorems fail when we restrict our attention to finite structures. From the Completeness Theorem it follows that the set of valid formulas for first-order logic is recursively enumerable (r.e.), and VALID is in fact r.e.-complete. Thus, the set of satisfiable formulas is not r.e. For finite structures, Trahtenbrot's Theorem says that the reverse is true: The set of formulas satisfiable in a finite structure is r.e.-complete, so the set of formulas valid in all finite structures is not r.e., and thus not axiomitizable [Tra50].

For a finite alphabet, $\Sigma = \{\sigma_1, \ldots, \sigma_r\}$, consider the vocabulary $\tau_\Sigma = \langle S^1_{\sigma_1}, \ldots, S^1_{\sigma_r}\rangle$. The set STRUC[$\tau_\Sigma$] consists of the set of non-empty words of vocabulary Σ. Languages without BIT are well-studied for these vocabularies. The following two theorems are fundamental:

Theorem 1.36. *The set of boolean queries expressible in second-order, monadic logic, without* BIT, *over the vocabularies τ_Σ consist exactly of the regular languages. In symbols,*

$$\text{SO(monadic)(wo BIT)} = \text{Regular}.$$

Theorem 1.37. *The set of boolean queries expressible in first-order logic, without* BIT, *over the vocabularies τ_Σ consist exactly of the star-free regular languages. In symbols,*

$$\text{FO(wo BIT)} = \text{star-free Regular}.$$

Theorem 1.36 is due to Büchi [Büc60] and Theorem 1.37 is due to McNaughton and Papert, [MP71]. From the point of view of this book, the languages without BIT are slightly too weak to provide a robust view of computation. A good reference for these results and other relations between logic and automata theory is [Str94] by Straubing.

2
Background in Complexity

Computational Complexity measures the amount of computational resources, such as time and space, that are needed, as a function of the size of the input, to compute a query. This chapter introduces the reader to complexity theory. We define the complexity measures and complexity classes that we study in the rest of the book. We also explain some of their basic properties, complete problems, and interrelationships.

2.1 Introduction

In the 1930's many models of computation were invented, including Church's lambda calculus, Gödel's recursive functions, Markov algorithms and Turing machines. It is very striking that these interesting and apparently different models — all independent efforts to precisely define the intuitive notion of "mechanical procedure" — were proved equivalent. This leads to the universally accepted "Church's thesis", which states that the intuitive concept of what can be "automatically computed" is appropriately captured by the Turing machine (and all its variants).

If one appropriately measures the complexity of computation in a Markov algorithm, a lambda expression, a recursive function, or a Turing machine, one obtains equivalent values. A consequence of this is that efficiency is not model-dependent, but is in fact a fundamental concept.

We get the same theory of complexity whether we approach it via Turing machines or any of these other models. As we will see in this book, descriptive complexity gives definitions of complexity that are equivalent to those of all the above models. Different formalisms lend themselves to different ways of thinking and working. We find that the insights gained from the descriptive approach to complexity offer a different point of view from more traditional machine-based complexity. In particular, there are well understood methods in logic for ascertaining what can and cannot be expressed in a given language. We will introduce some of these methods in Chapter 6.

In Descriptive complexity, we measure the difficulty of describing queries. We will see that natural measures of descriptive complexity such as depth of nesting of quantifiers and number of variables correspond closely to natural notions of complexity in Turing machines.

2.2 Preliminary Definitions

We assume that the reader is familiar with the Turing machine. We start from there and present a survey of computational complexity theory.

We write $M(w){\downarrow}$ to mean that Turing machine M accepts input w, and we write $L(M)$ to denote the language accepted by M,

$$L(M) = \{w \in \{0,1\}^* \mid M(w){\downarrow}\}.$$

Instead of just accepting or rejecting, Turing machines may compute functions from binary strings to binary strings. We use $T(w)$ to denote binary string that Turing machine T leaves on its write-only output tape when it is started with the binary string w on its input tape. If T does not halt on input w, then $T(w)$ is undefined.

Everything that a Turing machine does may be thought of as a query from binary strings to binary strings. In order to make Descriptive complexity rich and flexible it is useful to consider queries that use other vocabularies. To relate such queries to Turing machine complexity, we fix a scheme that encodes the structures of vocabulary τ as boolean strings. To do this, for each τ, we define an encoding query,

$$\text{bin}_\tau : \text{STRUC}[\tau] \to \text{STRUC}[\tau_s]$$

Recall that $\tau_s = \langle S^1 \rangle$ is the vocabulary of boolean strings. The details of the encoding are not important, but it is useful to know that for each τ, bin_τ and its inverse are first-order queries (Exercise 2.3).

Definition 2.1 (The binary encoding of structures: bin($\mathcal{A}$)) Let $\tau = \langle R_1^{a_1}, \ldots, R_r^{a_r}, c_1, \ldots, c_s \rangle$ be a vocabulary, and let $\mathcal{A} = \langle \{0, 1, \ldots, n-1\}, R_1^\mathcal{A} \ldots R_r^\mathcal{A}, c_1^\mathcal{A} \ldots c_s^\mathcal{A} \rangle$ be an ordered structure of vocabulary τ. The relation $R_i^\mathcal{A}$ is a subset of $|\mathcal{A}|^{a_i}$. We encode this relation as a binary string $\text{bin}^\mathcal{A}(R_i)$ of length n^{a_i} where "1" in a given position indicates that the corresponding tuple is in $R_i^\mathcal{A}$. Similarly, for each constant $c_j^\mathcal{A}$, its number is encoded as a binary string $\text{bin}^\mathcal{A}(c_j)$ of length $\lceil \log n \rceil$.

The binary encoding of the structure $\mathcal{A}$ is then just the concatenation of the bit strings coding its relations and constants,

$$\text{bin}_\tau(\mathcal{A}) \;=\; \text{bin}^\mathcal{A}(R_1)\text{bin}^\mathcal{A}(R_2)\cdots \text{bin}^\mathcal{A}(R_r)\text{bin}^\mathcal{A}(c_1)\cdots\text{bin}^\mathcal{A}(c_s)$$

We do not need any separators between the various relations and constants because the vocabulary τ and the length of $\text{bin}_\tau(\mathcal{A})$ determines where each section belongs. Observe that the length of $\text{bin}_\tau(\mathcal{A})$ is given by

$$\hat{n}_\tau \;=\; \|\text{bin}_\tau(\mathcal{A})\| \;=\; n^{a_1} + \cdots + n^{a_r} + s\lceil \log n \rceil \qquad (2.2)$$

Note: We do not bother to include any numeric predicates or constants in $\text{bin}_\tau(\mathcal{A})$ since they can be easily recomputed. However, the coding $\text{bin}_\tau(\mathcal{A})$ does presuppose an ordering on the universe. There is no way to code a structure as a string without an ordering. Since a structure determines its vocabulary, in the sequel we usually write $\text{bin}(\mathcal{A})$ instead of $\text{bin}_\tau(\mathcal{A})$ for the binary encoding of $\mathcal{A} \in \text{STRUC}[\tau]$. Here bin is the union of bin_τ over all vocabularies τ. In the special case where τ includes no input relations symbols, we pretend that there is a unary relation symbol that is always false. For example, if $\tau = \emptyset$, then $\text{bin}(\mathcal{A}) = 0^{|\mathcal{A}|}$. We do this to insure that the size of $\text{bin}(\mathcal{A})$ is at least as large as $\|\mathcal{A}\|$. $\square$

When $\tau = \tau_s$, the map bin_{τ_s} maps strings to strings. The reader should check from the above definition that in this case, bin_{τ_s} is the identity map and thus $\hat{n}_{\tau_s} = n$.

In complexity theory, n is usually reserved for the length of the input. However, in this book, n is used to denote the size of the input structure, $n = \|\mathcal{A}\|$. When the inputs are structures of vocabulary τ, the length of the input is $\hat{n}_\tau$. For the case of binary strings, these two sizes coincide because $\hat{n}_{\tau_s} = n$. When τ is understood, we write $\hat{n}$ for $\hat{n}_\tau$. Observe that n and $\hat{n}$ are always polynomially related.

There are two requirements of a coding function such as "bin". First, it must be computationally very easy to encode and decode. Secondly, the coding must be fairly space efficient, e.g., coding in unary would not be acceptable. In the next exercise, the reader is asked to show that both the encoding and decoding of bin are first-order expressible.

Exercise 2.3 Show that for any vocabulary τ, the queries $\text{bin}_\tau : \text{STRUC}[\tau] \to \text{STRUC}[\tau_s]$ and its inverse $\text{bin}_\tau^{-1} : \text{STRUC}[\tau_s] \to \text{STRUC}[\tau]$ are first-order queries. More explicitly, let $\tau = \langle R_1^{a_1}, \ldots, R_r^{a_r}, c_1, \ldots, c_s \rangle$.

1. Construct a first-order query β_τ that is equal to the mapping bin_τ.
2. Construct a first-order query $\delta_\sigma : \text{STRUC}[\tau_s] \to \text{STRUC}[\tau]$, such that for all $\mathcal{A} \in \text{STRUC}[\tau], \delta_\sigma(\text{bin}_\tau(\mathcal{A})) = \mathcal{A}$. The query δ should be unary, that is, $k = 1$ in the definition of k-ary first-order query (Definition 1.26). $\square$

Using the encoding bin, we define what it means for a Turing machine to compute a query:

Definition 2.4 Let $I : \text{STRUC}[\sigma] \to \text{STRUC}[\tau]$ be a query. Let T be a Turing machine. Suppose that for all $\mathcal{A} \in \text{STRUC}[\sigma]$, $T(\text{bin}(\mathcal{A})) = \text{bin}(I(\mathcal{A}))$. Then we say that *T computes I*. $\square$

We use the notation DTIME[$t(n)$] to denote the set of boolean queries that are computable by a deterministic, multi-tape Turing machine in $O(t(n))$ steps for inputs of universe size n.[1] Similarly we use NTIME[$t(n)$], DSPACE[$s(n)$], and NSPACE[$s(n)$] to denote nondeterministic time, deterministic space and nondeterministic space, respectively. We assume that the reader is familiar with the following classical complexity classes: L = DSPACE[$\log n$], NL = NSPACE[$\log n$], P = polynomial time = $\bigcup_{k=1}^{\infty}$ DTIME[n^k], NP = nondeterministic polynomial time = $\bigcup_{k=1}^{\infty}$ NTIME[n^k], PSPACE = polynomial space = $\bigcup_{k=1}^{\infty}$ DSPACE[n^k] = $\bigcup_{k=1}^{\infty}$ NSPACE[n^k], and EXPTIME = exponential time = $\bigcup_{k=1}^{\infty}$ DTIME[2^{n^k}]. To talk about space $s(n)$, for $s(n) < \hat{n}$, the Turing machine is assumed to have a read-only input tape of length $\hat{n}$ and some number of work tapes of total length $O(s(n))$.

In the definition of complexity classes, we consider only boolean queries. This is in order to be consistent with the standard definitions of complexity classes as sets of decision problems, i.e., boolean queries. For any complexity class $\mathcal{C}$, we use the notation $Q(\mathcal{C})$ to denote the set of all queries that are computable in the complexity class $\mathcal{C}$. Since $\mathcal{C}$ just consists of boolean queries, what does it mean for a general query to be "computable in $\mathcal{C}$"? It means that each bit of bin($I(\mathcal{A})$) is uniformly computable in $\mathcal{C}$ from bin($\mathcal{A}$). In other words,

Definition 2.5 ($Q(\mathcal{C})$, **the Queries Computable in** $\mathcal{C}$) Let I : STRUC[σ] → STRUC[τ] be a query. We say that I *is computable in* $\mathcal{C}$ iff the boolean query I_b is an element of $\mathcal{C}$, where

$$I_b = \{(\mathcal{A}, i, a) \mid \text{The } i^{\text{th}} \text{ bit of bin}(I(\mathcal{A})) \text{ is "}a\text{"}\}. \quad (2.6)$$

Let $Q(\mathcal{C})$ be the set of all queries computable in $\mathcal{C}$:

$$Q(\mathcal{C}) = \mathcal{C} \cup \{I \mid I_b \in \mathcal{C}\} \qquad \square$$

For each of the above resources (deterministic and nondeterministic time and space) there is a hierarchy theorem saying that more of the given resource enables us to compute more boolean queries (see Exercise 2.8). These theorems are proved by diagonalization arguments.

We say that a function s : **N** → **N** is *space constructible* (resp. *time constructible*) iff there is a deterministic Turing machine running in space $O(s(n))$, (resp. time $O(s(n))$) that on input 0^n, i.e., n in unary, computes $s(n)$ in binary. This is the same thing as saying that $s' \in Q(\text{DSPACE}[s(n)])$, resp. $s' \in Q(\text{DTIME}[s(n)])$ where s' is the function that on input 0^n computes $s(n)$ in binary.

Every reasonable function is constructible, as is every function one finds in this book. Many theorems in this book need to assume that the time and space bounds in question are constructible.

---

[1] The usual definition in complexity theory writes $t(n)$ as the function $t'(\hat{n})$, a polynomially-related function of the length of the encoding of the input. We use n to be the size of the universe of the input structure and measure all sizes in this uniform way.

Exercise 2.7 1. Show that the following functions are time constructible: n, n^2, $\lceil n \log n \rceil$, 2^n.
2. Show that the following are space constructible: n, n^2, $\lceil n \log n \rceil$, 2^n, $\lceil \log n \rceil$.
□

Exercise 2.8 Prove the **Space Hierarchy Theorem:** For all space constructible $s(n) \geq \log n$, if $\lim_{n \to \infty}(t(n)/s(n)) = 0$, then DSPACE$[t(n)]$ is strictly contained in DSPACE$[s(n)]$.

[Hint: this is a diagonalization argument, but you have to be careful. On input M, the diagonalization program marks off $s(|M|)$ tape cells and then simulates machine M on input M. If $M(M)$ exceeds the given space or loops, then it should accept. Otherwise, do the opposite of what M would do.]   □

When comparing different resources, we are able to prove much less. For example, by Savitch's Theorem (Theorem 2.32), for $s(n) \geq \log n$,

$$\text{DSPACE}[s(n)] \subseteq \text{NSPACE}[s(n)] \subseteq \text{DSPACE}[(s(n))^2] \;;$$

However, we know only the trivial relationships between nondeterministic and deterministic time:

$$\text{DTIME}[t(n)] \subseteq \text{NTIME}[t(n)] \subseteq \text{DTIME}[2^{O(t(n))}] \;.$$

Consider the following series of containments. It follows from Savitch's Theorem and the Space Hierarchy Theorem that NL is not equal to PSPACE; but even now, more than twenty-five years after the introduction of the classes P and NP, no other inequalities, including that L is not equal to NP, are known.

$$\text{L} \subseteq \text{NL} \subseteq \text{P} \subseteq \text{NP} \subseteq \text{PSPACE}$$

2.3 Reductions and Complete Problems

There are several ways to compare the complexity of boolean queries. Perhaps the most natural is the Turing reduction. Complexity in this model is defined via oracles. Let A and B be boolean queries that may be difficult to compute. An *oracle* for B is a mythical device that when given a structure $\mathcal{B}$ will answer in unit time whether or not $\mathcal{B}$ satisfies query B.

Turing gave his original definition of oracles for the case of unsolvable problems. After proving that the halting problem, K, is undecidable, he considered the question of what can be decided by a Turing machine that had an oracle for K. In complexity theory, we use oracles for computable but difficult sets. Such an oracle can speed up some computations.

We say that A is *Turing reducible* to B if it is easy to compute query A given an oracle for B. The following makes this definition more precise.

Definition 2.9 An *oracle Turing machine* is a Turing machine equipped with an extra tape called the query tape. Let M be an oracle Turing machine and B be any boolean query. We write M^B to denote the oracle Turing machine M equipped with an oracle for set B. M^B may write on its query tape like any other tape. At any time, M^B may enter the "query state". Assume that the string $w = \text{bin}(\mathcal{A})$ is written on the query tape when M^B enters the query state. At the next step, "1" will appear on the query tape if $\mathcal{A} \in B$ and "0" otherwise. Thus, M^B may answer any membership question "Does $\mathcal{A}$ satisfy B?" in linear time: the time to copy the string $\text{bin}(\mathcal{A})$ to its query tape.

Given two boolean queries A, B and a complexity class $\mathcal{C}$, we say that A is $\mathcal{C}$-*Turing reducible* to B iff there exists an oracle Turing machine M such that M^B runs in complexity class $\mathcal{C}$ and $\mathcal{L}(M^B) = A$. We denote this by $A \leq_{\mathcal{C}}^T B$. The superscript "T" stands for Turing reduction. An important example is the polynomial-time Turing reduction, $\leq_{\text{p}}^T$. □

Example 2.10 As an example of a Turing reduction, define the boolean query CLIQUE to be the set of pairs $\langle G, k \rangle$ such that G is a graph having a complete subgraph of size k. The vocabulary for CLIQUE is $\tau_{gk} = \langle E^2, k \rangle$. As usual, we can identify the universe of a structure $\mathcal{A} \in \text{STRUC}[\tau_{gk}]$ with the set $\{0, 1, \ldots, n-1\}$, where $n = \|\mathcal{A}\|$ is the number of vertices of $\mathcal{A}$, and the constant k thus represents not only a vertex, but a number between 0 and $n-1$. (We will see later that CLIQUE is an NP-complete problem.)

Define the query MAX-CLIQUE(G) to be the size of a largest clique in graph G. We show that the boolean version of MAX-CLIQUE is polynomial-time Turing reducible to CLIQUE. In symbols, this would be written,

$$\text{MAX-CLIQUE}_b \leq_{\text{p}}^T \text{CLIQUE} .$$

Where MAX-CLIQUE$_b$ is defined as in Equation (2.6),

$$\text{MAX-CLIQUE}_b = \left\{ (G, i, a) \mid \text{bit } i \text{ of } \text{bin}(I(G)) \text{ is "a"} \right\} .$$

The reduction is as follows. Given (G, i, a), perform binary search using an oracle for CLIQUE to determine the size s of the maximum clique for G. This is done by asking if $(G, n/2) \in$ CLIQUE. If yes, ask about $(G, 3n/4)$, if no, ask about $(G, n/4)$. After $\log n$ queries to the oracle, s has been computed. Now accept iff bit i of s is "a". □

A simpler and more popular kind of reduction in complexity theory is the many-one reduction. (In descriptive complexity, we use first-order reductions. These are first-order queries that are at the same time many-one reductions.)

Definition 2.11 (**Many-One Reduction**) Let $\mathcal{C}$ be a complexity class, and let $A \subseteq \text{STRUC}[\sigma]$ and $B \subseteq \text{STRUC}[\tau]$ be boolean queries. Suppose that the query $I : \text{STRUC}[\sigma] \to \text{STRUC}[\tau]$ is an element of $Q(\mathcal{C})$ with the property that for all $\mathcal{A} \in \text{STRUC}[\sigma]$,

$$\mathcal{A} \in A \quad \Leftrightarrow \quad I(\mathcal{A}) \in B$$

Then I is a $\mathcal{C}$-*many-one reduction* from A to B. We say that A is $\mathcal{C}$-*many-one reducible* to B, in symbols, $A \leq_{\mathcal{C}} B$. For example, when I is a first-order query

(Definition 1.26), it is a first-order reduction ($\leq_{\text{fo}}$), when $I \in Q(L)$, it is a logspace reduction ($\leq_{\log}$); and when $I \in Q(P)$, it is a polynomial-time reduction ($\leq_p$). □

Observe that a many-one reduction is a particularly simple kind of Turing reduction: To decide whether $\mathcal{A}$ is an element of A, compute $I(\mathcal{A})$ and ask the oracle whether $I(\mathcal{A})$ is an element of B. Many-one reductions are simpler than Turing reductions, and they seem to suffice in most situations.

Example 2.12 We give a first-order reduction from PARITY to MULT_b. PARITY is the boolean query on binary strings that is true iff the string has an odd number of ones. We will see later that PARITY is not first-order (Theorem 13.1). MULT, the multiplication query, maps a pair of boolean strings of length n to their product: a boolean string of length $2n$. Let $\tau_{ab} = \langle A^1, B^1 \rangle$ be the vocabulary of structures that are a pair A, B of boolean strings. Then MULT: $\text{STRUC}[\tau_{ab}] \to \text{STRUC}[\tau_s]$. Since reductions map boolean queries to boolean queries, we actually deal with the boolean version of MULT. MULT_b is a boolean query on structures of vocabulary $\tau_{abcd} = \langle A^1, B^1, c, d \rangle$ that is true iff bit c of the product of A and B is "d".

Recall that $\tau_s = \langle S^1 \rangle$ is the vocabulary of boolean strings. The first-order reduction $I_{PM} : \text{STRUC}[\tau_s] \to \text{STRUC}[\tau_{abcd}]$ is given by the following formulas:

$$\varphi_A(x, y) \equiv y = \mathit{max} \wedge S(x)$$
$$\varphi_B(x, y) \equiv y = \mathit{max}$$
$$I_{PM} \equiv \lambda_{xy} \langle \mathbf{true}, \varphi_A, \varphi_B, \langle 0, \mathit{max} \rangle, \langle 0, 1 \rangle \rangle$$

Observe that the effect of this reduction is to line up all the bits of string $\mathcal{A}$ into column $n - 1$ of the generated product. (See Figure 2.13.) It follows that

$$\mathcal{A} \in \text{PARITY} \quad \Leftrightarrow \quad I_{PM}(\mathcal{A}) \in \text{MULT}_b$$

as desired. Thus, PARITY $\leq_{\text{fo}}$ MULT_b. It follows that if MULT were first-order, then PARITY would be as well. We will see later that PARITY is not first-order (Theorem 13.1), so we can conclude that MULT is not first order. □

When $\mathcal{C}$ is a weak complexity class such as FO or L, the intuitive meaning of $A \leq_\mathcal{C} B$ is that the complexity of problem A is less than or equal to the complexity of problem B. The intuitive meaning of A being complete for $\mathcal{C}$ is that A is a

A		0	...	0	s_0	0	...	0	s_1	...	0	...	0	s_{n-1}
B	×	0	...	0	1	0	...	0	1	...	0	...	0	1
		0	...	0	s_0	0	...	0				...		
		0	...	0	s_1	0	...	0				...		
		0	...	0	⋮	0	...	0				...		
		0	...	0	s_{n-1}	0	...	0				...		
			...		P							...		

Figure 2.13: First-order reduction of PARITY to MULT

hardest query in $\mathcal{C}$ and in fact every query in $\mathcal{C}$ can be rephrased as an instance of A; more precisely,

Definition 2.14 (**Completeness for a Complexity Class**) Let A be a boolean query, let $\mathcal{C}$ be a complexity class, and let $\leq_r$ be a reducibility relation. We say that A is *complete for $\mathcal{C}$ via $\leq_r$* iff

1. $A \in \mathcal{C}$, and,
2. for all $B \in \mathcal{C}$, $B \leq_r A$.

In this book, when we say that a problem is complete for a complexity class and we do not say under what reduction, then we mean via first-order reductions $\leq_{fo}$. It will follow from Theorem 3.1 that if a problem is complete via first-order reductions, then it is complete via logspace and polynomial-time reductions. □

For reasons that are not well understood, naturally occurring problems tend to be complete for important complexity classes such as P, NP, and NL. Completeness was originally defined via reductions such as polynomial-time many-one reductions ($\leq_p$) and later, logspace reductions ($\leq_{\log}$). However, problems complete via $\leq_p$ and $\leq_{\log}$ tend to remain complete via $\leq_{fo}$.

Most natural complexity classes include a large number of interesting complete problems. Here are a few boolean queries that are complete for their respective complexity classes. We state these problems very informally here, just to give the reader an idea. More details on these problems and completeness proofs or references are be provided later in the text.

Complete for L:

- CYCLE: Given an undirected graph, does it contain a cycle?
- REACH$_d$: Given a directed graph, is there a deterministic path from vertex s to vertex t? (A deterministic path is such that for every edge (u, v) on the path, there is only one edge in the graph from u.)

Complete for NL:

- REACH: Given a directed graph, is there a path from vertex s to vertex t?
- 2-SAT: Given a boolean formula in conjunctive normal form with only two literals per clause, is it satisfiable?

Complete for P:

- CIRCUIT-VALUE-PROBLEM (CVP): Given an acyclic boolean circuit, with inputs specified, does its output gate have value one?
- NETWORK-FLOW: Given a directed graph, with capacities on its edges, and a value V, is it possible to achieve a steady-state flow of value V through the graph?

Complete for NP:

- SAT: Given a boolean formula, is it satisfiable?
- 3-SAT: Given a boolean formula in conjunctive normal form with only three literals per clause, is it satisfiable?
- CLIQUE: Given an undirected graph and a value k, does the graph have a complete subgraph with k vertices?

Complete for PSPACE:

- QSAT: Given a quantified boolean formula, is it satisfiable?
- HEX, GEOGRAPHY, GO: Given a position in the generalized versions of the games hex, geography, or go, is there a forced win for the player whose move it is?

Exercise 2.15 1. Show that the relations $\leq_{fo}$, $\leq_{log}$, and $\leq_p$ are transitive.

2. Let $\leq_r$ be a transitive, many-one reduction, and let A be complete for complexity class $\mathcal{C}$ via $\leq_r$. Let T be any boolean query. Show that T is complete for $\mathcal{C}$ via $\leq_r$ iff the following two conditions hold:

$$T \in \mathcal{C} \quad \text{and} \quad A \leq_r T$$

□

Exercise 2.16 1. Show the the value $\lceil \log n \rceil$ is first-order expressible.
2. Define the majority query as follows:

$$\text{MAJ} \;=\; \{\mathcal{A} \in \text{STRUC}[\tau_s] \mid \text{string } \mathcal{A} \text{ contains more than } \|\mathcal{A}\|/2 \text{ "1"s}\}$$

Modify the reduction in Example 2.12 to show that $\text{MAJ} \leq_{fo} \text{MULT}$.

□

To prove a natural problem complete for a complexity class, one usually reduces another natural complete problem to it as in Exercise 2.15, part (2). Proving that the first natural complete problem is complete is more subtle, and we defer these proofs to later chapters. The following exercise, however, introduces an unnatural but universal kind of complete problem.

Exercise 2.17 Consider the following boolean query over strings from $\{0, 1, \#\}^*$.

$$U_{\text{time}} \;=\; \{M\#w\#^r \mid M(w)\!\downarrow \text{ in } r \text{ steps}\}$$

Show that U_{time} is complete for P via first-order, many-one reductions ($\leq_{fo}$). Hint: you must show that $U_{\text{time}} \in$ P and that for any problem $B \in$ P, $B \leq_{fo} U_{\text{time}}$. The latter is the easier part in this case: let M_B be a polynomial-time Turing machine that accepts B; then the first-order reduction maps w to $M_b\#w\#^r$, where r is sufficiently large. □

We next describe the problem SAT, which we will see later is NP-complete via first-order reductions (Theorem 7.16).

Example 2.18 SAT is the set of boolean formulas in conjunctive normal form (CNF) that admit a satisfying assignment, i.e., a way to set each boolean variable to **true** or **false** so that the whole formula evaluates to **true**. For example, consider the following boolean formulas:

$$\varphi_0 = (x_1 \vee \overline{x_2} \vee x_3) \wedge (\overline{x_1} \vee \overline{x_2} \vee x_4) \wedge (x_2 \vee \overline{x_3} \vee x_5)$$

$$\varphi_1 = (x_1 \vee \overline{x_2} \vee x_3) \wedge (x_1 \vee x_2 \vee \overline{x_3}) \wedge (x_1 \vee x_2 \vee x_3) \wedge (x_1 \vee \overline{x_2} \vee \overline{x_3})$$
$$\wedge (\overline{x_1} \vee \overline{x_4} \vee x_5) \wedge (\overline{x_1} \vee x_4 \vee \overline{x_5}) \wedge (\overline{x_1} \vee x_4 \vee x_5) \wedge (\overline{x_1} \vee \overline{x_4} \vee \overline{x_5})$$

Here the notation $\bar{v}$ means $\neg v$. The reader should verify that $\varphi_0 \in$ SAT and $\varphi_1 \notin$ SAT.

It is easy to see that SAT is in NP. In linear time, a nondeterministic algorithm can write down a "0" or a "1" for each boolean variable. Then it can deterministically check that each clause has been assigned at least one "1", and if so, accept.

A boolean formula φ that is in CNF may be thought of as a set of clauses, each of which is a disjunction of literals. Recall that a literal is an atomic formula — in this case a boolean variable — or its negation. Thus, a natural way to encode φ is via the structure $\mathcal{A}_\varphi = \langle A, P, N \rangle$.

The universe A is a set of clauses and variables. The relation $P(c, v)$ means that variable v occurs positively in clause c and $N(c, v)$ means that v occurs negatively in c. We can think of every element of the universe as a variable and a clause. Thus, $n = \|\mathcal{A}_\varphi\|$ is equal to the maximum of the number of variables and the number of clauses occurring in φ. If v is really a variable but not a clause, we can harmlessly make it the clause $(v \vee \bar{v})$ by adding the pair (v, v) to the relations P and N. For example, a structure coding φ_0 in this way is:

$$\mathcal{A}_{\varphi_0} = \langle \{1, 2, 3, 4, 5\}, P, N \rangle$$
$$P = \{(1, 1), (1, 3), (2, 4), (3, 2), (3, 5), (4, 4), (5, 5)\}$$
$$N = \{(1, 2), (2, 1), (2, 2), (3, 3), (4, 4), (5, 5)\}$$

□

We next show that SAT $\leq_{\text{fo}}$ CLIQUE. It then follows from Theorem 7.16 that CLIQUE is also NP-complete.

Example 2.19 We show that SAT is first-order reducible to CLIQUE. Let $\mathcal{A}$ be a boolean formula in CNF with clauses $C = \{c_1, \ldots, c_n\}$ and variables $V = \{v_1, \ldots, v_n\}$.

Let $L = \{v_1, \ldots, v_n, \bar{v}_1, \ldots, \bar{v}_n\}$. Define the instance of the clique problem $g(\mathcal{A}) = (V^{g(\mathcal{A})}, E^{g(\mathcal{A})}, k)$ as follows:

$$V^{g(\mathcal{A})} = (C \times L) \cup \{w_0\}$$
$$E^{g(\mathcal{A})} = \{(\langle c_1, \ell_1 \rangle, \langle c_2, \ell_2 \rangle) \mid c_1 \neq c_2 \text{ and } \overline{\ell}_1 \neq \ell_2\} \cup$$
$$\{(w_0, \langle c, \ell \rangle), (\langle c, \ell \rangle, w_0) \mid \ell \text{ occurs in } c\} \quad (2.20)$$
$$k^{g(\mathcal{A})} = n + 1 = \|\mathcal{A}\| + 1$$

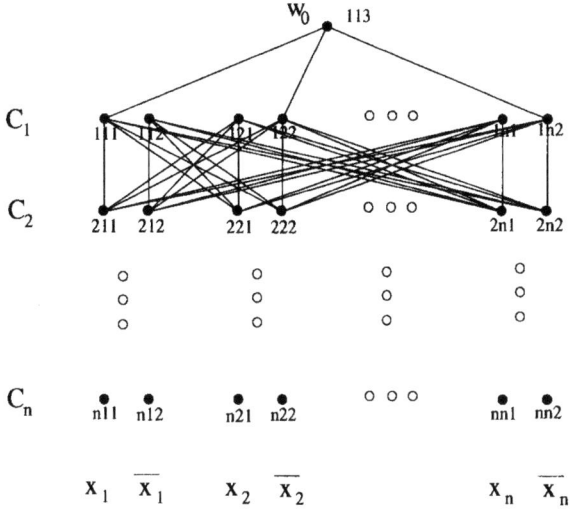

Figure 2.21: SAT $\leq_{\text{fo}}$ CLIQUE, $C_1 = (x_1 \vee \overline{x_2} \vee \overline{x_n})$

The graph $g(\mathcal{A})$ is an $n \times n$ array of vertices containing a row for every clause in $\mathcal{A}$ and a column for every literal in L, plus a top vertex w_o. (See Figure 2.21). There are edges between vertices $\langle c_1, \ell_1 \rangle$ and $\langle c_2, \ell_2 \rangle$ iff $c_1 \neq c_2$, i.e., the points come from different clauses; and, $\overline{\ell_1} \neq \ell_2$, i.e., literals ℓ_1 and ℓ_2 are not the negations of each other. The other edges in the graph are between w_0 and those $\langle c, \ell \rangle$ such that literal ℓ occurs in clause c.

Observe that a clique of size $n + 1$ must involve w_0 and one vertex from each clause. This corresponds to a satisfying assignment, because no literal and its negation can be in a clique. Conversely, any satisfying assignment to $\mathcal{A}$ determines an $n+1$-clique consisting of w_0 together with one literal per clause that is assigned "true". It follows that mapping g is indeed a many-one reduction,

$$(\mathcal{A} \in \text{SAT}) \quad \Leftrightarrow \quad (g(\mathcal{A}) \in \text{CLIQUE}) \tag{2.22}$$

We now give the rather technical details of writing g as a first-order query, $g = \lambda_{x^1 x^2 x^3 y^1 y^2 y^3} \langle \varphi_0, \varphi_1, \psi_1 \rangle$. We encode the vertices as triples $\langle x^1, x^2, x^3 \rangle$, where x^1 corresponds to the clause, x^2 to the variable, and $x^3 = 1$ means the variable is positive and $x^3 = 2$ means the variable is negative. Vertex w_0 is $\langle 1, 1, 3 \rangle$, the only triple with $x^3 > 2$. See Figure 2.21, which shows part of $g(\mathcal{A})$ for a formula $\mathcal{A}$ whose first clause is $(x_1 \vee \overline{x_2} \vee \overline{x_n})$. The numeric formula φ_0, which describes the universe of $g(\mathcal{A})$, is the following,

$$\varphi_0 \equiv (x^3 \leq 2) \vee (x^1 x^2 x^3 = 113)$$

To define the edge relation, let

$$\varphi_1'(\bar{x}, \bar{y}) \equiv \alpha_1 \vee (\alpha_2 \wedge P(y^1, y^2)) \vee (\alpha_3 \wedge N(y^1, y^2))$$

$$\alpha_1 \equiv x^1 \neq y^1 \wedge x^3 < 3 \wedge y^3 < 3 \wedge (x^2 = y^2 \rightarrow x^3 = y^3)$$

$$\alpha_2 \equiv x^3 = 3 \wedge y^3 = 1$$
$$\alpha_3 \equiv x^3 = 3 \wedge y^3 = 2$$

Next, let φ_1 be the symmetric closure of φ'_1,

$$\varphi_1(x^1, x^2, x^3, y^1, y^2, y^3) \equiv \varphi'_1(x^1, x^2, x^3, y^1, y^2, y^3) \vee \varphi'_1(y^1, y^2, y^3, x^1, x^2, x^3)$$

Notice that φ_1 is a direct translation of Equation (2.20). We are thinking of the elements of the ordered universe as $1, 2, \ldots, n$ instead of the usual $0, 1, \ldots n-1$. For this reason, the number $n+1$ which would usually by 011 in lexicographic order, is instead 122. Formula ψ_1 identifies k as $n+1$:

$$\psi_1(x^1, x^2, x^3) \equiv x^1 x^2 x^3 = 122$$

It follows that we have correctly encoded the desired first-order reduction g, and Equation (2.22) holds as desired. □

Exercise 2.23 Construct sets S and T such that S is polynomial-time, Turing reducible to T, but S is not polynomial-time, many-one reducible to T, i.e., $S \leq_p^T T$, $S \not\leq_p T$.

[Hint: construct S and T systematically so that for all w, $w \in S$ iff $(0w \in T) \oplus (1w \in T)$ and systematically satisfy the conditions,

$$C_i \equiv \left(\begin{array}{c} \text{Turing machine } M_i \text{ running in time } i \cdot n^i \text{ does} \\ \text{not compute a many-one reduction from } S \text{ to } T. \end{array} \right)$$

If you are careful, you can carry out this construction so that S and T are in ETIME = DTIME$[2^{O(n)}]$.] □

2.4 Alternation

The concept of a nondeterministic acceptor of a boolean query has a long and rich history, going back to various kinds of nondeterministic automata. On the other hand, it is important to remember that these are fictitious machines and suspect that they cannot be built. We elaborate on the unreasonableness of nondeterministic machines.

For $A \subseteq \text{STRUC}[\tau]$ a boolean query, define its complement $\overline{A} = \text{STRUC}[\tau] - A$. Given a complexity class $\mathcal{C}$, one can define the complementary class as follows,

$$\text{co-}\mathcal{C} = \{\overline{A} \mid A \in \mathcal{C}\}$$

For example, since SAT is in NP, its complementary problem $\overline{\text{SAT}} = \text{UNSAT}$ is in co-NP. The question whether NP is closed under complementation, i.e., is NP equal to co-NP? is open. Most people believe that these classes are different. Notice that if one could really build an NP machine, then one could also build a co-NP machine: All that is needed is a single gate to invert the former machine's answer. Thus from a very practical point of view, the complexity of a problem A and its complement, $\overline{A}$ are identical.

One way to imagine a realization of an NP machine is via a parallel or biological machine with many processors. At each step, each processor p_i creates two copies of itself and sets them to work on two slightly different problems. If either of these offspring ever accepts, i.e., says "yes" to p_i, then p_i in turn says "yes" to its parent. These "yes"es travel up a binary tree to the root and the whole nondeterministic process accepts. In such a view of nondeterminism, in time $t(n)$ we can build about $2^{t(n)}$ processors. However, we do not make very good use of them. Their pattern of communication is very weak. Each processor can compute only the "or" of its children. Thus, the whole computation is one big "or" of its leaves.

There is a natural way to generalize the concept of nondeterminism so that it is closed under complementation and makes better use of its processors. Namely, we can let the processors compute either the "or" or the "and" of their children. This leads to the notion of *alternation*. An alternating Turing machine has both "or" states like a nondeterministic Turing machine and "and" states.

We now formally define and study alternating machines. We will see that in many ways the concept of alternation is more robust than the concept of nondeterminism.

Definition 2.24 An *alternating Turing machine* is a Turing machine whose states are divided into two groups: the existential states and the universal states. Recall that a *configuration* of any Turing machine — also called an *instantaneous description* (ID) — consists of the machine's state, work-tape contents, and head positions. The notion of when such a machine accepts an input is defined by induction: The alternating Turing machine in a given configuration c *accepts* iff

1. c is in a final accepting state, or
2. c is in an existential state and there exists a next configuration c' that accepts, or
3. c is in a universal state, there is at least one next configuration, and all next configurations accept.

Note that this is a generalization of the notion of acceptance for a nondeterministic Turing machine, which is an alternating Turing machine all of whose states are existential.

Turing machines have an awkward way of accessing their tapes, which makes it difficult for them to do anything in sublinear time. Since alternating Turing machines can sensibly use sublinear time, it is more reasonable to use machines that have a more random access nature. As a compromise, from now on we assume that our Turing machines have a *random access* read-only input. This works as follows: there is an *index tape*, which can be written and read like other tapes. Whenever the value v, written in binary, appears on the index tape, the read head automatically scans bit v of the input. □

Define the complexity classes ASPACE[$s(n)$] and ATIME[$t(n)$] to be the set of boolean queries accepted by alternating Turing machines using a maximum of

$O(s(n))$ tape cells, respectively a maximum of $O(t(n))$ time steps in any computation path on an input of length n. The main relationships between alternating and deterministic complexity are given by the following theorem.

Theorem 2.25. *For $s(n) \geq \log n$, and for $t(n) \geq n$,*

$$\bigcup_{k=1}^{\infty} \text{ATIME}[(t(n))^k] = \bigcup_{k=1}^{\infty} \text{DSPACE}[(t(n))^k]$$

$$\text{ASPACE}[s(n)] = \bigcup_{k=1}^{\infty} \text{DTIME}[k^{s(n)}]$$

In particular, $\text{ASPACE}[\log n] = P$, and alternating polynomial time is equal to PSPACE.

Figure 2.26 shows the computation graph of an alternating machine. We assume for convenience that such machines have a unique accepting and a unique rejecting configuration and that each configuration has at most two possible next moves. The start configuration is labeled "s" and the accept configuration is labeled "t". We also assume that these machines have clocks that uniformly cause the machines to shut off at a fixed time that is a function of the length of the input. Shutting off means entering the reject configuration unless the machine is already in the accept configuration.

Observe Figure 2.26. The letters "E" and "A" below the vertices indicate whether the corresponding configurations are existential or universal. If they were all existential, then this would be a nondeterministic computation. The time $t(n)$ measures the depth of the computation graph. It is convenient to think of alternating Turing machines as a parallel model in which at each branching move an extra processor is created, and these two processors take the two branches. Eventually these two processors complete their tasks and report their answers to their parent. If the parent was existential, then it reports "accept" iff either of its children accept. If the

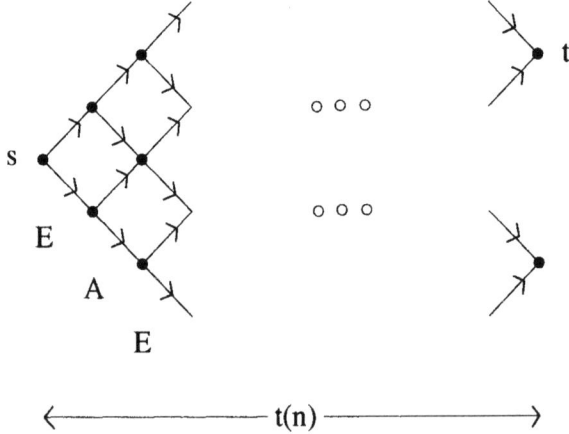

Figure 2.26: Computation graph of an alternating Turing machine

parent is universal, then it reports "accept" iff both of its children accept. Notice that in time $t(n)$ potentially $2^{O(t(n))}$ processors are created. For purely nondeterministic machines these processors have a very poor system of communication: each parent can perform only the "or" of its children. The ability to perform "and"s as well as "or"s lets alternating machines make more extensive use of their extra processors. We will see in Chapter 4 that alternating time $t(n)$ corresponds to a reasonable notion of parallel time $t(n)$ when $2^{O(t(n))}$ processors are available. As we will see, we can allow at most polynomially many processors by restricting the alternating machines to use no more than logarithmic space. The space used by an alternating machine is the maximum amount of space used in any path through its computation graph.

Before we prove some simulation results concerning alternating Turing machines, we give some examples of their use. The first example involves the circuit value problem (CVP). We will see in Exercise 3.28 that CVP is complete for P.

Definition 2.27 A *boolean circuit* is a directed, acyclic graph (DAG),

$$C = (V, E, G_\wedge, G_\vee, G_\neg, I, r) \in \text{STRUC}[\tau_c]; \quad \tau_c = \langle E^2, G_\wedge^1, G_\vee^1, G_\neg^1, I^1, r \rangle$$

Internal node w is an and-gate if $G_\wedge(w)$ holds, an or-gate if $G_\vee(w)$ holds, and a not-gate if $G_\neg(w)$ holds. The nodes v with no edges entering them are called leaves, and the input relation $I(v)$ represents the fact that the leaf v is on. Often we will be given a circuit C, and separately we will be given its input relation I.

Define the Circuit Value Problem (CVP) to consist of those circuits whose root gate r evaluates to one. Define the monotone, circuit value problem (MCVP) to be the subset of CVP in which no negation gates occur. □

Exercise 2.28 Show that a boolean circuit can be evaluated in linear time.

[Hint: do this bottom up: from the leaves to the root. Each edge should be processed only once. By "linear time" we mean linear time on a random-access machine (RAM). Such machines correspond better to real computers than do multi-tape Turing machines. There is a polynomial-size memory and an $O(\log n)$-bit word size.] □

Proposition 2.29. MCVP *is recognizable in* ASPACE[$\log n$].

Proof Let G be a monotone boolean circuit as in Definition 2.27. Define the procedure "EVAL(a)", where a is a vertex of G, as follows:

1. **if** $I(a)$ **then** accept
2.   **else if** a has no outgoing edges **then** reject
3. **if** $G_\wedge(a)$ **then** in a universal state choose a child b of a
4.   **else** in an existential state choose a child b of a
5. Return (EVAL(b))

The machine M simply calls EVAL(r). Observe that EVAL(a) returns "accept" iff gate a evaluates to one. Furthermore, the space used by EVAL is just the space to name two vertices a, b. Thus, M is an ASPACE[$\log n$] machine accepting MCVP,

38   2. Background in Complexity

as desired. Observe that the alternating time required for this computation is the depth of circuit G — the length of the longest path in G starting at r. Recall that we have said that all alternating machines have timers. In this case, an appropriate time limit would be $n = \|G\|$, which is an upper bound on the length of the longest path. □

Another boolean query that is well suited for alternating computation is the quantified satisfiability problem:

Definition 2.30 The *quantified satisfiability problem* (QSAT) is the set of true formulas of the following form:

$$\Psi \;=\; (Q_1 x_1)(Q_2 x_2) \cdots (Q_r x_r) \varphi$$

where φ is a boolean formula and each Q_i's is each either $\forall$ or $\exists$, and $x_1, \ldots x_r$ are the boolean variables occurring in φ. □

Observe that for any boolean formula φ on variables $\overline{x}$,

$$\varphi \in \text{SAT} \quad \Leftrightarrow \quad (\exists \overline{x}) \varphi \in \text{QSAT} \quad \text{and} \quad \varphi \notin \text{SAT} \quad \Leftrightarrow \quad (\forall \overline{x}) \neg \varphi \in \text{QSAT}$$

Thus QSAT logically contains SAT and $\overline{\text{SAT}}$.

Proposition 2.31. *QSAT is recognizable in* ATIME$[n]$.

Proof Construct an alternating machine A that works as follows. To evaluate the sentence

$$\Phi \equiv (\exists x_1)(\forall x_2) \cdots (Q_r x_r) \alpha(\overline{x})$$

in an existential state, A writes down a boolean value for x_1, in a universal state it writes a bit for x_2, and so on. Next A must evaluate the quantifier-free boolean formula α on these chosen values. This is especially easy for an alternating machine: for each "$\wedge$" in α, A universally chooses which side to evaluate and for each "$\vee$", A existentially chooses. Thus A only has to read one of the chosen bits x_i and accept iff it is true and occurs positively, or false and occurs negatively. Observe that A runs in linear time and accepts the sentence Φ iff Φ is true. □

The next theorem explains the relationship between alternating time and deterministic and nondeterministic space.

Theorem 2.32. *Let $s(n) \geq \log n$ be space constructible. Then,*

$$\text{NSPACE}[s(n)] \subseteq \text{ATIME}[s(n)^2] \subseteq \text{DSPACE}[s(n)^2]$$

Proof For the first inclusion, let N be an NSPACE$[s(n)]$ Turing machine. Let w be an input to N. Let G_w denote the computation graph of N on input w. Note that N accepts w iff there is a path from s to t in G_w. We construct an ATIME$[s(n)^2]$ machine A that accepts the same language as N. A does this by calling the subroutine, $P(d, x, y)$, which accepts iff there is a path in G_w of length at most 2^d from x to y. For $d > 0$, P is defined as follows:

$$P(d, x, y) \equiv (\exists z)(P(d - 1, x, z) \wedge P(d - 1, z, y))$$

P works by existentially choosing a middle configuration z, universally choosing the first half or the second half, and then checking that the appropriate path of length 2^{d-1} exists. Thus, the time $T(d)$ taken to compute $P(d, x, y)$ is the time to write down a new, middle configuration plus the time to compute $P(d - 1, x', y')$. The number of bits in a configuration of G_w is $O(s(n))$ where $n = |w|$. Thus,

$$T(d) = O(s(n)) + T(d - 1) = O(d \cdot s(n))$$

The length of the maximum useful path in G_w is bounded by the number of configurations of N on input w, i.e. $2^{cs(n)}$ for an appropriate value of c. Thus, on input w, A calls $P(cs(n), s, t)$ and receives its answer in time $O(cs(n)s(n)) = O(s(n)^2)$, as desired.

For the second inclusion, let A be an ATIME[$t(n)$] machine. On input w, A's computation graph — pictured in Figure 2.26 — has depth $O(t(n))$ and size $2^{O(t(n))}$. A deterministic Turing machine can systematically search this entire and-or graph using space $O(t(n))$. This is done by keeping a string of length $O(t(n))$: $c_1 c_2 \ldots c_r \star \ldots \star$ denoting that we are currently simulating step r of A's computation, having made choices $c_1 \ldots c_r$ on all of the existential and universal branches up until this point. The rest of the simulation will report an **answer** as to whether choices $c_1 \ldots c_r$ will lead to acceptance. This is done as follows:

If one of the following conditions holds:

1. $c_r = 1$, or
2. **answer** = "yes" and step r was existential, or
3. **answer** = "no" and step the r was universal,

then let $c_r = \star$ and report **answer** back to step $r - 1$. Otherwise, set $c_r = 1$ and continue. Note, that we do not have to store intermediate configurations of the simulation because the sequence $c_1 c_2 \ldots c_r \star \ldots \star$ uniquely determines which configuration of A to go to next. □

An immediate corollary is that NSPACE[$s(n)$] is contained in DSPACE[$s(n)^2$]. This is Savitch's theorem (Theorem 2.32), and it is the best known simulation of nondeterministic space by deterministic space. It is unknown whether equality holds in either or both of the inclusions of Theorem 2.32. Another corollary of Theorem 2.32 is the first part of Theorem 2.25.

We complete the proof of Theorem 2.25 by showing that ASPACE[$s(n)$] is DTIME[$O(1)^{s(n)}$]. One direction of this is obvious: an ASPACE[$s(n)$] machine has $O(1)^{s(n)}$ possible configurations. Thus, its entire computation graph is of size $O(1)^{s(n)}$ and thus may be traversed in DTIME[$O(1)^{s(n)}$]. The same traversal algorithm as in the second half of the proof of Theorem 2.32 works in this case.

In the other direction, we are given a DTIME[$k^{s(n)}$] machine M. Let w be an input to M and let $n = |w|$. We can view M's computation as a $k^{s(n)} \times k^{s(n)}$ table — see Figure 2.34. Cell (t, p) of this table contains the symbol that is in position p

of M's tape at time t of the computation. Furthermore, if M's head was at position p at time t, then this cell should also include M's state at time t.

Define an alternating procedure $C(t, p, a)$ that accepts iff the contents of cell p at time t in M's computation on input w is symbol a. $C(0, p, a)$ holds iff a is the correct symbol in position p of M's initial configuration on input w. This means that position 1 contains $\langle q_0, w_1 \rangle$ where q_0 is M's start state, and w_1 is the first symbol of w. Similarly, for $2 \leq p \leq n$, $C(0, p, a)$ holds iff $a = w_p$.

Inductively, $C(t+1, p, a)$ holds iff the three symbols a_{-1}, a_0, a_1 in tape positions $p-1, p, p+1$ lead to an "a" in position p in one step of M's computation. We denote this symbolically as $(a_{-1}, a_0, a_1) \stackrel{M}{\to} a$. This condition can be read directly from M's transition table.

$$C(t+1, p, a) \equiv (\exists a_{-1}, a_0, a_1)\Big((a_{-1}, a_0, a_1) \stackrel{M}{\to} a \ \wedge$$
$$(\forall i \in \{-1, 0, 1\})(C(t, p+i, a_i))\Big) \qquad (2.33)$$

See Figure 2.34 which shows values of $C(t, p, \star)$ for a DTIME$[T(n)]$ Turing machine. In the present case, $T(n) = k^{s(n)}$.

Formula 2.33 can be evaluated by an alternating machine using the space to hold the values of t and p. This is $O(\log k^{s(n)}) = O(s(n))$. Note that M accepts w iff $C(k^{s(n)}, 1, a_f)$ holds, where a_f is the contents of the first cell of M's accept configuration. For example, we can use $a_f = \langle q_f, 0 \rangle$, where q_f is M's accept state.

This completes the proof of Theorem 2.25.

2.5 Simultaneous Resource Classes

Let the classes ASPACE-TIME$[s(n), t(n)]$ (resp. ATIME-ALT$[t(n), a(n)]$) be the sets of boolean queries accepted by alternating machines simultaneously using space $s(n)$ and time $t(n)$ (resp. time $t(n)$ and making at most $a(n)$ alterna-

	Space 1	2	p		n		$T(n)$
Time 0	$\langle q_0, w_1 \rangle$	w_2	$\cdots$		w_n	b $\cdots$	b
1	w_1	$\langle q_1, w_2 \rangle$	$\cdots$		w_n	b $\cdots$	b
	$\vdots$	$\vdots$	$\vdots$		$\vdots$		
t			$a_{-1} a_0 a_1$				
$t+1$			a				
	$\vdots$	$\vdots$	$\vdots$		$\vdots$		
$T(n)$	$\langle q_f, 0 \rangle$	$\cdots$	$\cdots$		$\cdots$		

Figure 2.34: A DTIME[T(n)] computation on input $w_1 w_2 \cdots w_n$

tions between existential and universal states, and starting with existential. Thus ATIME-ALT[$n^{O(1)}$, 1] = NP). Two more important complexity classes may now be defined using these simultaneous alternating classes. Define the polynomial-time hierarchy (PH) to be the set of boolean queries accepted in polynomial time by alternating Turing machines making a bounded number of alternations between existential and universal states:

$$\text{PH} = \bigcup_{k=1}^{\infty} \text{ATIME-ALT}[n^k, k] . \qquad (2.35)$$

Also define NC (Nick's Class) to be the set of boolean queries accepted by alternating Turing machines in log n space and poly log time:

$$\text{NC} = \bigcup_{k=1}^{\infty} \text{ASPACE-TIME}[\log n, \log^k n] \qquad (2.36)$$

See Theorem 5.33 for the more usual definition of NC as the class of boolean queries accepted by a parallel random access machine using polynomially much hardware in poly log parallel time.

2.6 Summary

We conclude this section with a list of the complexity classes defined so far. These will be a main focus for much of what follows:

$$\text{L} \subseteq \text{NL} \subseteq \text{NC} \subseteq \text{P} \subseteq \text{NP} \subseteq \text{PH} \subseteq \text{PSPACE} \qquad (2.37)$$

The above containments are easy to prove. (It is a good exercise for the reader to now show how to simulate each of these complexity classes by the next larger one.)

On the other hand, despite intense effort, there is very little known about the strictness of the above inclusions. It has not yet been proved that L is not equal to PH, or that P is not equal to PSPACE. It would probably be safe to assume the strictness of each of the inequalities in Equation (2.37) as an axiom and go on with the rest of one's life.

The fact that we cannot prove these inequalities reveals just the tip of the iceberg of what we do not know concerning the computational complexity of important computational problems. As an example, for the thousands of known NP complete problems, the best known algorithms to get an exact solution are all exponential time in the worst case. However, we have no proof that these are not computable in linear time.

See Figure 2.38 for a view of the computability and complexity world. The classes in the diagram that have not yet been defined will be described later in the text. The intuitive idea behind this diagram is that there is a set of boolean queries called "truly feasible". These are the queries that can be computed exactly with an "affordable" amount of time and hardware, on all "reasonably sized" instances.

2. Background in Complexity

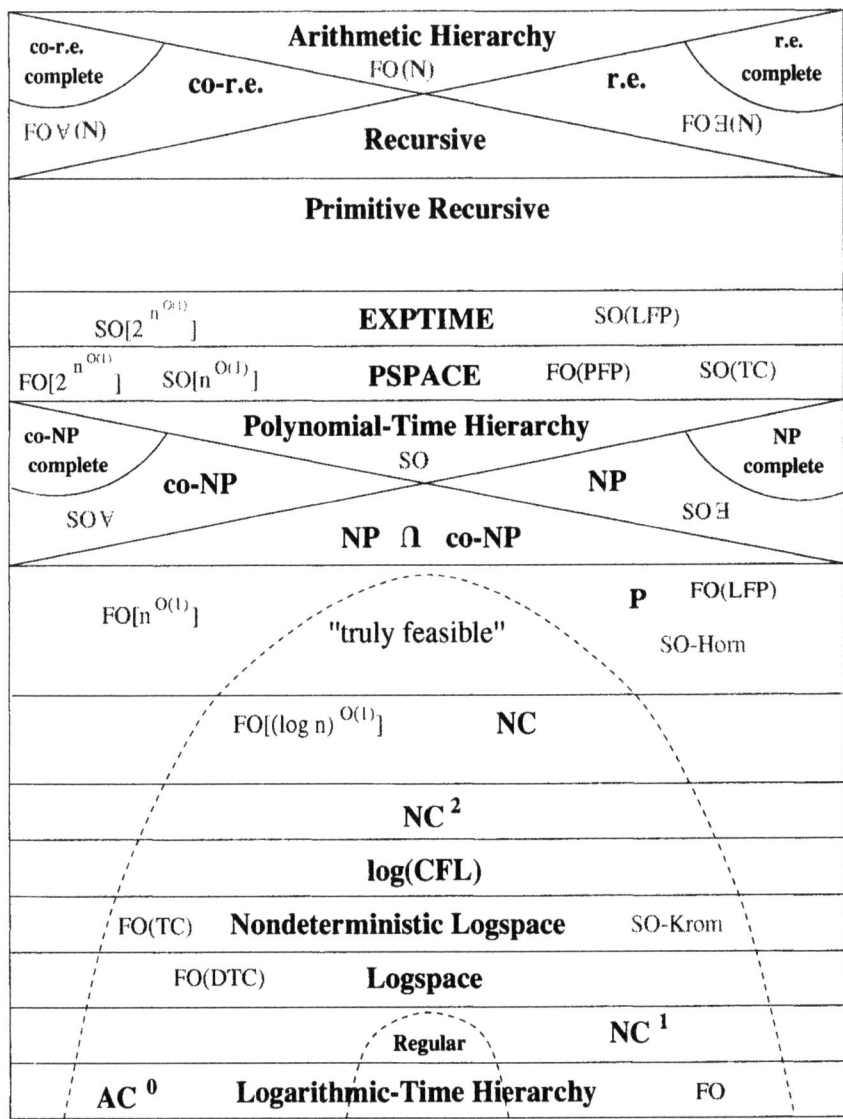

Figure 2.38: The World of Descriptive and Computational Complexity.

The truly feasible queries are a proper subset of P^2. Many important problems that we need to solve are not truly feasible. For example, they may be NP-complete, EXPTIME-complete, or even r.e.-complete. The theory of algorithms and complexity helps us determine whether the problem we need to solve is feasible, and if

---

[2] Those readers who believe that the class BPP properly contains P should change this sentence to, "... a proper subset of BPP."

not, how to choose a limited set of instances of the problem — or easier versions of them — that are feasible.

As the reader will learn, the theory of complexity via Turing machines is isomorphic to descriptive complexity, i.e., the theory of complexity via logic formulas. We will give descriptive characterizations of almost all of the classes in Figure 2.38. For example, the logarithmic-time hierarchy is equal to the set of first-order boolean queries (LH = FO, Theorem 5.30). The polynomial-time hierarchy is the set of second-order boolean queries (PH = SO, Theorem 7.21). The arithmetic hierarchy is an analogous but much larger class defined to be the set of boolean queries that are describable in the first-order theory of the natural numbers.

The descriptive characterizations of complexity classes in this book are all constructive. Efficient algorithms can be automatically translated into efficient descriptions and vice versa. The descriptive approach has added significant new insights and brought new methods to bear on the basic problems in complexity. Descriptive complexity affords an elegant and simple view of complexity. We hope that the reader will use this and perhaps other approaches to add a few more pieces to the puzzle.

Historical Notes and Suggestions for Further Reading

The reader is referred to [Pap94] for much more information on computational complexity. See also [BDG] for more detail on "structural complexity" and [Str94] for a combination of algebraic, logical, and automata-theoretic approaches to complexity theory.

Chandra and Harel [CH80a] introduced the general notion of queries as in Definition 1.24. We got the idea for using queries as the general paradigm of computation from a lecture by Phokion Kolaitis.

Alternating machines were defined and Theorem 2.25 was proved independently by Kozen and by Chandra and Stockmeyer [CKS81]. Kozen titled his description of these machines, "Parallelism in Turing Machines", which is very apt (cf. Theorem 5.33); but Chandra and Stockmeyer coined the name "alternating Turing machine".

The QSAT boolean query was so named in [Pap94] because it is a natural generalization of SAT. Previously, it had been called the quantified boolean formula problem (QBF).

Many-one reductions have their name for an historical reason: to contrast them with 1:1 reductions in which the corresponding mapping is a 1:1 function.

Exercise 2.16 was suggested by Jose Balcázar, cf. [CSV84].

Nick's class (NC) was originally defined by Nick Pippenger; see [Coo85].

3
First-Order Reductions

First-order reductions are simple translations that have very little computational power of their own. Thus it is surprising that the vast majority of natural complete problems remain complete via first-order reductions. We provide examples of such complete problems for many complexity classes. All the complexity classes and descriptive languages studied in this book are closed under first-order reductions. Once we express a complete problem for some complexity class C in a language $\mathcal{L}$, it will follow that $\mathcal{L}$ contains all queries computable in C.

3.1 FO ⊆ L

Recall that FO is the set of first-order definable boolean queries (Definition 1.26). It may be expected that the computational complexity of easy-to-describe queries is low. It was very surprising to us at first that descriptive classes like FO are identical to natural complexity classes.

We will see in Chapter 5 that FO is equal to the set of boolean queries computable in constant parallel time. The following theorem will get us started comparing descriptive versus machine characterizations of complexity. We show that first-order definable queries are all computable in logspace. We will see in Chapter 13 that this containment is strict.

Theorem 3.1. *The set of first-order boolean queries is contained in the set of boolean queries computable in deterministic logspace:* FO ⊆ L.

Proof Let $\sigma = \langle R_1^{a_1}, \ldots, R_r^{a_r}, c_1, \ldots, c_s \rangle$. Let $\varphi \in \mathcal{L}(\sigma)$ determine a first-order boolean query, $I_\varphi : \text{STRUC}[\sigma] \to \{0, 1\}$,

$$\varphi \equiv (\exists x_1)(\forall x_2)\ldots(Q_k x_k)\alpha(\bar{x})$$

where α is quantifier-free. We must construct a logspace Turing machine M such that for all $\mathcal{A} \in \text{STRUC}[\sigma]$, $\mathcal{A}$ satisfies φ iff M accepts the binary encoding of $\mathcal{A}$. In symbols,

$$\mathcal{A} \models \varphi \quad \Leftrightarrow \quad M(\text{bin}(\mathcal{A}))\downarrow \qquad (3.2)$$

We construct the logspace Turing machine M inductively on k, the number of quantifiers occurring in φ. If $k = 0$, then $\varphi = \alpha$ is a quantifier-free sentence. Thus, α is a fixed, finite boolean combination of atomic formulas. The atomic formulas are either occurrences of input relations $R_i(p_1, \ldots, p_{a_i})$ or numeric relations $p_1 = p_2$, $p_1 \leq p_2$, or $\text{BIT}(p_1, p_2)$, and the p_i's are members of $\{c_1, \ldots, c_s, 0, 1, max\}$. Once we know that M can determine, on input $\mathcal{A}$, whether $\mathcal{A}$ satisfies each of these atomic formulas, M can then determine whether $\mathcal{A} \models \alpha$, by performing the fixed, finite boolean combination using its finite control.

The reader should convince herself that a logspace machine that knows its input is of the form $\text{bin}(\mathcal{A})$, for some $\mathcal{A} \in \text{STRUC}[\sigma]$, can calculate the values n and $\lceil \log n \rceil$. Then, by counting, the machine can go to the appropriate constants and copy the p_i's that it needs onto its worktape. To calculate one of the input predicates, M can just look up the appropriate bit of its input.

For example, to calculate $R_3(c_2, max, c_1)$, M first copies the values $c_2, n-1, c_1$ to its worktape. Next it moves its read head to location $n^{a_1} + n^{a_2} + 1$, which is the beginning of the array encoding R_3. Finally, it moves its read head $n^2 \cdot c_2 + n \cdot (n-1) + c_1$ spaces to the right. The bit now being read is "1" iff $\mathcal{A} \models R_3(c_2, max, c_1)$.

It is easy to see that a logspace Turing machine may test the numeric predicates. This completes the construction of M in the base case.

Inductively, assume that all first-order queries with $k-1$ quantifiers are logspace computable. Let

$$\psi(x_1) = (\forall x_2)\ldots(Q_k x_k)\alpha(\bar{x}) \ .$$

Let M_0 be the logspace Turing machine that computes the query $\psi(c)$. Note that c is a new constant symbol substituted for the free variable x_1. To compute the query $\varphi \equiv (\exists x_1)(\psi(x_1))$ we build the logspace machine that cycles through all possible values of x_1, substitutes each of these for c, and runs M_0. If any of these lead M_0 to accept, then we accept, else we reject. Note that the extra space needed is just $\log n$ bits to store the possible values of x_1. Simulating a universal quantifier is similar. □

3.2 Dual of a First-Order Query

A first-order query I from $\text{STRUC}[\sigma]$ to $\text{STRUC}[\tau]$ maps any $\mathcal{A} \in \text{STRUC}[\sigma]$ to $I(\mathcal{A}) \in \text{STRUC}[\tau]$. It does this by defining the relations of $I(\mathcal{A})$ via first-order formulas. In a similar way, I has a natural dual $\widehat{I}$, which translates any formula in $\mathcal{L}(\tau)$ to a formula in $\mathcal{L}(\sigma)$.

3.2 Dual of a First-Order Query

In this section we define and characterize this dual mapping. The dual is useful in showing that relevant languages and complexity classes are closed under first-order reductions.

Let I be a k-ary first-order query. For each formula $\varphi \in \mathcal{L}(\tau)$, the formula $\widehat{I}(\varphi) \in \mathcal{L}(\sigma)$ is constructed as follows: Replace each variable by a k-tuple of variables. Replace each symbol of τ by its definition in I. It follows that the length of $\widehat{I}(\varphi)$ is linear in the length of φ. In the following, we give the details.

Definition 3.3 (Dual of I) Let $I = \lambda_{x_1 \ldots x_d} \langle \varphi_0, \ldots, \psi_s \rangle$ be a k-ary first-order query from STRUC[σ] to STRUC[τ]. Then I also defines a dual map, which we call $\widehat{I} : \mathcal{L}(\tau) \to \mathcal{L}(\sigma)$, as follows:

Let $\tau = \langle R_1^{a_1}, \ldots, R_r^{a_r}, c_1, \ldots, c_s \rangle$. For $\varphi \in \mathcal{L}(\tau)$, $\widehat{I}(\varphi)$ is the result of replacing all relation and constant symbols in φ by the corresponding formulas in I, using a map f_I defined as follows:

- Each variable is mapped to a k-tuple of variables: $f_I(v) = v^1, \ldots, v^k$
- Input relations are replaced by their corresponding formulas:

$$f_I(R_i(v_1, \ldots, v_{a_i})) = \varphi_i(f_I(v_1), \ldots, f_I(v_{a_i}))$$

- Constant c_i is replaced by a special k-tuple of variables[1]:

$$f_I(c_i) = z_i^1, \ldots, z_i^k$$

- Quantifiers are replaced by restricted quantifiers:

$$f_I(\exists v) = (\exists f_I(v) \cdot \varphi_0(f_I(v)))$$

- The equality relation and the other numeric relations are replaced by their appropriate formulas as in Exercise 1.33.
- Second-order quantifiers — which we will need in Chapter 7 — have the arities of the relations being quantified multiplied by k:

$$f_I(\exists R^a) = (\exists R^{ka})$$

- On boolean connectives, f_I is the identity.

The only thing to add is that the variables $z_i^1, \ldots, z_i^k$ corresponding to the constant symbol c_i must be quantified before they are used. It does not matter where these quantifiers go because the values are uniquely defined. Typically, we can place these quantifiers at the beginning of a first-order formula. (For a second-order formula, they would be placed just after the second-order quantifiers.)

Thus, the mapping $\widehat{I}$ is defined as follows, for $\theta \in \mathcal{L}(\tau)$,

$$\widehat{I}(\theta) = (\exists z_1^1 \ldots z_1^k \cdot \psi_1(z_1^1 \ldots z_1^k)) \cdots (\exists z_s^1 \ldots z_s^k \cdot \psi_s(z_s^1 \ldots z_s^k))(f_I(\theta)) \quad \square$$

Exercise 3.4 To get the idea of what the dual mapping does, consider the query I_{PM} from Example 2.12. Here is one sample value of the map $\widehat{I}_{PM}$:

---

[1] For many applications, each constant from τ may be replaced by a corresponding k-tuple of constants from σ.

$$\widehat{I}_{PM}(A(c)) \equiv (\exists z_1 z_2 \, . \, z_1 = 0 \land z_2 = max)(z_2 = max \land S(z_1))$$
$$\equiv S(0)$$

Compute the value of $\widehat{I}_{PM}$ on the following,

1. $(\forall v)(A(v) \leftrightarrow B(v))$
2. $A(max)$
3. $A(0)$ □

It follows immediately from Definition 3.3 that:

Proposition 3.5. *Let σ, τ, and I be as in Definition 3.3. Then for all sentences $\theta \in \mathcal{L}(\tau)$ and all structures $\mathcal{A} \in \text{STRUC}[\sigma]$,*

$$\mathcal{A} \models \widehat{I}(\theta) \quad \Leftrightarrow \quad I(\mathcal{A}) \models \theta$$

Remark 3.6. Proposition 3.5 goes through for formulas with free variables as well. In this case, I must behave appropriately on interpretations of variables. That is, $I(\mathcal{A}, i) = (I(\mathcal{A}), i')$, where $i'(x)$ is defined iff all of $i(x^1), \ldots, i(x^k)$ are defined. In this case, $i'(x) = \langle i(x^1), \ldots, i(x^k) \rangle$.

In the following exercise you are asked to show that structures of any vocabulary σ may be transformed to graphs via a first-order, invertible query. It then follows from Proposition 3.5 that every formula in $\mathcal{L}(\sigma)$ may be translated into the language of graphs. This is true even without the ordering relation. Exercise 2.3 shows the same thing about binary strings, i.e., that everything can be coded in a first-order way as a binary string; but in the case of strings, ordering and arithmetic are required.

Exercise 3.7 (**Everything is a Graph**) Let σ be any vocabulary, and let $\tau_e = \langle E^2 \rangle$ be the vocabulary with one binary relation symbol, i.e., the vocabulary of graphs with no specified points. In this exercise, you will show that every structure may be thought of as a graph.

Show that there exist first-order queries $I_\sigma : \text{STRUC}[\sigma] \to \text{STRUC}[\tau_e]$ and $I_\sigma^{-1} : \text{STRUC}[\tau_e] \to \text{STRUC}[\sigma]$ with the following property,

$$\text{for all } \mathcal{A} \in \text{STRUC}[\sigma], \quad I_\sigma^{-1}(I_\sigma(\mathcal{A})) \cong \mathcal{A} \quad (3.8)$$

[Hint: to build the graph $I_\sigma(\mathcal{A})$, you can construct "gadgets", i.e., small recognizable graphs to label different sorts of vertices, e.g., those corresponding to elements of $|\mathcal{A}|$, those corresponding to tuples from each relation $R_i^\mathcal{A}$, etc.] Note that this exercise does not require ordering, which is why Equation (3.8) says only that the two objects are isomorphic. If we include ordering, we can require that equality holds. □

If A and B are boolean queries and A is first-order reducible to B ($A \leq_{\text{fo}} B$), then intuitively the complexity of A is not greater than the complexity of B. The following definition makes this intuitive idea explicit.

Definition 3.9 (**Closure Under First-Order Reductions**) A set of boolean queries S is *closed under first-order reductions* iff whenever there are boolean queries A and B such that $B \in \mathcal{S}$ and $A \leq_{fo} B$, we have that $A \in \mathcal{S}$. We say that a language $\mathcal{L}$ is closed under first-order reductions iff the set of boolean queries definable in $\mathcal{L}$ is closed. □

The following proposition follows immediately from Theorem 3.1.

Proposition 3.10. *Let $\mathcal{S}$ be any set of boolean queries that is closed under logspace reductions. Then $\mathcal{S}$ is also closed under first-order reductions.*

The next proposition cannot be proved immediately. It is a global exercise. It is striking that with the exception of the dynamic-complexity classes, all complexity classes we discuss in the book are closed under first-order reductions. Every time a new language or complexity class is introduced, the reader should check that it is closed under first-order reductions. For languages, we mean that the set of boolean queries definable in that languages is closed under first-order reductions. This is immediate if the language is closed under quantification and boolean operations. Most languages we consider are so closed. Some languages such as SO∃ — see Theorem 7.8 — do not seem to be closed under negation. However, they still allow full use of first-order logic at their bottom levels and are thus closed under first-order reductions.

Meta-Proposition 3.11. *With the exception of the dynamic complexity classes defined in Chapter 14, all the complexity classes $\mathcal{C}$ that we discuss in this book are closed under first-order reductions. All the languages $\mathcal{L}$ that we discuss in this book are closed under first-order reductions.*

Framework 3.12 Here is a method for proving this proposition whenever a new complexity class or logical language is encountered. For complexity classes we can usually use Proposition 3.10 as most complexity classes are closed under logspace reductions.

For languages, let $A \leq_{fo} B$ be two boolean queries, where B is expressible as the formula φ_B in language $\mathcal{L}$. Let I_{AB} be the first-order reduction from A to B. We know that for all structures $\mathcal{S}$,

$$\mathcal{S} \in A \quad \Leftrightarrow \quad I_{AB}(\mathcal{S}) \in B$$

It follows from Proposition 3.10 that

$$\mathcal{S} \in A \quad \Leftrightarrow \quad \mathcal{S} \models \widehat{I}_{AB}(\varphi_B)$$

So if $\widehat{I}_{AB}(\varphi_B)$ is in $\mathcal{L}$ then the proof is complete. Since the definition of $\widehat{I}(\varphi)$ (Definition 3.3) is a simple substitution that doesn't change the structure of φ very much, we will find that for the languages we consider $I_{AB}(\varphi_B)$ will be in $\mathcal{L}$ as desired. □

First-order reductions are simple and natural reductions. It is very surprising that they seem to suffice in almost all complexity theoretic settings. We will see that "natural" problems that are complete via polynomial-time reductions for some complexity class tend to remain complete via first-order reductions.

Suppose that we know that a boolean query A is complete via first-order reductions for a complexity class C. Suppose further that A is expressible in a language $\mathcal{L}$ which is closed under first-order reductions. It follows immediately that $\mathcal{L}$ expresses everything in C.

Suppose that $\mathcal{L}$ is a set of boolean queries describable in some language and that C is a complexity class, that is, a set of boolean queries computable in some complexity bound. In the sequel our paradigm for proving that $\mathcal{L} = C$ will be the following four steps:

1. Show that $\mathcal{L} \subseteq C$ by producing for each formula φ from the language an algorithm in C that computes the boolean query,
$$\text{MOD}[\varphi] \;=\; \{\mathcal{A} \mid \mathcal{A} \models \varphi\}$$
2. Produce a boolean query T that is complete for C via first-order reductions.
3. Show that $\mathcal{L}$ is closed under first-order reductions.
4. Express T in the language, thus showing that $T \in \mathcal{L}$.

A typical example is in a proof of Theorem 7.8, which says that NP = SO∃. We can show: (1) Each SO∃ formula can be checked by an NP machine; (2) The problem SAT is complete for NP via $\leq_{\text{fo}}$; (3) SO∃ is closed under first-order reductions; and finally, (4) SAT is expressible in SO∃. (Actually, in Chapter 7 we present a different proof for Theorem 7.8 that provides more information.)

In the remainder of this chapter, we give several examples of first-order reductions as we produce natural complete problems for the complexity classes L, NL, and P. The proofs encode Turing machine computations using first-order formulas. The proofs are quite intricate. For this reason, it would be fine to skim the remainder of this chapter on first reading. On the other hand, since many of the results on capturing complexity classes by logics depend on this material, at some point this material should be read.

3.3 Complete problems for L and NL

Natural complete problems for L and NL are the REACH$_d$ and REACH problems.

Definition 3.13 Define REACH to be the set of directed graphs G such that there is a path in G from s to t. Define REACH$_d$ to be the subset of REACH such that the path from s to t is deterministic. This means that for each edge (u, x) on the path, this is the unique edge in G leaving u. See Figure 3.14 for a directed graph that is in REACH but not REACH$_d$. Note that there is a directed path in this figure from p to t. Also, define REACH$_u$ — the undirected graph reachability problem — to be the restriction of REACH to undirected graphs,

$$\text{REACH}_u \;=\; \{G \in \text{REACH} \mid G \models (\forall xy)(E(x, y) \to E(y, x))\} \qquad \square$$

The following NSPACE[$\log n$] algorithm recognizes REACH. Note that the space used is just the $O(\log n)$ bits needed to name the two vertices a and b.

3.3 Complete problems for L and NL

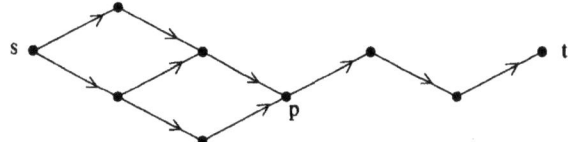

Figure 3.14: A graph that is in REACH but not $REACH_d$

Algorithm 3.15. Recognizing REACH in NL

1. $b := s$
2. **while** $(b \neq t)$ **do** {
3. $a := b$
4.   nondeterministically choose new b
5.   **if** $(\neg E(a, b))$ **then** reject }
6. accept

Theorem 3.16. REACH *is complete for* NL *via first-order reductions.*

Proof Let $S \subseteq \text{STRUC}[\sigma]$ be a boolean query in NL. Let N be the nondeterministic logspace Turing machine that accepts S. We construct a first-order reduction $I : \text{STRUC}[\sigma] \to \text{STRUC}[\tau_g]$ such that for all $\mathcal{A} \in \text{STRUC}[\sigma]$,

$$N(\text{bin}(\mathcal{A}))\!\downarrow \quad \Leftrightarrow \quad I(\mathcal{A}) \in \text{REACH} \qquad (3.17)$$

Let c be such that N uses at most $c \log n$ bits of worktape for inputs $\text{bin}(\mathcal{A})$, with $n = \|\mathcal{A}\|$. Let $\sigma = \langle R_1^{a_1}, \ldots, R_r^{a_r}, c_1, \ldots, c_s \rangle$ and let $a = \max\{a_i \mid 1 \leq i \leq r\}$. Let $k = 1 + a + c$. Consider a run of N on input $\text{bin}(\mathcal{A})$. We code an instantaneous description (ID) of N's computation as a k-tuple of variables:

$$\text{ID} \;=\; (p, r_1, \ldots r_a, w_1, \ldots, w_c)$$

The idea is that variables $r_1, \ldots, r_a$ encode where in one of the input relations the read head of N is looking. If for example it is looking at relation R_i, then,

$$N\text{'s read head is looking at a "1"} \quad \Leftrightarrow \quad \mathcal{A} \models R_i(r_1, \ldots, r_{a_i}) \qquad (3.18)$$

Variables $w_1, \ldots, w_c$ encode the contents of N's work tape. Remember that each variable represents an element of $\mathcal{A}$'s n-element universe, so it corresponds to a $\log n$-bit number. (We are assuming the presence of the numerical relations $\leq$ and BIT. Of these, $\leq$ is necessary (see Proposition 6.14), but BIT is merely convenient (see Proposition 9.16). Finally, we need $O(\log \log n)$ bits of further information to encode: (1) the state of N, (2) which input relation or constant symbol the read head is currently scanning, and (3) the position of the work head. We assume that n is sufficiently large that all of this information can be encoded into a single variable, p.

Now we start to build the desired k-ary first-order query I and show that it satisfies Equation (3.17). I will be constructed as follows:

$$I \;=\; \lambda_{\text{ID, ID}'} \langle \textbf{true}, \varphi_N, \alpha, \omega \rangle$$

where

1. The universe relation being "**true**" indicates that for any $\mathcal{A} \in \mathrm{STRUC}[\sigma]$, the universe of $I(\mathcal{A})$ consists of *all* k-tuples from the universe of $\mathcal{A}$, $|I(\mathcal{A})| = |\mathcal{A}|^k$.
2. $\mathcal{A} \models \varphi_N(\mathrm{ID}, \mathrm{ID}')$ iff $(\mathrm{ID}, \mathrm{ID}')$ is a valid move of N on input $\mathrm{bin}(\mathcal{A})$,
3. $\mathcal{A} \models \alpha(\mathrm{ID}_i)$ iff ID_i is the unique initial ID of N, for inputs of size $\|\mathcal{A}\|$, and,
4. $\mathcal{A} \models \omega(\mathrm{ID}_f)$ iff ID_f is the unique accept ID of N for inputs of size $\|\mathcal{A}\|$.

Formulas α and ω are the following,

$$\begin{aligned} \alpha(x_1,\ldots,x_k) &\equiv& x_1 = x_2 = \ldots = x_k = 0 \\ \omega(x_1,\ldots,x_k) &\equiv& x_1 = x_2 = \ldots = x_k = max \end{aligned} \quad (3.19)$$

Formula φ_N is not hard, but it is more tedious. It is essentially a disjunction over N's finite transition table.

A typical entry in the transition table is $(\langle q, b, w \rangle, \langle q', i_d, w', w_d \rangle)$. This says that in state q, looking at bit b with the input head and bit w with the work head, N may go to state q', move its input head one step in direction i_d, write bit w' on its work tape and move its work head one step in direction w_d. The corresponding disjunct in φ_N must decode the old state from variable p and must decode from p which input relation is being read. Say it is R_i. Then the bit b is "1" iff $R_i(r_1,\ldots,r_{a_i})$ holds. Similarly, we must extract from p the segment j of the work tape that is currently being scanned together with the position s on that worktape. Thus, bit w is "1" iff $\mathrm{BIT}(w_j, s)$ holds.

With these details completed, it now follows that for any $\mathcal{A} \in \mathrm{STRUC}[\sigma]$, $I(\mathcal{A})$ is the computation graph of N on input $\mathrm{bin}(\mathcal{A})$. It follows that N accepts $\mathrm{bin}(\mathcal{A})$ iff there is a path in $I(\mathcal{A})$ from s to t, i.e., Equation (3.17) holds. □

Exercise 3.20 There are several gaps left in the proof of Theorem 3.16 that the reader should now fill in.

1. Using numeric relation BIT, write first-order formulas to uniquely identify elements $l_1 = \lceil \log n \rceil$ and $l_2 = \lceil \log \log n \rceil$ of the universe.
2. Show that since the coding is somewhat arbitrary, we may use Equation (3.19) as our definitions of α and ω.
3. Assume that the first l_2 bits of p encode the work head's position, s. Write a formula to uniquely identify element s.
4. Do the same problem as (3) but this time assume that the bits of s are encoded in the last l_2 bits of p. In order to do this you will need addition, which is available (Theorem 1.17). □

We next show that REACH_d is complete for L via first-order reductions. We first must show that REACH_d is in L. A modification of Algorithm 3.15 recognizes REACH_d in logspace. Since a deterministic path has at most one edge leaving each vertex, nondeterminism is no longer needed. We add a counter to detect cycles:

Algorithm 3.22. Recognizing REACH_d in L

1. $b := s$; $i := 0$; $n := \|G\|$

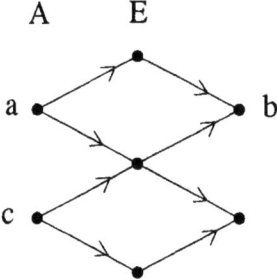

Figure 3.21: An alternating graph with two universal nodes: a, c.

2. **while** $b \neq t \wedge i < n \wedge (\exists ! a)(E(b, a))$ **do** {
3. $b :=$ the unique a for which $E(b, a)$
4. $i := i + 1$ }
5. **if** $b = t$ **then** accept **else** reject

The definition of REACH$_d$ was made just so that the following theorem would be true:

Theorem 3.23. REACH$_d$ *is complete for* L *via first-order reductions.*

Proof This proof is similar to the proof of Theorem 3.16. In fact, we copy the whole construction with $S \subseteq \text{STRUC}[\sigma]$ an arbitrary boolean query from L. The only difference is that now N is a deterministic logspace Turing machine that computes S. Since N is deterministic, for any $\mathcal{A} \in \text{STRUC}[\sigma]$, the graph $I(\mathcal{A})$ has at most one edge leaving any vertex. It follows that $I(\mathcal{A})$ is in REACH iff it is in REACH$_d$. Thus,

$$N(\text{bin}(\mathcal{A}))\downarrow \quad \Leftrightarrow \quad I(\mathcal{A}) \in \text{REACH}_d \qquad \square$$

3.4 Complete Problems for P

Alternation provides a nice way to characterize P, namely, P = ASPACE[$\log n$] (Theorem 2.25). This leads to a natural analogue of the reachability problem that is complete problem for P.

Definition 3.24 Let an *alternating graph* $G = (V, E, A, s, t)$ be a directed graph whose vertices are labeled universal or existential. $A \subseteq V$ is the set of universal vertices. Let $\tau_{ag} = \langle E^2, A^1, s, t \rangle$ be the vocabulary of alternating graphs.

Alternating graphs have a different notion of accessibility. Let $P_a^G(x, y)$ be the smallest relation on vertices of G such that:

1. $P_a^G(x, x)$
2. If x is existential and $P_a^G(z, y)$ holds for some edge (x, z) then $P_a^G(x, y)$.
3. If x is universal, there is at least one edge leaving x, and $P_a^G(z, y)$ holds for all edges (x, z) then $P_a^G(x, y)$.

See Figure 3.21 where $P_a^G(a, b)$ holds but $P_a^G(c, b)$ does not. Let

$$\text{REACH}_a = \{G \mid P_a^G(s, t)\} \qquad \square$$

Observe that the following marking algorithm computes REACH_a in linear time.

Algorithm 3.25. Recognizing REACH_a in linear time on a RAM.

1. make QUEUE empty; mark(t); insert t into QUEUE
2. **while** QUEUE not empty **do** {
3.   remove first element, x, from QUEUE
4.   **for** each unmarked vertex y such that $E(y, x)$ **do** {
5.   delete edge (y, x)
6.   **if** y is existential or y has no outgoing edges
7.   **then** mark(y); insert y into QUEUE } }
8. **if** s is marked then **accept else reject**

Not surprisingly we have,

Theorem 3.26. REACH_a *is complete for* P *via first-order reductions.*

Proof This proof is similar to the proof of Theorem 3.16. Let $S \subseteq \text{STRUC}[\sigma]$ be an arbitrary boolean query. Assume that $S \in P$ and let T be the alternating, logspace Turing machine that computes S. We construct a first-order reduction $I_a : \text{STRUC}[\sigma] \to \text{STRUC}[\tau_{ag}]$ such that for all $\mathcal{A} \in \text{STRUC}[\sigma]$,

$$T(\text{bin}(\mathcal{A}))\downarrow \quad \Leftrightarrow \quad I_a(\mathcal{A}) \in \text{REACH}_a \qquad (3.27)$$

Indeed, the only difference between I, the query from the proof of Theorem 3.16, and I_a is that I_a must also describe the relation A that identifies the universal states of T. Assume for simplicity that the universal states are exactly the odd-numbered states. Assume further that the variable p in an ID encodes its state in its low-order bits. Thus the state of an ID is universal iff the corresponding p is odd. This occurs iff $\text{BIT}(p, 0)$ holds. Thus, let

$$\psi_A = \text{BIT}(p, 0), \qquad I = \lambda_{\text{ID},\text{ID}'}\langle \text{true}, \varphi_T, \psi_A, \alpha, \omega \rangle$$

where φ_T, α, and ω are defined exactly as in the proof of Theorem 3.16.

It follows that,

$$T(\text{bin}(\mathcal{A}))\downarrow \quad \Leftrightarrow \quad I_a(\mathcal{A}) \in \text{REACH}_a \qquad \square$$

Exercise 3.28 Recall the circuit value problem (CVP) and the monotone circuit value problem (MCVP) from Definition 2.27.

1. Produce a first-order reduction from REACH_a to MCVP.
2. Conclude that MCVP is complete for P via first-order reductions.
3. Conclude that CVP is also complete via first-order reductions. Be slightly careful because it is certainly not true that for any two boolean queries S, T, if S is complete for $\mathcal{C}$ and $T \in \mathcal{C}$ and $S \subseteq T$ then T is complete for $\mathcal{C}$. $\qquad \square$

Historical Notes and Suggestions for Further Reading

Savitch proved that REACH is complete for NL via logspace reductions, [Sav73]. Hartmanis, Immerman, and Mahaney REACH$_d$ showed that REACH$_d$ is complete for L via one-way logspace reductions, [HIM78]. The completeness of these problems via first-order reductions was proved in [I83]. In these references, REACH was called "GAP" and REACH$_d$ was called "1GAP".

The complexity of REACH$_u$ is also a rich subject. Aleliunas, Karp, Lipton, Lovász, and Rackoff, proved that taking a random walk in an undirected graph will — with very high probability — quickly reach all reachable vertices [AKL79]. It follows that boolean query REACH$_u$ is computable in "random logspace". Lewis and Papadimitriou define a restriction of nondeterministic space called "symmetric space" and prove that REACH$_u$ is complete via logspace reductions for symmetric logspace [LP82].

REACH$_a$ was shown to be complete for P via logspace reductions in [I80] and via first-order reductions (and in fact quantifier-free projections) in [I83].

Exercise 3.28 shows that the monotone, circuit value problem is complete for P via first-order reductions. The completeness of CVP for P via logspace reductions was first shown by Ladner in [L75]. The completeness of MCVP via logspace reductions was originally shown by Goldschlager in [Go77].

4
Inductive Definitions

First-order logic is not rich enough to express most interesting computations. A useful and natural way to increase the expressive power of first-order logic is to add the power to define new relations by induction. In this chapter we formalize the notion of inductive definitions via the least fixed point operator (LFP). We prove that first-order logic extended by the least fixed point operator captures exactly polynomial-time.

4.1 Least Fixed Point

A useful way to increase the power of first-order logic without jumping all the way up to second order logic is to add the power to define new relations by induction. Many such inductively defined relations are not first-order expressible.

One useful example of a relation that is not first-order expressible, but can be defined inductively, is transitive closure. Recall the vocabulary $\tau_g = \langle E^2, s, t \rangle$ of graphs. We can define the reflexive, transitive closure E^* of E as follows. Let R be a binary relation variable and consider the formula,

$$\varphi_{4.1}(R, x, y) \equiv x = y \vee \exists z(E(x, z) \wedge R(z, y)) \qquad (4.1)$$

The formula $\varphi_{4.1}$ formalizes an inductive definition of E^*. This may be more suggestively written as follows,

$$E^*(x, y) \equiv x = y \vee \exists z(E(x, z) \wedge E^*(z, y)) \qquad (4.2)$$

For any structure $\mathcal{A}$ with vocabulary τ_g, $\varphi_{4.1}$ induces a map from binary relations on the universe of $\mathcal{A}$ to binary relations on the universe of $\mathcal{A}$,

4. Inductive Definitions

$$\varphi_{4.1}^{\mathcal{A}}(R) = \{\langle a, b\rangle \mid \mathcal{A} \models \varphi_{4.1}(R, a, b)\}$$

Such a map is called *monotone* if for all R, S,

$$R \subseteq S \Rightarrow \varphi^{\mathcal{A}}(R) \subseteq \varphi^{\mathcal{A}}(S).$$

The relation symbol R appears only positively in $\varphi_{4.1}$, i.e., within an even number of negation symbols. It follows that $\varphi_{4.1}^{\mathcal{A}}$ is monotone. Let $(\varphi_{4.1}^{\mathcal{A}})^r$ denote $\varphi_{4.1}^{\mathcal{A}}$ iterated r times. With $\varphi_{4.1}$ defined as in Equation (4.1), $\mathcal{A}$ any graph, and $r \geq 0$, observe that,

$$(\varphi_{4.1}^{\mathcal{A}})(\emptyset) = \{\langle a, b\rangle \in |\mathcal{A}|^2 \mid \text{distance}(a, b) \leq 0\}$$
$$(\varphi_{4.1}^{\mathcal{A}})^2(\emptyset) = \{\langle a, b\rangle \in |\mathcal{A}|^2 \mid \text{distance}(a, b) \leq 1\}$$

and in general,

$$(\varphi_{4.1}^{\mathcal{A}})^r(\emptyset) = \{\langle a, b\rangle \in |\mathcal{A}|^2 \mid \text{distance}(a, b) \leq r - 1\}$$

Thus, for $n = \|\mathcal{A}\|$, then $(\varphi_{4.1}^{\mathcal{A}})^n(\emptyset) = E^*$ = the least fixed point of $\varphi_{4.1}^{\mathcal{A}}$, i.e., the minimal relation T such that $\varphi_{4.1}^{\mathcal{A}}(T) = T$. This is a general situation, as we now show in the finite version of the Knaster-Tarski Theorem.

Theorem 4.3. *Let R be a new relation symbol of arity k, and let $\varphi(R, x_1, \ldots, x_k)$ be a monotone first-order formula. Then for any finite structure $\mathcal{A}$, the least fixed point of $\varphi^{\mathcal{A}}$ exists. It is equal to $(\varphi^{\mathcal{A}})^r(\emptyset)$ where r is minimal so that $(\varphi^{\mathcal{A}})^r(\emptyset) = (\varphi^{\mathcal{A}})^{r+1}(\emptyset)$. Furthermore, letting $n = \|\mathcal{A}\|$, we have $r \leq n^k$.*

Proof Consider the sequence

$$\emptyset \subseteq (\varphi^{\mathcal{A}})(\emptyset) \subseteq (\varphi^{\mathcal{A}})^2(\emptyset) \subseteq (\varphi^{\mathcal{A}})^3(\emptyset) \subseteq \cdots \qquad (4.4)$$

The containment follows because $\varphi^{\mathcal{A}}$ is monotone. If $(\varphi^{\mathcal{A}})^{i+1}(\emptyset)$ strictly contains $(\varphi^{\mathcal{A}})^i(\emptyset)$, then it must contain at least one new k-tuple from $|\mathcal{A}|$. Since there are at most n^k such k-tuples, for some $r \leq n^k$, $(\varphi^{\mathcal{A}})^r(\emptyset) = (\varphi^{\mathcal{A}})^{r+1}(\emptyset)$, i.e, $(\varphi^{\mathcal{A}})^r(\emptyset)$ is a fixed point of $\varphi^{\mathcal{A}}$.

Let S be any other fixed point of $\varphi^{\mathcal{A}}$. We show by induction that $(\varphi^{\mathcal{A}})^i(\emptyset) \subseteq S$ for all i. The base case is that,

$$(\varphi^{\mathcal{A}})^0(\emptyset) = \emptyset \subseteq S.$$

Inductively, suppose that $(\varphi^{\mathcal{A}})^i(\emptyset) \subseteq S$. Since $\varphi^{\mathcal{A}}$ is monotone,

$$(\varphi^{\mathcal{A}})^{i+1}(\emptyset) = \varphi^{\mathcal{A}}((\varphi^{\mathcal{A}})^i(\emptyset)) \subseteq \varphi^{\mathcal{A}}(S) = S.$$

Thus, $(\varphi^{\mathcal{A}})^r(\emptyset) \subseteq S$ and $(\varphi^{\mathcal{A}})^r(\emptyset)$ is the least fixed point of $\varphi^{\mathcal{A}}$ as claimed. □

If R occurs only positively in φ, i.e., within an even number of negation symbols, then φ is monotone. Theorem 4.3 tells us that any R-positive formula $\varphi(R^k, x_1, \ldots, x_k)$ determines a least fixed point relation. We write $(\text{LFP}_{R^k x_1 \ldots x_k} \varphi)$ to denote this least fixed point. The least fixed point operator (LFP) thus formalizes the definition of new relations by induction. The subscript "$R^k x_1 \ldots x_k$" explicitly tells us which relation and domain variables we are taking the fixed point with respect to. When the choice of variables is clear, these subscripts may be omitted. As an example, (LFP$_{Rxy}\varphi_{4.1}$) denotes the reflexive, transitive closure of the edge relation E. Thus boolean query REACH is expressible as:

$$\text{REACH} \equiv (\text{LFP}_{Rxy}\varphi_{4.1})(s,t)$$

Definition 4.5 Define FO(LFP), the language of first-order inductive definitions, by adding a least fixed point operator (LFP) to first-order logic. If $\varphi(R^k, x_1, \ldots, x_k)$ is an R^k-positive formula in FO(LFP), then (LFP$_{R^k x_1 \ldots x_k} \varphi$) may be used as a new k-ary relation symbol denoting the least fixed point of φ. □

Example 4.6 In Definition 3.24 we gave an inductive definition of alternating paths. We then defined boolean query REACH$_a$ to be the set of graphs having an alternating path from s to t. (See Figure 4.8 which shows a graph that satisfies REACH$_a$.)

We now give a first-order inductive definition of the alternating path property, P_a,

$$\varphi_{ap}(P, x, y) \equiv x = y \vee [(\exists z)(E(x,z) \wedge P(z,y)) \wedge$$
$$(A(x) \rightarrow (\forall z)(E(x,z) \rightarrow P(z,y)))]$$

Thus,

$$P_a = (\text{LFP}_{Pxy}\varphi_{ap}) \quad \text{and} \quad \text{REACH}_a = (\text{LFP}_{Pxy}\varphi_{ap})(s,t) \qquad \square$$

Recall that REACH$_a$ is complete for P via first-order reductions, Theorem 3.26. It follows from Example 4.6 and from the following exercise that FO(LFP) contains all the polynomial-time boolean queries.

Exercise 4.7 Show that FO(LFP) is closed under first-order reductions.

Hint: this is a special case of Meta-Proposition 3.11. You have to show that if Q is a k-ary first-order query and $\Phi \in$ FO(LFP), then $\widehat{Q}(\Phi) \in$ FO(LFP). This is clear once you observe that

$$\widehat{Q}(\text{LFP}_{R^a, x_1, \ldots, x_a} \alpha) \equiv (\text{LFP}_{R^{ka}, x_1^1 \ldots x_1^k, \ldots, x_a^1 \ldots x_a^k} \widehat{Q}(\alpha)) \qquad \square$$

Now that we have formalized inductive definitions, we feel free to write them in the intuitive form of Equation (4.2) rather than as (LFP$_{Rx,y} \varphi_{4.1}$). It is often convenient to define several relations by simultaneous induction. The following exercise shows that there is no harm in doing this.

Exercise 4.9 Suppose $\psi(\bar{y}, S, T)$ and $\varphi(\bar{x}, S, T)$ are first order formulas that are positive in S and T. Let $r_0 = \text{arity}(S) = |\bar{y}|$ and $r_1 = \text{arity}(T) = |\bar{x}|$. For any structure $\mathcal{A}$ define the relations I_0^ω and I_1^ω by simultaneous induction:

4. Inductive Definitions

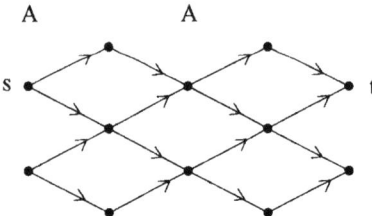

Figure 4.8: A graph satisfying the boolean query REACH_a.

$$I_0^0 = I_1^0 = \emptyset$$
$$\bar{a} \in I_0^n \Leftrightarrow \mathcal{A} \models \psi(\bar{a}, I_0^{n-1}, I_1^{n-1})$$
$$\bar{b} \in I_1^n \Leftrightarrow \mathcal{A} \models \varphi(\bar{b}, I_0^{n-1}, I_1^{n-1})$$
$$I_b^\omega = \bigcup_{n=1}^{\infty} I_b^n, \quad b = 0, 1$$

Show that both I_0^ω and I_1^ω are expressible in FO(LFP).

Hint: Assume that there are distinct constants $c_0 \neq c_1$.[1] Assume that $r_0 \leq r_1$. Define a single new relation U of arity $1 + r_1$ so that $U(c_0, \bar{y}, \bar{u})$ refers to $S(\bar{y})$, with $\bar{u}$ an $(r_1 - r_0)$-tuple of dummy variables and $U(c_1, \bar{x})$ refers to $T(\bar{x})$. □

We next show that FO(LFP) — the closure of first-order logic under the power to inductively define new relations — describes exactly the set of all polynomial-time computable boolean queries.

Theorem 4.10. *Over finite, ordered structures,*

$$\text{FO(LFP)} = \text{P}$$

Proof ($\subseteq$): Let $\mathcal{A}$ be an input structure, let $n = \|\mathcal{A}\|$, and let $(\text{LFP}_{Rx_1...x_k} \varphi)$ be a fixed-point formula. By Theorem 4.3, we know that this fixed point evaluated on $\mathcal{A}$ is $(\varphi^{\mathcal{A}})^{n^k}(\emptyset)$. This amounts to evaluating the first-order query φ at most n^k times. We have seen in Theorem 3.1 that first-order queries may be evaluated in L, thus easily in P.

($\supseteq$): Since FO(LFP) includes query REACH_a, which is complete for P via first-order reductions, and FO(LFP) is closed under first-order reductions, FO(LFP) includes all polynomial-time queries. □

Theorem 4.10 equates polynomial time — one of the most important complexity classes — with FO(LFP) — the closure of first-order logic under the power to make inductive definitions. The latter is very natural from a descriptive point of view.

---

[1] For example, 0 and *max* would do for ordered structures of size greater than one. If no constants are available, then one can quantify two distinct elements u, v and use u in place of c_0 and v in place of c_1. If the universe has size only one, then any formula is trivial, and that case can be dealt with separately; see Proviso 1.15.

This theorem thus increases our intuition that polynomial time is a class whose fundamental nature goes beyond the machine models with which it is usually defined.

The use of ordering in Theorem 4.10 is required in the proof that $REACH_a$ is complete via $\leq_{fo}$. Stripped of its numeric relations including ordering, FO(LFP) does not describe all polynomial-time properties. For example, we will see that it cannot even express the parity of its universe. (The rôle of ordering is described extensively in Chapter 12.)

To conclude this section we note that the above proof leads to the following normal form theorem for language FO(LFP) over ordered structures. The language FO(LFP) allows the application of a complicated series of nested fixed points including extra quantifications and negations. The normal form theorem says that any such formula is equivalent to a single fixed point applied to a first-order formula. Such a normal form theorem makes it easier to understand exactly what can be expressed in language FO(LFP). The same result holds without ordering, but the proof is more subtle (Theorem 9.6).

Corollary 4.11. *Let φ be any formula in language* FO(LFP). *There exists a first-order formula ψ and a tuple of constants $\bar{c}$ such that over finite, ordered structures,*

$$\varphi \equiv (\text{LFP } \psi)(\bar{c})$$

Proof The completeness of $REACH_a$ for P means that every polynomial-time query is expressible as $\widehat{Q}(REACH_a)$ for some first-order query Q. Now, in Example 4.6 we saw that,

$$REACH_a = (\text{LFP}\varphi_{ap})(s, t)$$

Thus, an arbitrary polynomial-time query is expressible as,

$$\widehat{Q}(REACH_a) = (\text{LFP } \widehat{Q}(\varphi_{ap}))\widehat{Q}(s, t) \qquad (4.12)$$

Since, the first-order reductions used in Theorem 3.26 replace the constants s and t by k-tuples of the constants 0 and *max* — see Equation (3.19) — the form of Equation (4.12) is as desired. □

4.2 The Depth of Inductive Definitions

The number of iterations until an inductive definition closes is called its *depth*[2]. We will see that inductive depth is an important complexity measure, corresponding to parallel time (Theorem 5.2).

---

[2]In the logic literature where structures are usually infinite this is called the *closure ordinal*.

4. Inductive Definitions

Definition 4.13 Let $\varphi(R, x_1, \ldots, x_k)$ be an R-positive formula, where R is a relation symbol of arity k, and let $\mathcal{A}$ be a structure of size n. Define the *depth* of φ in $\mathcal{A}$, in symbols $|\varphi^{\mathcal{A}}|$, to be the minimum r such that

$$\mathcal{A} \models \left(\varphi^r(\emptyset) \leftrightarrow \varphi^{r+1}(\emptyset) \right)$$

As we saw in the proof of Theorem 4.3, $|\varphi^{\mathcal{A}}| \leq n^k$. Define the depth of φ as a function of n equal to the maximum depth of φ in $\mathcal{A}$ for any structure $\mathcal{A}$ of size n:

$$|\varphi|(n) = \max_{|\mathcal{A}|=n} \{|\varphi^{\mathcal{A}}|\} \qquad \square$$

Remark 4.14. The inductive definition $\varphi_{4.1}$ given in Equation (4.1) has depth $|\varphi|(n) = n$. However, the following alternate inductive definition of E^* has depth $|\varphi_{tc}|(n) = \lceil \log n \rceil + 1$.

$$\varphi_{tc}(R, x, y) \equiv x = y \lor E(x, y) \lor \exists z (R(x, z) \land R(z, y)) \qquad (4.15)$$

In computer science, the depth of an inductive definition corresponds to the depth of the stack needed to evaluate a recursive definition, that is, the depth of nesting of recursive calls. We will see in Theorem 5.2 that this also corresponds to the parallel time needed to evaluate such a recursive definition.

Definition 4.16 Let $\text{IND}[f(n)]$ be the sublanguage of FO(LFP) in which only fixed points of first-order formulas φ for which $|\varphi|$ is $O[f(n)]$ are included. For example, REACH is expressible as $(\text{LFP}_{Rxy}\,\varphi_{tc})$ and is thus in $\text{IND}[\log n]$. Note also that,

$$\text{FO(LFP)} = \bigcup_{k=1}^{\infty} \text{IND}[n^k] \; . \qquad \square$$

The facts that REACH $\in \text{IND}[\log n]$, REACH is complete for NL via first-order reductions (Theorem 3.16), and $\text{IND}[\log n]$ is closed under first-order reductions imply

Proposition 4.17. $\text{NL} \subseteq \text{IND}[\log n]$

We will see in Theorem 5.22 that $\text{IND}[\log n]$ is equal to AC^1, a well-studied complexity class. As another example, the inductive definition of REACH_a in Example 4.6 has depth equal to the length of the longest path from s to t in the graph.

In the following exercise you are asked to show that the numeric relations BIT, PLUS, and $\times$, are all definable in $\text{IND}(wo\text{BIT})[\log n]$, that is, via first-order inductive definitions that use only the numeric relation $\leq$. This shows that the descriptive class $\text{IND}[\log n]$ is somewhat more robust than $\text{IND}[0] = \text{FO}$ and has a more general definition. This is a general pattern: the more powerful the language, the less important exactly which numeric relations are included.

Exercise 4.18 Show that $\text{BIT} \in \text{IND}(wo\text{BIT})[\log n]$, i.e., relation BIT is definable by a depth $\log n$ induction just from $\leq$.

[Hint: by successive inductive definitions, show that PLUS and then MULT are definable in IND(woBIT)[$\log n$]. The result will then follow by Theorem 1.17.] □

Exercise 4.19 Recall that PARITY $\subset$ STRUC[τ_s] is the set of binary strings with an odd number of 1's (Example 2.12).

1. Show that PARITY $\in$ IND[$\log n$].
2. For a more challenging problem, show that PARITY $\in$ IND[$\log n / \log \log n$]. This requires BIT. [Hint: divide the n bit input string into $\lfloor \log n \rfloor$ pieces. The string has odd parity iff an odd number of the pieces have odd parity. You may use Lemma 1.18 for the fact that the parity of a $\log n$ bit string is first-order.]

We see in Chapter 6 that PARITY $\notin$ IND(woBIT)[$o(\log n)$]. We also see in Corollary 13.8 that with BIT, [$\log n / \log \log n$] is optimal. □

4.3 Iterating First-Order Formulas

Theorem 4.3 shows that for any first-order inductive definition $\varphi(R, x_1, \ldots, x_k)$, the least fixed point of φ amounts to iterating φ at most n^k times. Thus, in some sense, LFP is a polynomial iteration operator. This is even more apparent when we put the inductive definitions into the following normal form. Then the effect of the least-fixed-point operator is to iterate a certain block of quantifiers a polynomial number of times. (Recall that the notation $(\forall x . M)\psi$ means $(\forall x) M \rightarrow \psi$ and that $(\exists x . M)\psi$ means $(\exists x) M \wedge \psi$.)

Lemma 4.20. *Let φ be an R-positive first-order formula. Then φ can be written in the following form,*

$$\varphi(R, x_1, \ldots, x_k) \equiv (Q_1 z_1 . M_1) \ldots (Q_s z_s . M_s)(\exists x_1 \ldots x_k . M_{s+1}) R(x_1, \ldots, x_k) . \tag{4.21}$$

where the M_i's are quantifier-free formulas in which R does not occur.

Proof By induction on the complexity of φ. We assume that all negations have been pushed all the way inside. There are two base cases: If $\varphi \equiv R(v_1, \ldots, v_k)$, then,

$$\varphi \equiv (\exists z_1, \ldots, z_k . M_1)(\exists x_1, \ldots, x_k . M_2) R(x_1, \ldots, x_k)$$
$$M_1 \equiv z_1 = v_1 \wedge \cdots \wedge z_k = v_k$$
$$M_2 \equiv x_1 = z_1 \wedge \cdots \wedge x_k = z_k$$

If φ is quantifier free and R does not occur in φ, then,

$$\varphi \equiv (\forall z . \neg \varphi)(\exists x_1, \ldots, x_k . x_1 \neq x_1) R(x_1, \ldots, x_k) .$$

In the inductive cases $\varphi = (\exists v)\psi$ and $\varphi = (\forall v)\psi$, we simply put the new quantifier $(\exists v)$ in front of the quantifier block for ψ.

The remaining cases for $\wedge$ and $\vee$ are similar to each other. Suppose that $\varphi = \alpha \wedge \beta$ and

$$\alpha \equiv (Q_1 y_1.N_1)\ldots(Q_t y_t.N_t)(\exists x_1\ldots x_k.N_{t+1})R(x_1,\ldots,x_k)$$
$$\beta \equiv (Q_1 z_1.M_1)\ldots(Q_s z_s.M_s)(\exists x_1\ldots x_k.M_{s+1})R(x_1,\ldots,x_k)$$

where we may assume that the y's and z's are disjoint. Let

$$\mathrm{QB}_1 \equiv (Q_1 y_1.N_1')\ldots(Q_t y_t.N_t'),$$
$$\mathrm{QB}_2 \equiv (Q_1 z_1.M_1')\ldots(Q_s z_s.M_s').$$

where,

$$N_i' \equiv N_i \vee b = 1; \quad \text{and } M_i' \equiv M_i \vee b = 0$$

Let $\psi(\bar{u}/\bar{x})$ denote the formula ψ with variables $u_1,\ldots,u_k$ substituted for $x_1,\ldots,x_k$ and define the quantifier-free formulas,

$$S \equiv (b = 0 \wedge N_{t+1}(\bar{u}/\bar{x})) \vee (b = 1 \wedge M_{s+1}(\bar{u}/\bar{x})),$$
$$T \equiv (u_1 = x_1 \wedge \ldots \wedge u_k = x_k).$$

Recall that bool(b) means that $b = 0$ or $b = 1$, that is, b is a boolean variable (Definition 1.16). We can now write φ in the desired form,

$$\varphi \equiv (\forall b.\mathrm{bool}(b))(\mathrm{QB}_1)(\mathrm{QB}_2)(\exists \bar{u}.S)(\exists \bar{x}.T)R(x_1,\ldots,x_k) \qquad \square$$

Note that in Equation (4.21), the requantification of the x_i's means that these variables may occur free in $M_1\ldots M_s$, but they are bound in M_{s+1} and $R(x_1,\ldots,x_k)$. The same variables may now be requantified. Let us write QB to denote the quantifier block $(Q_1 z_1.M_1)\ldots(Q_s z_s.M_s)(\exists x_1\ldots x_k.M_{s+1})$. Thus, in particular, for any structure $\mathcal{A}$, and any $r \in \mathbf{N}$,

$$\mathcal{A} \models ((\varphi^{\mathcal{A}})^r(\emptyset)) \leftrightarrow ([\mathrm{QB}]^r \text{ false}).$$

Here $[\mathrm{QB}]^r$ means QB literally repeated r times. It follows immediately that if $t = |\varphi|(n)$ and $\mathcal{A}$ is any structure of size n then

$$\mathcal{A} \models (\mathrm{LFP}\,\varphi) \leftrightarrow ([\mathrm{QB}]^t \text{ false}).$$

Example 4.22 We show how to write the inductive definition of transitive closure in the normal form of Lemma 4.20.

Recall the definition of transitive closure from Equation (4.15),

$$\varphi_{tc}(R, x, y) \equiv x = y \vee E(x, y) \vee (\exists z)(R(x, z) \wedge R(z, y))$$

First, code the base case using a dummy universal quantification,

$$\varphi_{tc}(R, x, y) \equiv (\forall z.M_1)(\exists z)(R(x, z) \wedge R(z, y))$$
$$M_1 \equiv \neg(x = y \vee E(x, y))$$

Note that there are no free occurrences of z within the scope of the $(\forall z.M_1)$ quantifier. Next, use universal quantification to replace the two occurrences of R

with a single one:

$$\varphi_{tc}(R, x, y) \equiv (\forall z.M_1)(\exists z)(\forall uv.M_2)R(u, v)$$
$$M_2 \equiv (u = x \wedge v = z) \vee (u = z \wedge v = y).$$

Finally, requantify x and y. We have transformed Equation (4.15), into the normal form of Lemma 4.20,

$$M_3 \equiv (x = u \wedge y = v)$$

$$\varphi_{tc}(R, x, y) \equiv (\forall z.M_1)(\exists z)(\forall uv.M_2)(\exists xy.M_3)R(x, y) \qquad (4.23)$$

□

Define the quantifier block,

$$QB_{tc} \equiv (\forall z.M_1)(\exists z)(\forall uv.M_2)(\forall xy.M_3).$$

Equation (4.23) tells us that an application of the operator φ_{tc} corresponds exactly to the writing of QB_{tc},

$$\varphi_{tc}(R, x, y) \equiv [QB_{tc}]R(x, y)$$

It follows that for any r,

$$\varphi_{tc}^r(\emptyset) \equiv [QB_{tc}]^r(\text{false}).$$

We have thus demonstrated a syntactic uniformity for the inductive definition of REACH. For any structure $\mathcal{A} \in \text{STRUC}[\tau_g]$,

$$\mathcal{A} \in \text{REACH} \Leftrightarrow \mathcal{A} \models (\text{LFP}\varphi_{tc})(s, t)$$
$$\Leftrightarrow \mathcal{A} \models \left([QB_{tc}]^{\lceil 1+\log |\mathcal{A}|\rceil}\right)\text{false}(s/x, t/y)$$

We now define $\text{FO}[t(n)]$ to be the set of properties defined by quantifier blocks iterated $t(n)$ times. (This is the same as being iterated $O(t)$ times since a quantifier block may be any constant size.) Even though such expressions grow as a function of the size of their inputs, they use only the variables in the quantifier block. That is, the number of variables is a fixed constant independent of the size of the input.

Definition 4.24 A set $S \subseteq \text{STRUC}[\tau]$ is a member of $\text{FO}[t(n)]$ iff there exist quantifier free formulas M_i, $0 \leq i \leq k$, from $\mathcal{L}(\tau)$, a tuple $\bar{c}$ of constants and a quantifier block,

$$QB = [(Q_1x_1.M_1)\ldots(Q_kx_k.M_k)]$$

such that for all $\mathcal{A} \in \text{STRUC}[\tau]$,

$$\mathcal{A} \in S \Leftrightarrow \mathcal{A} \models \left([QB]^{t(|\mathcal{A}|)}M_0\right)(\bar{c}/\bar{x})$$

□

The reason for the substitution of constants is that the quantifier block QB may contain some free variables that must be substituted for to build a sentence. See Example 4.22 which shows that REACH $\in$ FO[$\log n$].

Combining Lemma 4.20 and Definition 4.24, we see that,

Lemma 4.25. *For all $t(n)$ and all classes of finite structures,*

$$\text{IND}[t(n)] \subseteq \text{FO}[t(n)] \;.$$

A converse of Lemma 4.25 also holds, but we put off its proof until the next chapter.

Exercise 4.26 Write an inductive definition showing that CVP, the circuit value problem (Definition 2.27) is describable in FO(LFP). The depth of your inductive definition should be equal to the depth of the circuit, i.e., the length of the longest path from root to leaf. □

Exercise 4.27 Write a sentence in FO(wo$\leq$)(LFP) meaning that a graph is two-colorable, i.e., each vertex may be colored red or blue in such a way that no two adjacent vertices are the same color. [Hint: one way to do this is to first simultaneously define the relations OPath(x, y) and EPath(x, y) meaning that there is a path of odd length, respectively even length, from x to y. A graph is two-colorable iff it has no cycles of odd length.] □

Historical Notes and Suggestions for Further Reading

Moschovakis has written a thorough and excellent book on inductive definitions, [Mos74]. Although its main focus is infinite structures, our treatment of inductive definitions follows the approach set out in that book. In particular, Exercise 4.9 and Lemma 4.20 are from [Mos74].

The Knaster-Tarski Theorem (Theorem 4.3) originally appeared in [Tar55]. Ajtai and Gurevich proved that over finite structures, not every monotone formula has an equivalent positive formula [AG87]. Theorem 4.10 is due to Vardi and Immerman [I82] and [Var82].

5
Parallelism

Descriptive complexity is inherently parallel in nature. This is a particularly delightful dividend of applying this form of logic to computer science. The time to compute a query on a certain parallel computer corresponds exactly to the depth of a first-order induction needed to describe the query. There is also a close relationship between the amount of hardware used — memory and processors — and the number of variables in the inductive definition.

Quantification is a parallel operation. The query $(\exists x) S(x)$ can be executed using n processors in constant parallel time. Processor p_i checks whether $S(i)$ holds for $i = 0, 1, \ldots, n - 1$. Any p_i for which $S(i)$ does hold should write a one into a specified location in global memory that was originally zero.

The real world is inherently parallel. There are many atoms, molecules, cells, organisms, computers, factories, towns, countries, all working on their own. For a long time, however, computers were sequential devices having a single processor and thus executing one instruction at a time. Over recent years, we have vastly increased our ability to produce small, fast, inexpensive processors. It is thus possible to build large parallel computers as well as a large number of personal computers that can interact. One of the fundamental problems in computer science is how to make efficient and effective use of this dramatic proliferation of computing power, including many processors that we may use at once.

As researchers and practitioners struggle with the question of how to use many processors at once, numerous models of parallel computation have been developed. The main axis along which these models vary is how tightly coupled the processors are. One extreme is the parallel random access machine (PRAM) in which the inter-connection pattern is essentially a complete graph. In this model, a word of memory can be sent from any processor to any other processor in the time it takes to perform a single instruction. The other extreme is distributed computation, in which many personal computers or work stations are connected via a network, which might be fairly fast and local — or it might be the Internet.

Both of these models are important, and neither is well enough understood. For general applications, it is still very difficult to effectively use a tightly coupled parallel computer or a distributed network of computers and gain a large speed up compared to doing the computation at a single uni-processor. Of course there have been great successes. Some problems, such as linear algebra, are very easy to parallelize. Other problems may be inherently sequential, but this remains to be seen.

In this chapter we study highly coupled parallelism, that is, parallelism as on a PRAM. We show that this model corresponds very closely and nicely with descriptive complexity. We see in particular that the optimal depth of inductive definitions of a query corresponds exactly to the optimal parallel time needed to compute the query on a PRAM. There is also a close relationship between the number of processors needed by the PRAM and the number of variables used in the inductive definition.

This connection between parallelism and descriptive complexity sheds a great deal of light on the parallel nature of computation. It is very natural to describe our queries via inductive definitions. It is enlightening that inductive depth corresponds exactly to parallel time.

Later in this chapter we also study other models of parallel computation namely circuit complexity and alternating Turing machines. We tie these other two parallel models together with PRAMs and descriptive complexity.

The net result of this approach is that we can see what is basic about parallel computation and what is merely model dependent; and, if we choose, we can understand the main issues via the quantifier depth and the number of variables needed for first-order descriptions of the queries of interest.

5.1 Concurrent Random Access Machines

In this section we define a precise model of PRAMs called Concurrent Random Access Machines. This model is *synchronous*, that is the processors work in lock step, and it is *concurrent*, that is, several processors may read from the same location at the same time step and several processors may try to write the same location at the same time step.

A *concurrent random access machine* (CRAM) consists of a large number of processors, all connected to a common, global memory. See Figure 5.1. The processors are identical except that they each contain a unique processor number. At each step, any number of processors may read or write any word of global memory. If several processors try to write the same word at the same time, then the lowest numbered processor succeeds.

This is the "priority write" model. The results in this chapter remain true if instead we use the "common write" model, in which the program guarantees that different values will never be written to the same location at the same time. The common write model is the more natural model for logic: a formula such as $(\forall x)\varphi$ specifies a parallel program using n processors — one for each possible value of x. Any processor finding that φ is false for its value of x will write a zero into a location in global memory that was initially one. See Corollary 5.8.

The CRAM is a special case of the concurrent read, concurrent write, parallel random access machine (CRCW-PRAM). We now describe the precise instruction set for the CRAM. This model will be familiar to anyone who has ever programmed a computer using assembly language, and it may seem strange to anyone who has not. Whether or not it seems strange, we must describe our model in a level that is detailed enough so that we may prove its equivalence to the descriptive model. Once we have done this, we can from then on write our parallel programs using first-order logic.

In addition to assignments, the CRAM instruction set includes addition, subtraction, and branch-on-less. The CRAM instruction set also includes a Shift instruction. Shift(x, y) causes the word x to be shifted y bits to the right.

Some of our choices for the instruction set correspond to the choices we have made earlier concerning which numeric relations to include in our first-order language. The Shift operation for the CRAM allows each bit of global memory is available to every processor in constant time. Without Shift, CRAM$[t(n)]$ would be too weak to simulate FO$[t(n)]$ for $t(n) < \log n$. It was in working to prove Theorem 5.2 that we originally realized that BIT should be added as a numeric predicate to first-order logic. Without BIT, a first-order formula cannot access the individual bits in a given variable. In more powerful logics such as FO(LFP), or even IND[$\log n$], BIT is definable from ordering and does not need to be explicitly added (Exercise 4.18).

Each processor has a finite set of registers, including the following, Processor: containing the number between 1 and $p(n)$ of the processor; Address: containing an address of global memory; Contents: containing a word to be written or read from global memory; and ProgramCounter: containing the line number of the instruction to be executed next. The instructions of a CRAM consist of the following:

READ: Read the word of global memory specified by Address into Contents.
WRITE: Write the Contents register into the global memory location specified by Address.
OP R_a R_b: Perform OP on R_a and R_b and leave the result in R_b. Here OP may be Add, Subtract, or, Shift.
MOVE R_a R_b: Move R_a to R_b.
BLT R L: Branch to line L if the contents of R is less than zero.

The above instructions each increment the ProgramCounter, with the exception of BLT which replaces it by L, when R is less than 0.

We assume initially that the contents of the first $|\text{bin}(\mathcal{A})|$ words of global memory contain one bit each of the input string bin$(\mathcal{A})$. We will see in Corollary 5.8 that

70   5. Parallelism

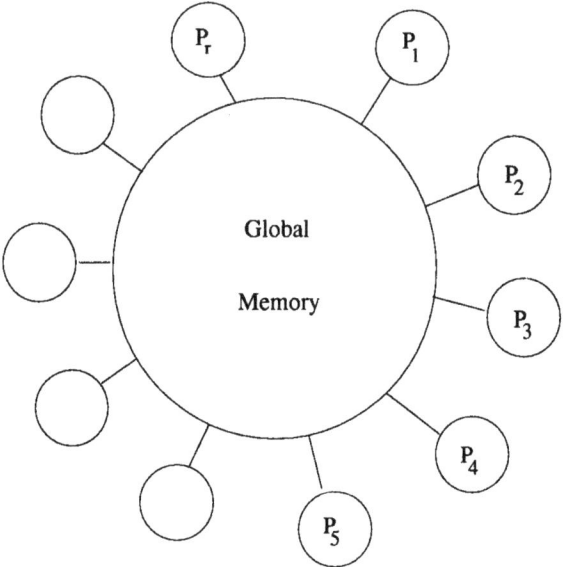

Figure 5.1: A concurrent random access machine (CRAM)

any other plausible setting of the input will work as well. We assume that a section of global memory is specified as the output. One of the bits of the output may serve as a flag indicating that the output is available.

Our measure of parallel time complexity will be time on a CRAM. Define CRAM[$t(n)$] to be the set of boolean queries computable in parallel time $t(n)$ on a CRAM that has at most polynomially many processors. When we want to measure how many processors are needed, we use the complexity classes CRAM-PROC[$t(n), p(n)$]. This is the set of boolean queries computable by a CRAM using at most $p(n)$ processors and time $O(t(n))$. Thus, CRAM[$t(n)$] = CRAM-PROC[$t(n), n^{O(1)}$].

We will see in Theorem 5.2 and Corollary 5.8 that the complexity class CRAM[$t(n)$] is quite robust. In particular, it is not affected by exactly how we place the input in the CRAM, by the size of the global memory word size, or by the size of the local registers, as long as these are both polynomially bounded.

5.2   Inductive Depth Equals Parallel Time

The following theorem says that parallel time is identical to inductive depth. In other words, a depth-optimal first-order inductive description of a query is a parallel-time-optimal algorithm to compute the query. The theorem also completes the circle and shows that number of quantifier-block iterations and inductive depth are equal.

5.2 Inductive Depth Equals Parallel Time

This is a very basic theorem, and its proof is not difficult. There are, however, many details to combine in showing that these rather different looking models are identical. For this reason, the proof is longer than it is deep.

Theorem 5.2. *Let S be a boolean query. For all polynomially bounded, parallel time constructible $t(n)$, the following are equivalent:*

1. S is computable by a CRAM in parallel time $t(n)$ using polynomially many processors and registers of polynomially bounded word size.

2. S is definable as a uniform first-order induction whose depth, for structures of size n, is at most $t(n)$.

3. There exists a first-order quantifier-block [QB], a quantifier-free formula M_0 and a tuple $\bar{c}$ of constants such that the query S for structures of size at most n is expressed as $[QB]^{t(n)} M_0(\bar{c}/\bar{x})$, i.e., the quantifier-block repeated $t(n)$ times followed by M_0.

In symbols, the equivalence of these three conditions can be written,

$$\text{CRAM}[t(n)] \;=\; \text{IND}[t(n)] \;=\; \text{FO}[t(n)]$$

Theorem 5.2 follows immediately from three containments: Lemmas 4.25, 5.3 and 5.4. The first of these — $\text{IND}[t(n)] \subseteq \text{FO}[t(n)]$ — has already been proved. We state and prove the remaining two now.

Lemma 5.3. *For any polynomially bounded $t(n)$ we have,*

$$\text{CRAM}[t(n)] \;\subseteq\; \text{IND}[t(n)]$$

Proof We want to simulate the computation of a CRAM M. On input $\mathcal{A}$, a structure of size n, M runs in $t(n)$ synchronous steps, using $p(n)$ processors, for some polynomial $p(n)$. Since the number of processors, the time and the memory word size are all polynomially bounded, we need only a constant number of variables $x_1, \ldots, x_k$, each ranging over the n element universe of $\mathcal{A}$, to name any bit in any register belonging to any processor at any step of the computation. We can thus define the contents of all the relevant registers for any processor of M by induction on the time step.

We now write a first-order inductive definition for the relation $\text{VALUE}(\bar{p}, \bar{t}, \bar{x}, r, b)$ meaning that bit $\bar{x}$ in register r of processor $\bar{p}$ just after step $\bar{t}$ is equal to b.

The base case is that if $\bar{t} = 0$, then memory is correctly loaded with $\text{bin}(\mathcal{A})$. This is first-order expressible (Exercise 2.3). We also need to say that the initial contents of each processor's register Processor is its processor number. This is easy, since we are given the processor number as the argument $\bar{p}$.

The inductive definition of the relation $\text{VALUE}(\bar{p}, \bar{t}, \bar{x}, r, b)$ is a disjunction depending on the value of $\bar{p}$'s program counter at time $\bar{t} - 1$. The most interesting case is when the instruction to be executed is READ. Here we simply find the most recent time $\bar{t}' < \bar{t}$ at which the word specified by $\bar{p}$'s register Address at time $\bar{t}$ was written into, and the lowest numbered processor $\bar{p}'$ that wrote into this address at time $\bar{t}'$. In this way we can access the answer, namely bit $\bar{x}$ of $\bar{p}'$s register Contents at time $\bar{t}'$. If there exists no such time $\bar{t}'$ then this memory location contains its input value. This is bit i of the input $\text{bin}(\mathcal{A})$ if $i < |\text{bin}(\mathcal{A})|$, and zero otherwise.

72   5. Parallelism

It remains to check that Addition, Subtraction, BLT, and Shift are first-order expressible. Addition was handled in Proposition 1.9. In a similar way we can express Subtraction and Less Than.

Relation BIT allows our first-order formulas to examine any of the $\log n$ bits of a domain variable. It follows that the addition relation on such variables is first-order expressible (Theorem 1.17). Using addition, we can specify the Shift operation.

Thus we have described an inductive definition of relation VALUE, coding M's entire computation. Furthermore, one iteration of the definition occurs for each step of M. □

Notice that the above proof is simple because first-order inductive definitions are very general and easy to use. Notice also that we did not write out the definition in its entirety. This would be a complete formal definition of the CRAM which we do not need. All that we needed to show is that the contents of all the bits of all the registers at time $t + 1$ is first-order definable from this same information at time t or earlier.

Lemma 5.4. *For polynomially bounded and parallel time constructible $t(n)$,*

$$\text{FO}[t(n)] \subseteq \text{CRAM}[t(n)]$$

Proof Let the FO[$t(n)$] problem be determined by the following quantifier free formulas, quantifier block, and tuple of constants,

$$M_0, M_1, \ldots, M_k, \quad \text{QB} = (Q_1 x_1.M_1)\ldots(Q_k x_k.M_k), \quad \bar{c}, .$$

Our CRAM must test whether an input structure $\mathcal{A}$ satisfies the sentence,

$$\varphi_n \equiv [\text{QB}]^{t(n)} M_0(\bar{c}/\bar{x})$$

where $n = \|\mathcal{A}\|$. The CRAM will use n^k processors and n^{k-1} bits of global memory. Note that each processor has a number $a_1 \ldots a_k$ with $0 \leq a_i < n$. Using the Shift operation it can retrieve each of the a_i's in constant time.[1]

The CRAM will evaluate φ_n from right to left, simultaneously for all values of the variables $x_1, \ldots, x_k$. At its final step, it will output the bit $\varphi_n(\bar{c}/\bar{x})$.

For $0 \leq r \leq t(n) \cdot k$, let,

$$\varphi^r \equiv (Q_i x_i.M_i)\ldots(Q_k x_k.M_k)[\text{QB}]^q M_0 \tag{5.5}$$

where $r = k \cdot (q + 1) + 1 - i$. Let $x_1 \ldots \hat{x}_i \ldots x_k$ be the $k - 1$-tuple resulting from $x_1 \ldots x_k$ by removing x_i. We will now give a program for the CRAM which is broken into rounds each consisting of three processor steps such that:

Just after round r, the contents of memory location $a_1 \ldots \widehat{a_i} \ldots a_k$ is 1 or 0 according as whether $\mathcal{A} \models \varphi^r(a_1, \ldots, a_k)$ or not. (5.6)

---

[1] This is obvious if n is a power of 2. If not, we can just let each processor break its processor number into $k \lceil \log n \rceil$-tuples of bits. If any of these is greater than or equal to n, then the processor should do nothing during the entire computation.

Note that x_i does not occur free in φ^r. At round r, processor number $a_1 \ldots a_k$ executes the following three instructions according to whether Q_i is $\exists$ or Q_i is $\forall$:

$\{Q_i$ is $\exists\}$

1. $b := loc(a_1 \ldots \hat{a}_{i+1} \ldots a_k);$
2. $loc(a_1 \ldots \hat{a}_i \ldots a_k) := 0;$
3. **if** $M_i(a_1, \ldots, a_k)$ and b **then** $loc(a_1 \ldots \hat{a}_i \ldots a_k) := 1;$

$\{Q_i$ is $\forall\}$

1. $b := loc(a_1 \ldots \hat{a}_{i+1} \ldots a_k);$
2. $loc(a_1 \ldots \hat{a}_i \ldots a_k) := 1;$
3. **if** $M_i(a_1, \ldots, a_k)$ and $\neg b$ **then** $loc(a_1 \ldots \hat{a}_i \ldots a_k) := 0;$

It is not hard to prove by induction that Equation (5.6) holds and thus that the CRAM simulates the formula. The bit fetched into b tells us whether $\mathcal{A}$ satisfies the formula,

$$\varphi^{r-1} \equiv (Q_{i+1}x_{i+1}.M_{i+1})\cdots[QB]^q M_0 .$$

The effect of lines 2 and 3 is that in parallel for all values of x_i, the truth of φ^r (Equation (5.5)) is tested and recorded. This completes the inductive step.

In the base case, at step 1, processor $(a_1 \ldots a_k)$ must set $b = 1$ iff $\mathcal{A} \models M_0(a_1, \ldots, a_k)$. Note that $M_0(x_1, \ldots, x_k)$ is a quantifier-free formula. Observe that in constant time using its processor number, the shift operation, and addition, processor $(a_1 \ldots a_k)$ can access the appropriate bits of $bin(\mathcal{A})$, for example the bit corresponding to $R_3(a_2, a_1)$, cf. Exercise 2.3. Furthermore, in constant time it can compute the boolean combination of these bits indicated by M_0. □

Remark 5.7. The proof of Lemma 5.4 provides a very simple network for simulating an FO[$t(n)$] property. The network has n^{k-1} bits of global memory and kn^k gates, where k is the number of distinct variables in the quantifier block. Each gate of the network is connected to two bits of global memory in a simple connection pattern. The blowup of processors going from CRAM to FO to CRAM seems large (cf. Corollary 5.10). However, it is plausible to build first-order networks with billions of processing elements, i.e. gates, thus accommodating fairly large n. It is crucial that k is kept small.

An immediate corollary of Theorem 5.2 is that the complexity class CRAM[$t(n)$] is not affected by minor changes in how the input is arranged, nor in the global memory word size, nor even by a change in the convention on how write conflicts are resolved.

Corollary 5.8. *For any function $t(n)$, the complexity class* CRAM[$t(n)$] *is not changed if the definition of a* CRAM *is modified in any consistent combination of the following ways. (By consistent, we mean that input words larger than the global word size or larger than the allowable length of applications of Shift are not allowed.)*

1. *Change the input distribution so that either*

   a. *The entire input is placed in the first word of global memory.*
   b. *The $I_\tau(n)$ bits of input are placed $\log n$ bits at a time in the first $I_\tau(n)/\log n$ words of global memory.*

2. *Change the global memory word size so that either*

   a. *The global word size is 1, i.e. words are single bits. (Local registers do not have this restriction so that the processor's number may be stored and manipulated.)*
   b. *The global word size is bounded by $O(\log n)$.*

3. *Modify the Shift operation so that shifts are limited to the maximum of the input word size and of the log base two of the number of processors.*
4. *Remove the polynomial bound on the number of memory locations, thus allowing an unbounded global memory.*
5. *Instead of the priority rule for the resolution of write conflicts, adopt the "common write" rule in which different processors never write different values into the same memory location at a given time step.*

Proof The proof is that Lemmas 5.3 and 5.4 still hold with any consistent set of these modifications. This is immediate for Lemma 5.3. For Lemma 5.4, we must only show that processor number $a_1 \ldots a_k$ still has the power to evaluate the quantifier free formula $M_i(a_1, \ldots, a_k)$ and to name the global memory location $a_1 \ldots \hat{a}_i \ldots a_k$, for $1 \leq i \leq k$, in constant time. Recall that we are assuming that the input structure $\mathcal{A} = \langle \{0, 1, \ldots, n-1\}, R_1^{\mathcal{A}} \ldots R_p^{\mathcal{A}}, c_1^{\mathcal{A}} \ldots c_q^{\mathcal{A}} \rangle$ is coded as a bit string of length $I_\tau(n) = n^{r_1} + \cdots + n^{r_p} + q\lceil \log n \rceil$. It is clear that all of the consistent modifications above allow processor $a_1 \ldots a_k$ to test in constant time whether the relation $R(t_1, \ldots, t_r)$ holds, where R is an input or numeric relation, and $t_j \in \{a_1, \ldots, a_k\} \cup \{c_j | 1 \leq j \leq q\}$. □

Exercise 5.9 Show that Corollary 4.11 holds for $\text{IND}[t(n)]$, for any $t(n)$. That is, show that any problem $S \in \text{IND}[t(n)]$ may be expressed as a single, depth $t(n)$ first-order induction. [Hint: Use Theorem 5.2.] □

5.3 Number of Variables Versus Number of Processors

We now show that the number of variables in an inductive definition determines the number of processors needed in the corresponding CRAM computation. The intuitive idea is that using k $\log n$-bit variables, we can name approximately n^k different parts of the CRAM. Thus, very roughly, k variables corresponds to n^k processors. The correspondence is not exact because the CRAM has a somewhat different pattern of interconnection between its processors and memory than the first-order inductive definition "model of parallelism". Later, we prove a tight relationship between number of variables and deterministic space (Theorem 10.16).

5.3 Number of Variables Versus Number of Processors

We now carefully analyze the proof of Theorem 5.2 to give processor-versus-variable bounds for translating between CRAM and IND. The proofs in this section consist of rather detailed variable counting. This whole section may be omitted.

Corollary 5.10. *Let* CRAM-PROC$[t(n), p(n)]$ *be the complexity class* CRAM$[t(n)]$ *restricted to machines using at most* $O(p(n))$ *processors. Let* IND-VAR$[t(n), v(n)]$ *be the complexity class* IND$[t(n)]$ *restricted to inductive definitions using at most* $v(n)$ *distinct variables. Assume for simplicity that the maximum size of a register word and* $t(n)$ *are both* $o[\sqrt{n}]$ *and that* $\pi \geq 1$ *is a natural number. Then,*

$$\text{CRAM-PROC}[t(n), n^\pi]$$
$$\subseteq \text{IND-VAR}[t(n), 2\pi + 2]$$
$$\subseteq \text{CRAM-PROC}[t(n), n^{2\pi+2}]$$

Proof We prove these bounds using the following two lemmas. For the first lemma, recall that Lemma 5.3 simulated a CRAM using an inductive definition. We inductively defined relation VALUE, which encoded the entire CRAM computation.

Lemma 5.11. *If the maximum size of a register word and of* $t(n)$ *are both* $o[\sqrt{n}]$, *and if M is a* CRAM-PROC$[t(n), n^\pi]$ *machine, then the inductive definition of* VALUE *may be written using* $2\pi + 2$ *variables.*

Proof We write out the inductive definition of VALUE in enough detail to count the number of variables used:

$$\text{VALUE}(\overline{p}, t, x, r, b) \equiv Z \vee W \vee S \vee R \vee M \vee B \vee A,$$

where the disjuncts have the following intuitive meanings:

Z: $t = 0$ and the initial value of r is correct.

W: $t \neq 0$, the instruction just executed is WRITE, and the value of r is correct, i.e., unchanged, unless r is Program-Counter.

S, R, M, B, A: Similarly for SHIFT, READ, MOVE, BLT, and, ADD or SUBTRACT, respectively.

It suffices to show that each disjunct can be written using the number of variables claimed. First we consider the disjunct Z. The only interesting part of Z is the case where r is "Processor". In this case we use relation BIT to say that $b = 1$ iff bit x of $\overline{p}$ is 1. No extra variables are needed. Note that the number of free variables in the relation is $\pi + 1$ because the values t, x, r, and b may be combined into a single variable.

Next we consider the case of Addition. Recall that the main work is to express the carry bit:

$$C[A, B](x) \equiv (\exists y < x)[A(y) \wedge B(y) \wedge (\forall z. y < z < x) A(z) \vee B(z)].$$

This definition uses two extra variables. Thus $\pi + 3 \leq 2\pi + 2$ variables certainly suffice. The cases S, M, and B are simpler.

The last and most interesting case is R. Here we must say,

1. The instruction just executed is READ,

76   5. Parallelism

2. Register r is register Contents,
3. There exists a processor $\overline{p'}$ and a time t' such that:
    a. $t' < t$,
    b. Address($\overline{p'}, t'$) = Address($\overline{p}, t$),
    c. VALUE($\overline{p'}, t', x, r, b$),
    d. Processor $\overline{p'}$ wrote at time t',
    e. For all $\overline{p''} < \overline{p'}$, if $\overline{p''}$ wrote at time t', then Address($\overline{p''}, t'$) $\neq$ Address($\overline{p'}, t'$),
    f. For all t'' such that $t' < t'' < t$ and for all $\overline{p''}$, if $\overline{p''}$ wrote at time t'', then Address($\overline{p''}, t''$) $\neq$ Address($\overline{p'}, t'$).

We count variables. On its face, this formula uses three $\overline{p}$'s and three t's. However, we show that two copies of each suffice. Observe that where we quantify $\overline{p''}$ in lines 3e and 3f, we no longer need $\overline{p}$, so we may use these variables instead.

The most subtle case is 3f. We use the fact that t is $o[\sqrt{n}]$, so t' and t'' can be coded into a single variable. We use a variable from $\bar{p}$ to encode t and t'. Then we can use t to universally quantify $t = \langle t', t'' \rangle$. Now we can universally quantify $\overline{p}$ to act as $\overline{p''}$. To say that Address($\overline{p''}, t''$) $\neq$ Address($\overline{p'}, t'$), we use the extra variable (t') to assert that there exists a bit position i and a bit b such that b is the bit at position i of Address($\overline{p''}, t''$), and $1 - b$ is the bit at position i of Address($\overline{p'}, t'$). To help with expressing the first conjunct, we may use a variable from $\overline{p'}$ and to help in the second conjunct, we may use a variable from $\overline{p}$.

Thus $2\pi + 2$ variables suffice as claimed.   □

The second lemma we need (Lemma 5.12) is a refinement of Lemma 5.4.

Lemma 5.12. *Let $\varphi(R, \bar{x})$ be an inductive definition of depth $d(n)$. Let k be the number of distinct variables, including $\bar{x}$, occurring in φ. Then the relation defined by φ is also computable in* CRAM-PROC$[d(n), O(n^k)]$.

Proof This is very similar to the proof of Lemma 5.4. Let T be the parse tree of φ. The CRAM will have $n^k |T|$ processors: one for each value of the k variables and each node in T. Let δ be the depth of T. As in the proof of Lemma 5.4, in rounds consisting of 3δ steps, the CRAM will evaluate an iteration of φ. Let r = arity(R) = the number of variables in $\bar{x}$; so $r \leq k$. The CRAM will have n^r bits of global memory to hold the truth value of $R_t = \varphi^t(\emptyset)$. It will use an additional $n^k |T|$ bits of memory to store the truth values corresponding to nodes of T. Thus $R_{d(n)}$, the least fixed point of φ, is computed in time $O(d(n))$ using $O(n^k)$ processors, as claimed.   □

This completes the proof of Corollary 5.10.   □

The above proofs give us some information concerning the efficiency of our simulation of CRAMs with first-order inductive definitions. After these results, the questions is, "Why is the number of variables needed to express a computation of n^π processors $2\pi + 2$, instead of π?" We discuss the multiplicative factor of two, and the additional two variables, respectively in the next two paragraphs.

We need the term 2π for two reasons: we must specify $\overline{p}$ and $\overline{p'}$ at the same time in order to say that their address registers are equal; and we need to say that no lower numbered processor $\overline{p''}$ wrote into the same address as $\overline{p'}$. This term points out a difference between the CRAM model and the network described in Remark 5.7 that was used to simulate a FO[$t(n)$] property.

The factor of two would be eliminated if we adopted a weaker parallel machine model allowing only common writes and such that the memory location accessed by a processor at a given time could be determined by a very simple computation on the processor number and the time.

The additional two variables arise for various bookkeeping reasons. This term can be reduced if we make the following two changes:

1. Rather than keeping track of all previous times, we can assume that every bit of global memory is written into at least every T time steps for some constant T.

2. The register size can be restricted to $O(\log n)$, so we need only $O(\log \log n)$ bits to name a bit of a word.

Remark 5.13. The above observations show that the relation between the number of variables needed to give an inductive definition of a relation and the logarithm to the base n of the number of processors needed to quickly compute the relation are nearly identical. The cost of programming with first-order inductive definitions rather than CRAMs is theoretically very small.

The number of variables needed to describe a query is not a perfect measure of the amount of hardware needed in the parallel computation of the query. This is due to the difference in connection patterns of the parallel models CRAM[$t(n)$] and FO[$t(n)$]. For example, in the proof of Lemma 5.4, processor $\bar{a}$ at round t accesses only a fixed pair of bits of global memory: bits $loc(a_1 \ldots \hat{a}_i \ldots a_k)$ and $loc(a_1 \ldots \hat{a}_{i+1} \ldots a_k)$. Thus a "first-order parallel machine" has a more restrictive pattern of connections between processors and global memory than a CRAM (cf. Remark 5.7).

Compare this to Theorem 10.16 where the set of queries describable using $k+1$ variables is proved identical to the set of queries computable using deterministic space n^k.

5.4 Circuit Complexity

Real computers are built from many copies of small and simple components. Circuit complexity is the branch of computational complexity that uses circuits of boolean logic gates as its model of computation. The circuits that we consider are directed acyclic graphs, in which inputs are placed at the leaves and signals proceed up the circuit toward the root r. Thus, in this idealized model, a gate is never reused during a computation.

This simple and basic model admits many beautiful and deep combinatorial arguments, some of which we will see in Chapter 13. In this section, we define the major circuit complexity classes. It should be intuitively clear that the depth

78   5. Parallelism

of a circuit, that is, the length of a longest path from root to leaf, corresponds to parallel time. We demonstrate this and the related connections between circuits and the other models of parallel computation, i.e., CRAMs, alternating machines, and first-order inductive definitions.

Let $S \subseteq \text{STRUC}[\tau_s]$ be a boolean query on binary strings. In circuit complexity, S would be computed by an infinite sequence of circuits

$$\mathcal{C} = \{C_i \mid i = 1, 2, \ldots\}, \qquad (5.14)$$

where C_n is a circuit with n input bits and a single output bit r. For $w \in \{0, 1\}^n$, let $C_n(w)$ be the value at C_n's output gate, when the bits of w are placed in its n input gates. We say that $\mathcal{C}$ *computes* S iff for all n and for all $w \in \{0, 1\}^n$,

$$w \in S \quad \Leftrightarrow \quad C_n(w) = 1.$$

In this section we present an introduction to circuit complexity and relate complexity classes defined via uniform sequences of circuits to descriptive complexity. We also derive a completely syntactic definition for circuit uniformity. This definition is equivalent to the usual Turing machine-based definition in the range where the latter exists.

As seen in Definition 2.27, a circuit is a directed, acyclic graph. The leaves of the circuit are the input nodes. Every other vertex is an "and", "or", or "not" gate. The edges of the circuit indicate connections between nodes. Edge (a, b) would indicate that the output of gate a is an input to gate b.

It is convenient to assume that all the "not" gates in our circuits have been pushed down to the bottom. We can do this using De Morgan laws ($\neg(\alpha \wedge \beta) \equiv (\neg \alpha \vee \neg \beta)$; $\neg(\alpha \vee \beta) \equiv (\neg \alpha \wedge \neg \beta)$) without increasing the depth and without significantly increasing the size of the circuit.

Furthermore, we can assume that the levels alternate, with the top level being all "or" gates, the next level all "and" gates and so on. Such a normalized circuit is called a *layered circuit*. See Figure 5.15, in which a layered circuit of depth $t(n)$ is drawn.

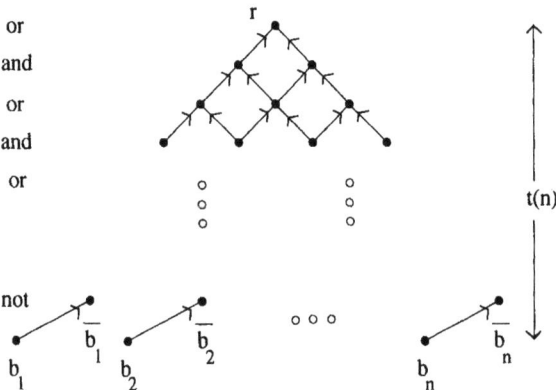

Figure 5.15: A layered circuit of depth $t(n)$.

5.4 Circuit Complexity

Below, we define three families of circuit complexity classes. They vary depending on whether all gates have bounded fan-in (NC), the "and" and "or" gates may have unbounded fan-in (AC), or there are threshold gates (ThC). A threshold gate with threshold value i has output one iff at least i of its inputs have value one. Note that threshold gates include as special cases "or" gates in which the threshold is one and "and" gates in which the threshold is equal to the number of inputs.

Now let us recall also from the earlier Definition 2.27 the vocabulary of circuits, $\tau_c = \langle E^2, G_\wedge^1, G_\vee^1, G_\neg^1, I^1, r \rangle$. Constant r refers to the root node, or output of the circuit. The gates that have no incoming edges are the leaves of the circuit. We use the following abbreviation,

$$L(x) \equiv (\forall y)(\neg E(y, x))$$

The leaves need to be ordered $1, 2, \ldots$ so that we know where to place input bits $b_1, b_2, \ldots, b_n$. We assume for simplicity that the leaves of a circuit are the initial elements of the universe of a circuit. That is, we assume that every circuit $\mathcal{C}$ satisfies the following formula:

$$\text{Leaves-Come-First} \equiv (\forall xy)(L(x) \wedge \neg L(y) \to x < y)$$

Input relation $I(v)$ represents the fact that leaf v contains value 1. Internal node w is an and-gate if $G_\wedge(w)$ holds, an or-gate if $G_\vee(w)$ holds, and a not-gate if $G_\neg(w)$ holds.

We generalize the vocabulary of circuits to the vocabulary of *threshold circuits*, $\tau_{thc} = \tau_c \cup \{G_t^2\}$. Relation $G_t(g, v)$ means that g is a threshold gate with threshold value v. Thus, g would take value 1 in a circuit iff at least v of its inputs have value one.

Let $\mathcal{A} \in \text{STRUC}[\tau]$ and let $n = \|\mathcal{A}\|$. A circuit C_n with $\hat{n}_\tau(n)$ leaves can take $\mathcal{A}$ as input by placing the binary string bin($\mathcal{A}$) into its leaves. We write $C(w)$ to denote the output of circuit C on input w, i.e., the value of the root node when w is placed at the leaves and C is then evaluated. We say that circuit C *accepts* structure $\mathcal{A}$ iff $C(\text{bin}(\mathcal{A})) = 1$.

In proving lower bounds on circuit complexity, one considers the size and structure of the circuits C_n, but one rarely needs to consider how the sequence of circuits relate for different values of n. To relate circuit complexity to machine or descriptive complexity, however, we must explain where these infinitely many circuits come from. The idea is that the circuits all come from unwindings of a particular program and architecture.

Formally we assume that there is a query of low complexity that on input 0^n produces C_n. We insist upon first-order uniformity. This means that there is a first-order query $I : \text{STRUC}[\tau_s] \to \text{STRUC}[\tau_c]$ with $C_n = I(0^n)$, $n = 1, 2, \ldots$. Here $0^n \in \text{STRUC}[\tau_s]$ is the string consisting of n zeros. Note that this uniformity condition implies that C_n has polynomially bounded size.

Definition 5.16 (**Uniform**) Let $\mathcal{C}$ be a sequence of circuits as in Equation (5.14). Let $\tau \in \{\tau_c, \tau_{thc}\}$ be the vocabulary of circuits or threshold circuits. Let $I : \text{STRUC}[\tau_s] \to \text{STRUC}[\tau]$ be a query such that for all $n \in \mathbf{N}$, $I(0^n) = C_n$. That is, on input a string of n zero's the query produces circuit n. If $I \in \text{FO}$, then $\mathcal{C}$ is a *first-order uniform* sequence of circuits. Similarly, if $I \in L$, then $\mathcal{C}$ is *logspace uniform*. If $I \in P$, then $\mathcal{C}$ is *polynomial-time uniform*, and so on. □

We now define the standard circuit complexity classes. The notion of uniformity that we use is first-order uniformity. Observe that whether we use first-order, logspace, or polynomial-time uniformity, any uniform sequence of circuits is polynomial-size. That is, there is a function $p(n)$ such that circuit C_n has size at most $p(n)$.

Definition 5.17 (Circuit Complexity) Let $t(n)$ be a polynomially bounded function and let $S \subseteq \text{STRUC}[\tau]$ be a boolean query. Then S is in the (first-order uniform) circuit complexity class $\text{NC}[t(n)]$, $\text{AC}[t(n)]$, $\text{ThC}[t(n)]$, respectively iff there exists a first-order query $I : \text{STRUC}[\tau_s] \to \text{STRUC}[\tau_{thc}]$ defining a uniform class of circuits $C_n = I(0^n)$ with the following properties:

1. For all $\mathcal{A} \in \text{STRUC}[\tau]$, $\mathcal{A} \in S \Leftrightarrow C_{\|\mathcal{A}\|}$ accepts $\mathcal{A}$.
2. The depth of C_n is $O(t(n))$.
3. The gates of C_n consist of binary "and" and "or" gates (NC), unbounded fan-in "and" and "or" gates (AC), and unbounded fan-in threshold gates (ThC), respectively.

For $i \in \mathbf{N}$, Let $\text{NC}^i = \text{NC}[(\log n)^i]$, $\text{AC}^i = \text{AC}[(\log n)^i]$, and $\text{ThC}^i = \text{ThC}[(\log n)^i]$. Finally, let

$$\text{NC} = \bigcup_{i=0}^{\infty} \text{NC}^i \qquad \square$$

The NC circuits correspond reasonably well to standard silicon-based hardware. The AC circuits are idealized hardware in that it is not known how to connect n inputs to a single gate with constant delay time. The practical way to do this is to connect them in a binary tree, causing an $O(\log n)$ time delay. On the other hand, once we have such a binary tree, we can also compute threshold functions. This explains what we rigorously prove below, namely:

$$\text{AC}[t(n)] \subseteq \text{ThC}[t(n)] \subseteq \text{NC}[t(n) \log n] \ .$$

We also see below that the unbounded fan-in gates in AC circuits correspond exactly to concurrent writing in the CRAM model. Similarly, threshold gates correspond to the tree connections in the NYU ultracomputer and to the "scan" operation on the connection machine cf. Theorem 5.27, Exercises 5.28, 5.29.

Recall that a *regular language* is a set of strings $S \subseteq \Sigma^*$ accepted by a *finite automaton*. A finite automaton is essentially a Turing machine with no work tapes. See [LP81] or [HU79] for details. As an example of computing with circuits, we prove the following,

Proposition 5.18. *Every regular language is in* NC^1.

Proof We are given a deterministic finite automaton $D = \langle \Sigma, Q, \delta, s, F \rangle$. We must construct a first-order query $I_D : \text{STRUC}[\tau_s] \to \text{STRUC}[\tau_c]$ such that, letting $C_n = I_D(0^n)$, for all strings $w \in \Sigma^*$,

$$w \in \mathcal{L}(D) \quad \Leftrightarrow \quad C_{|w|} \text{ accepts } w$$

Circuit C_n is a complete binary tree with n leaves. The input to leaf $L(i)$ is w_i, character i of the input string. Each such leaf contains the finite hardware to produce as output the transition function of D on reading input symbol w_i. That is, we store a table for $f_{L(i)} = \delta(\cdot, w_i) : Q \to Q$.

Each internal node v of the tree takes as input the transition functions f_{lc} and f_{rc} of its left child and right child, and computes their composition $f_v = f_{rc} \circ f_{lc}$. Thus, inductively, the output of every node v is the function $f_v = \delta^*(\cdot, w_v)$ where w_v is the subword of w that is sitting below v's subtree. In particular, w is in $L(D)$ iff $f_r(s) \in F$, where f_r is the mapping stored at the root.

Since D is a fixed, finite state automaton, the hardware at the leaves and at each internal node is a fixed, bounded size NC circuit. The first-order query I_D need only describe a complete binary tree with n leaves with these two fixed circuits placed at each leaf and each internal node, respectively. The height of the resulting circuits is $O(\log n)$ as desired. □

Exercise 5.19 Prove that the boolean majority query MAJ is in NC^1.

$$\text{MAJ} = \left\{ \mathcal{A} \in \text{STRUC}[\tau_s] \mid \text{string } \mathcal{A} \text{ contains more than } \|\mathcal{A}\|/2 \text{ ``1''s} \right\}$$

[Hint: The obvious way to try to build an NC^1 circuit for majority is to add the n input bits via a full binary tree of height $\log n$. The problem with this is that while the sums being added have more and more bits, they must be added in constant depth.

A solution to this problem uses ambiguous arithmetic notation. Consider a representation of natural numbers in binary, except that digits 0, 1, 2, 3 may be used. For example 3213 and 3221 are different representations of the decimal number 37 in this ambiguous notation,

$$3213 = 3 \cdot 2^3 + 2 \cdot 2^2 + 1 \cdot 2^1 + 3 \cdot 2^0 = 37 = 3221 = 3 \cdot 2^3 + 2 \cdot 2^2 + 2 \cdot 2^1 + 1 \cdot 2^0 .$$

Show that adding two n bit numbers in ambiguous notation can be done via an NC^0 circuit, i.e., with bounded depth.

See the following sample addition problem.

```
carries: 3  2  2  3

            3  2  1  3
         +  3  2  1  3
         ─────────────
         3  2  2  1  0
```

This is doable in NC^0 because the carry from column i can be computed by looking only at columns i and $i + 1$.

Translating from ambiguous notation back to binary, which must be done only once at the end, is just an addition problem. This is first-order, and thus AC^0, and thus NC^1.] □

82   5. Parallelism

Exercise 5.20 A good way to become familiar with circuit complexity classes is to prove the following containments. For all $i \in \mathbf{N}$,

$$\text{NC}^i \subseteq \text{AC}^i \subseteq \text{ThC}^i \subseteq \text{NC}^{i+1} \tag{5.21}$$

[Hint: the only subtle containment is the the last. For this, you should use Exercise 5.19.] □

The following theorem summarizes the relationships between all the parallel models that we have seen. Note that the equivalence of $\text{FO}[t(n)]$ and $\text{AC}[t(n)]$ shows that the uniformity of AC circuits can be defined in a completely syntactic way: circuit C_n is constructed by writing down a quantifier block $t(n)$ times.

Theorem 5.22. *For all polynomially bounded and first-order constructible $t(n)$, the following classes are equal:*

$$\text{CRAM}[t(n)] = \text{IND}[t(n)] = \text{FO}[t(n)] = \text{AC}[t(n)]$$

Proof The equality of the first three classes was already proved in Theorem 5.2. The proof of $\text{FO}[t(n)] \subseteq \text{AC}[t(n)]$ is similar to the proof of Lemma 5.4. Let S be a $\text{FO}[t(n)]$ boolean query given by the quantifier block, $\text{QB} = [(Q_1 x_1.M_1)\ldots(Q_k x_k.M_k)]$, initial formula, M_0, and tuple of constants, $\bar{c}$. We must write a first-order query, I, to generate circuit $C_n = I(0^n)$, so that for all $\mathcal{A} \in \text{STRUC}[\tau]$,

$$\mathcal{A} \models (\text{QB}^{t(|\mathcal{A}|)} M_0)(\bar{c}/\bar{x}) \quad \Leftrightarrow \quad C_{|\mathcal{A}|} \text{ accepts } \mathcal{A} \tag{5.23}$$

Initially the circuit evaluates the quantifier-free formulas M_i, $i = 0, 1, \ldots, k$. The nodes $\langle M_i, b_1, \ldots, b_k \rangle$ will be the gates that have evaluated these formulas, i.e.,

$$\langle M_i, b_1, \ldots, b_k \rangle (\text{bin}(\mathcal{A})) = 1 \quad \Leftrightarrow \quad \mathcal{A} \models M_i(b_1, \ldots, b_k)$$

As in Equation (5.5), let φ^r be the inside r quantifiers of $\text{QB}^{t(|\mathcal{A}|)} M_0$. The first of these quantifiers is Q_i, where $i \equiv 1 - r \pmod{k}$.

We construct the gate $\langle 2r, b_1, \ldots, \widehat{b_i}, \ldots, b_k \rangle$ so that

$$\langle 2r, b_1, \ldots, \widehat{b_i}, \ldots, b_k \rangle (\text{bin}(\mathcal{A})) = 1 \quad \Leftrightarrow \quad \mathcal{A} \models \varphi^r(b_1, \ldots, b_k)$$

This is achieved inductively by letting gate $\langle 2r, b_1, \ldots, \widehat{b_i}, \ldots, b_k \rangle$ be an "and"-gate, or "or"-gate according as $Q_i = \forall$ or $\exists$. This gate has inputs gates $\langle 2r - 1, b_1, \ldots, b_i, \widehat{b_{i+1}}, \ldots, b_k \rangle$ for b_i ranging over $|\mathcal{A}|$. Each $\langle 2r - 1, b_1, \ldots, b_i, \widehat{b_{i+1}}, \ldots, b_k \rangle$ is a binary "and"-gate whoses inputs are $\langle M_i, b_1, \ldots, b_k \rangle$ and $\langle 2r - 2, b_1, \ldots, b_i, \widehat{b_{i+1}}, \ldots, b_k \rangle$. See Figure 5.24 for a diagram of this construction.

The circuit we have described may be constructed via a first-order query I, and it satisfies Equation (5.23).

To prove that $\text{AC}[t(n)] \subseteq \text{IND}[t(n)]$, let $C_n = I(0^n)$, $n = 1, 2, \ldots$, be a uniform sequence of $\text{AC}[t(n)]$ circuits, with $I : \text{STRUC}[\tau_s] \to \text{STRUC}[\tau_c]$ a first-order query. We must write an inductive formula,

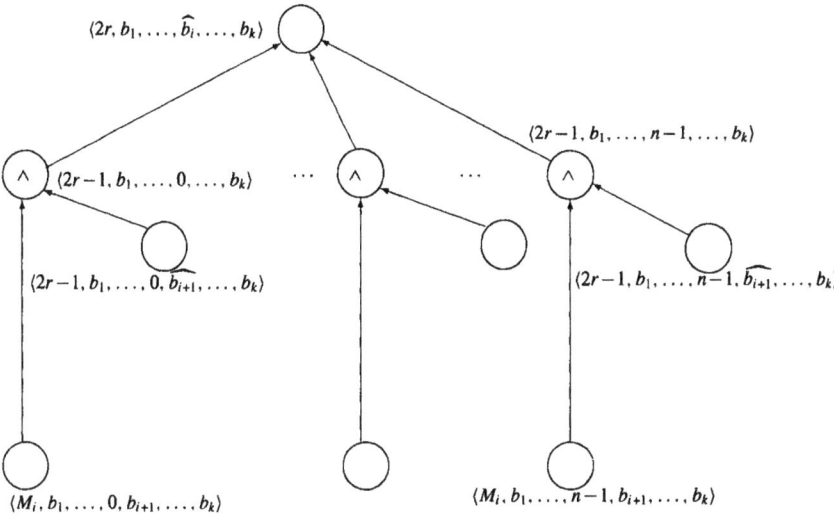

Figure 5.24: An AC[$t(n)$] circuit simulating an FO[$t(n)$] formula

$$\Phi \equiv (\text{LFP}\varphi)(\bar{c})$$

so that for all $\mathcal{A} \in \text{STRUC}[\tau]$,

$$\mathcal{A} \models \Phi \quad \Leftrightarrow \quad C_{|\mathcal{A}|} \text{ accepts } \mathcal{A}.$$

From $\mathcal{A}$ we can get the circuit $C_{|\mathcal{A}|} = \langle E, G_\wedge, G_\vee, G_\neg, \text{bin}(\mathcal{A}), r\rangle$ via the first-order query I. Note that the input string $I = \text{bin}(\mathcal{A})$ is first-order describable from $\mathcal{A}$ (Exercise 2.3). The following is a first-order inductive definition of the relation $V(x, b)$ meaning that gate x has boolean value b,

$$\begin{aligned}
V(x, b) \equiv{}& \text{DEFINED}(x) \wedge \big[L(x) \wedge (I(x) \leftrightarrow b) \vee \\
& G_\wedge(x) \wedge (C(x) \leftrightarrow b) \vee G_\vee(x) \wedge (D(x) \leftrightarrow b) \\
& \vee\ G_\neg(x) \wedge (N(x) \leftrightarrow b) \big]
\end{aligned}$$
(5.25)

Here, DEFINED(x), meaning that x is ready to be defined, is an abbreviation for $(\forall y)(\exists c)(E(y, x) \to V(y, c))$.

Predicates $C(x)$ says that all of x's inputs are true, $D(x)$ says that some of x's inputs are true, and $N(x)$ says that its input is false:

$$\begin{aligned}
C(x) &\equiv (\forall y)(E(y, x) \to V(y, 1)) \\
D(x) &\equiv (\exists y)(E(y, x) \wedge V(y, 1)) \\
N(x) &\equiv (\exists! y)(E(y, x)) \wedge (\exists y)(E(y, x) \wedge V(y, 0))
\end{aligned}$$

The inductive definition of V closes in exactly the depth of C_n, which is $O(t(n))$ iterations. Once it closes, $\Phi \equiv V(r, 1)$ expresses the acceptance condition in IND[$t(n)$], as desired. □

A corollary of Theorem 5.22 is the following characterization of the class NC,

Corollary 5.26.

$$\text{NC} = \bigcup_{k=1}^{\infty} \text{FO}[(\log n)^k] = \bigcup_{k=1}^{\infty} \text{CRAM}[(\log n)^k]$$

In the NYU ultracomputer, processors are connected using a complete binary tree. At each node v of this tree, there is enough logic to compute the sum of the words at v's children and send this value to v's parent. Thus, very quickly — and it is convenient to call this unit time — we can compute the "or" or the sum of n locations.

Thus, an ultracomputer is an extension of a CRAM in which when several processors try to write into the same location of global memory at the same time, that location contains at the next time step the sum of all the values written. Define the complexity class ULTRA$[t(n)]$ to be the set of boolean queries computable on an ultracomputer using polynomially much hardware and parallel time $t(n)$. The Connection Machine has an operation called "scan", which is equivalent to this n-ary sum. Thus, the Connection Machine is another example of an ultracomputer.

The following generalization of Theorem 5.22 is not difficult to prove. The main subtlety is in working with the first-order analogue of the circuit class ThC^0. Define the majority quantifier $(Mx)\varphi(x)$ to mean that more than half of the elements of the universe satisfy φ. Let FO(M), be first-order logic extended by the majority quantifier. It is shown in [BIS88] that FO(M) = ThC^0. Later, we define FO(COUNT) — a robust way to add counting to first-order logic (Definition 12.11). It is also shown in [BIS88] that FO(M) = FO(COUNT).

Theorem 5.27. *For all polynomially bounded and constructible $t(n)$, the following classes are equal:*

$$\text{ULTRA}[t(n)] = \text{FO}(M)[t(n)] = \text{ThC}[t(n)]$$

The following exercises give some sense of the complexity class ThC^0 = FO(M) = FO(COUNT).

Exercise 5.28 Let $\mathcal{A}$ be an ordered structure with $n = \|\mathcal{A}\|$. As usual, we can think of any element $i \in |\mathcal{A}|$ as being a number from 0 to $n-1$. Let the formula $C_\varphi(i)$ mean that the number of $a \in |\mathcal{A}|$ such that $\mathcal{A} \models \varphi(a)$ is at least i. Show that if $\varphi(x)$ is expressible in FO(M), then so is C_φ.

[Hint: use the ordering relation. Split the problem into showing that there are j a's no greater than $n/2$ satisfying φ and that there are $i - j$ a's greater than $n/2$. These statements can be written with one majority quantifier each.] □

Exercise 5.29 The class ThC^0 = FO(M) is a rather interesting complexity class. Show that the following arithmetic operations are computable in ThC^0:

1. The sum of n-bit natural numbers.
2. Multiplication of two n-bit integers.

3. Multiplication of two $n \times n$ integer matrices, each of whose entries is an integer of at most n bits.

[Hint: (3) follows easily from (2) which follows easily from (1). To do (1), let $S = \sum_{i=0}^{n-1} A_i$ be the sum we are trying to compute. Observe that we can count the number of 1s in each column. Thus, we just have to add the relevant column-counts, i.e., we have to add n $\log n$-bit numbers. To make this easier still, split each A_i into two parts: $A_i = B_i + C_i$, where the bits of B_i and C_i are split into blocks of size $\log n$, and B_i is zero on the even blocks and C_i is zero on the odd blocks. Thus $S = \sum B_i + \sum C_i$, and we have reduced the problem to computing $\sum B_i$ and $\sum C_i$. You have to figure out why this is easier than the original problem! (For a similar argument, see the part of the proof of Theorem 1.17 in which it is shown that TIMES is first-order definable using BIT.)] □

5.5 Alternating Complexity

Alternating Turing machines are very closely tied with quantifiers. In this section we establish the precise relationship between descriptive and alternating complexity. We begin by examining the relationship at the lowest level. Let the *logarithmic-time hierarchy* (LH) be ATIME-ALT[$\log n$, $O(1)$], i.e., the set of boolean queries computed by alternating Turing machines in $O[\log n]$ time, making a bounded number of alternations. The following theorem says that LH = FO.[2]

Theorem 5.30. *The logarithmic-time hierarchy is exactly the set of first-order expressible boolean queries.*

Proof The most delicate part of the proof is the following:

Lemma 5.31. DTIME[$\log n$] $\subseteq$ FO .

Proof Let T be a DTIME[$\log n$] machine. We must write a first-order sentence φ such that for all inputs $\mathcal{A}$,

$$T(\text{bin}(\mathcal{A}))\downarrow \quad \Leftrightarrow \quad \mathcal{A} \models \varphi$$

The sentence φ will begin with existential quantifiers, $\varphi \equiv (\exists x_1 \ldots x_c)\psi(\bar{x})$. The variables $\bar{x}$ will code the $\log n$ steps of T's computation including, for each time step t, the values q_t, w_t, d_t, I_t representing T's state, the symbol it writes, the direction its head moves, and the value of the input being scanned by the index-tape-controlled input head at time t, respectively. (It is important to remember that each variable is a $\lceil \log n \rceil$ bit number and that the numeric predicate BIT allows these bits to be specified.)

The formula ψ must now assert that the information in $\bar{x}$ meshes together to form a valid accepting computation of T. The work we must do to accomplish

---

[2]This is analogous to a result we will see later: The polynomial-time hierarchy (PH) is equal to the set of boolean queries expressible in second-order logic (SO), (Corollary 7.22).

this is to define the first-order relations $C(p, t, a)$ and $P(p, t)$ meaning that for the computation determined by $\bar{x}$, the contents of cell p at time t is a; and the work head is at position p at time t. Given C and P we can assert that $\bar{x}$ is self-consistent. Note, for example, that we can guess the contents y of the index tape, and then use C to verify that y is correct. Next, using y we can verify that the input symbol I_t is correct.

Next, note that using P we can write C because the contents of cell p at time t is just w_{t_1} where t_1 is the most recent time that the head was at position p.

Finally observe that to write the relation P it suffices to take the sum of $O[\log n]$ values each of which is either -1, or 1. We can do this in FO by Lemma 1.18. □

To prove LH $\subseteq$ FO, we need only note that an alternating logarithmic-time machine may be assumed to write its guesses on a work tape and then deterministically check for acceptance. Since there are a bounded number of alternations and the total time is $O(\log n)$, these guesses may be simulated by a bounded number of first-order quantifiers. The remaining work is in DTIME[$\log n$] and thus in FO by Lemma 5.31.

The other direction of Theorem 5.30 is fairly easy. We have to show that for every first-order sentence,

$$\varphi \equiv (\exists x_1)(\forall x_2)\ldots(Q_k x_k)M(\bar{x})$$

there exists an ATIME-ALT[$\log n$, $O(1)$] machine T such that for all input strings $\mathcal{A}$,

$$T(\text{bin}(\mathcal{A}))\downarrow \quad \Leftrightarrow \quad \mathcal{A} \models \varphi$$

Since M is a constant size quantifier-free formula, it is easy to build a DTIME[$\log n$] Turing machine which on input $\mathcal{A}$ and with values $a_1, \ldots, a_k$ on its tape, tests whether or not $\mathcal{A} \models M(\bar{a})$. (The most complicated part of this is to verify the BIT predicate, which requires counting in binary up to $O(\log n)$ on a work tape — this is straightforward.) Thus using $k - 1$ alternations between existential and universal states, a Σ_k logarithmic-time machine can guess $a_1, \ldots, a_k$ and then deterministically verify $M(\bar{a})$.

□

Notice that from Theorems 5.2, 5.22 and 5.30, we now have three interesting characterizations of the class FO.

Corollary 5.32.

$$\text{FO} = \text{AC}^0 = \text{CRAM}[1] = \text{LH}$$

Notice that the truth of Corollary 5.32 depends on our choice of including BIT as a numeric predicate, and the SHIFT operation in the CRAM, and our definition of uniformity for AC^0. In Chapter 11 we will obtain a non-uniform version of Corollary 5.32 in which we allow arbitrary numeric relations (Proposition 11.19).

5.5.1 Alternation as Parallelism

AC and NC circuits also have elegant characterizations via alternating machines.

Theorem 5.33. *For $t(n) \geq \log n$,*

$$\text{ASPACE-ALT}[\log n, t(n)] = \text{AC}[t(n)] = \text{FO}[t(n)].$$

Proof We have already seen the second equality in Theorem 5.22.

(ASPACE-ALT$[\log n, t(n)] \supseteq$ AC$[t(n)]$): Let $t(n) \geq \log n$ and consider the same AC$[t(n)]$ boolean query as in the proof that AC$[t(n)] \subseteq$ IND$[t(n)]$ in Theorem 5.22. Since ASPACE-ALT$[\log n, t(n)] \supseteq$ ATIME-ALT$[\log n, 1]$ = LH, we know from Theorem 5.30 that ASPACE-ALT$[\log n, t(n)] \supseteq$ FO. Thus, the circuit C_n is available in ASPACE-ALT$[\log n, t(n)]$.

Recall that Equation (5.25) simulates an AC$[t(n)]$ circuit via an IND$[t(n)]$ definition. Looking at Equation (5.25) we see that it makes at most $O(t(n))$ alternations between existential and universal quantifiers. This inductive definition can thus be directly simulated by an ASPACE-ALT$[\log n, t(n)]$ machine: Each universal quantifier is simulated by $\log n$ universal moves and each existential quantifier is simulated by $\log n$ existential moves. The space needed to hold the variables is $O(\log n)$. Furthermore, there are only a bounded number of alternations per iteration of the inductive definition.

(ASPACE-ALT$[\log n, t(n)] \subseteq$ AC$[t(n)]$): Conversely, let M be an ASPACE-ALT$[\log n, t(n)]$ machine. As in Theorem 3.16, an ID of M can be coded using a bounded number of variables. The acceptance condition of M can then be expressed via an inductive definition of depth $\log n + t(n)$ as follows. Let EPATH$_M$(ID$_1$, ID$_2$) mean that there is a computation path of M from ID$_1$ to ID$_2$ all of whose states except perhaps the last is existential. Let APATH mean the same thing for universal paths. It is easy to see that EPATH and APATH are expressible in IND$[\log n]$ (cf. Proposition 4.17). Thus, the following simultaneous induction has depth $O(\log n + t(n))$ and expresses the acceptance condition for M as desired.

$$\text{ACCEPT}_M(\text{ID}_1) \equiv \text{ID}_1 \text{ is the accept ID} \lor (\exists \text{ID}_2)[\text{ACCEPT}_M(\text{ID}_2) \land$$
$$(\text{EPATH}(\text{ID}_1, \text{ID}_2) \lor \text{APATH}(\text{ID}_1, \text{ID}_2))]$$

□

We leave a similar characterization of NC$[t(n)]$ to the reader:

Exercise 5.34 Prove that for $t(n) \geq \log n$,

$$\text{NC}[t(n)] = \text{ASPACE-TIME}[\log n, t(n)]$$

[Hint: this is similar to the proof of Theorem 5.33, the difference being that the definitions of C and D in Equation (5.25) now involve binary "and"s and "or"s rather than universal and existential quantifiers.] □

For $i \geq 1$, the bound NC$^i \subseteq$ AC^i in Equation (5.21) is not optimal. The following improvement is known to be optimal because the NC^1 query PARITY requires depth $\log n / \log \log n$ (Corollary 13.8).

Theorem 5.35. *For $t(n) \geq \log n$, the following containment holds,*

$$\text{NC}[t(n)] \subseteq \text{AC}[t(n)/\log\log n]$$

Proof We prove the equivalent containment,

$$\text{ASPACE-TIME}[\log n, t(n)] \subseteq \text{IND}[t(n)/\log\log n]$$

Suppose that we are given an ASPACE-TIME$[\log n, t(n)]$ machine M. As in Theorem 5.33, we can write an inductive definition for ACCEPT$_M$ — the acceptance condition of M. The straight forward way to do this is in IND$[t(n)]$ with one alternation of quantifiers per move of M.

As usual, we assume that M alternates at each step between existential and universal states. To improve this simulation by a $\log\log n$ factor, observe that a list of which existential moves to make in the event of each possible sequence of $(\log\log n)/2$ universal moves can be given in $\log n$ bits.

Let e be such a $\log n$-bit table of which existential move to make in the event of any sequence of $(\log\log n)/2$ universal moves, and let a be such a sequence of universal moves. Then we can inductively define the relation MOVES$_M(e, u, \text{ID}_1, \text{ID}_2)$ meaning that ID$_2$ follows from ID$_1$ in the $\log\log n$ moves of M determined by the universal moves u and the existential moves given by e indexed by u. It is easy to write such an inductive definition in depth $\log\log n$.

Our definition of ACCEPT$_M$ is then a simultaneous inductive definition with MOVES$_M$,

$$\text{ACCEPT}_M(\text{ID}_1) \equiv \text{ID}_1 \text{ is the accept ID } \vee \quad (\exists e)(\forall u)(\exists \text{ID}_2)$$
$$(\text{MOVES}_M(e, u, \text{ID}_1, \text{ID}_2) \wedge \text{ACCEPT}_M(\text{ID}_2))$$

The depth of this simultaneous induction is $t(n)/\log\log n$ as desired. □

Historical Notes and Suggestions for Further Reading

Cook has written a useful survey of parallel complexity classes and problems, [Coo85]. Stockmeyer and Vishkin proved the equality of non-uniform versions of CRAM$[t(n)]$ and AC$[t(n)]$ in [SV84]. Most of the uniform results in this chapter appeared first in [I88]. There was much previous work comparing different versions of PRAM's. See for example [FW78] or an early treatment of PRAM's and [FRW84] for the first proof that a common write machine can simulate a CRAM with a linear increase in time and a squaring of the number of processors.

Theorem 5.33 is due to Ruzzo and Tompa in [SV84]. The equality NC$[t(n)]$ = ASPACE-TIME$[\log n, t(n)]$ in Exercise 5.34 is due to Ruzzo [Ruz81].

Lindell has made some interesting observations about the definition FO = uniform AC^0, [L92]. The nonuniform version of Theorem 5.35 is due to Chandra, Stockmeyer, and Vishkin, [CSV84]. The uniform version appears in [I89a].

The ambiguous arithmetic notation used in Exercise 5.19 is from Borodin, Cook, and Pippenger, [BCP83].

More information about the NYU Ultracomputer and the Connection Machine may be found in [AG94, Hil85].

6
Ehrenfeucht-Fraïssé Games

We introduce combinatorial games that are useful for determining what can be expressed in various logics. Ehrenfeucht-Fraïssé games offer a semantics for first-order logic that is equivalent to, but more directly applicable than, the standard definitions.

6.1 Definition of the Games

Suppose we assert that structure $\mathcal{A}$ satisfies the formula $(\forall x)\varphi(x)$. This can be understood in a game-theoretic way: an opponent may choose any element $a \in |\mathcal{A}|$. We are then obliged to show that $\mathcal{A} \models \varphi(a)$. Similarly, if we assert $(\exists y)\psi(y)$, then we move first, choosing $b \in |\mathcal{A}|$ and asserting that $\mathcal{A} \models \psi(b)$.

This operational, game-theoretic view of logical assertions is the subject of the present chapter. Ehrenfeucht-Fraïssé games offer a convenient, model-theoretic approach to logic. These games have been used extensively for proving that certain queries are not expressible in certain logics.

Using games, we present a complete methodology for proving that a boolean query is not expressible in first-order logic. We provide many examples. In later chapters we will show how to modify the games to provide complete methodologies for other languages, stronger than first-order logic.

We begin by defining the games, which were invented by Ehrenfeucht and Fraïssé.

Definition 6.2 The game $\mathcal{G}_m^k$ is played by two players: Samson and Delilah on a pair of structures $\mathcal{A}$ and $\mathcal{B}$ of the same vocabulary, τ. $\mathcal{G}_m^k$ is played for m rounds, using k pairs of pebbles. Samson tries to point out a difference between the two

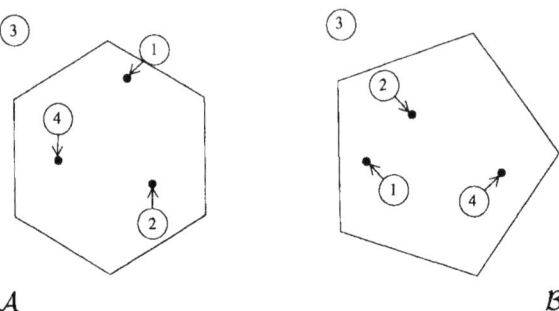

Figure 6.1: A four-pebble game.

structures, and Delilah tries to match his moves so that the differences between them are hidden.

At each move, Samson places one of the pebbles on an element of the universe of one of the two structures, i.e., he places pebble i on an element of $|\mathcal{A}|$, or $|\mathcal{B}|$. Delilah then responds by placing the other pebble i on an element of the other structure.

The position of the game immediately after move r is denoted by (α_r, β_r). Such a k-configuration of $\mathcal{A}, \mathcal{B}$ is a pair of partial functions

$$\begin{aligned} \alpha &: (\text{const}(\tau) \cup \{x_1, x_2, \ldots, x_k\}) \to |\mathcal{A}| \\ \beta &: (\text{const}(\tau) \cup \{x_1, x_2, \ldots, x_k\}) \to |\mathcal{B}| \end{aligned} \quad (6.3)$$

where we require that the domains of the functions α and β be equal, $\text{dom}(\alpha) = \text{dom}(\beta)$, and for all $c \in \text{const}(\tau)$, $\alpha(c) = c^{\mathcal{A}}$ and $\beta(c) = c^{\mathcal{B}}$.

The meaning of $\alpha_r(x_i) = a$ and $\beta_r(x_i) = b$ is that just after move r, the ith pebbles are sitting on $a \in |\mathcal{A}|$ and $b \in |\mathcal{B}|$. Some variable x_i is not in the domain of α_r iff just after move r, the ith pebbles are off the board. The valid positions of game $\mathcal{G}_m^k$ on $\mathcal{A}, \mathcal{B}$ consist of any possible k-configuration on $\mathcal{A}$ and $\mathcal{B}$. See Figure 6.1 in which the current configuration has $\text{dom}(\alpha) = \text{dom}(\beta) = \{1, 2, 4\} \cup \text{const}(\tau)$, indicating that pebbles 1, 2, and 4 are currently placed on elements of $|\mathcal{A}|$ and $|\mathcal{B}|$, and both pebbles numbered 3 are off the board.

Write $\mathcal{G}_m^k(\mathcal{A}, \alpha_0, \mathcal{B}, \beta_0)$ to denote the k-pebble, m-move game on $\mathcal{A}, \mathcal{B}$, with initial configuration (α_0, β_0). $\mathcal{G}_m^k(\mathcal{A}, \mathcal{B})$ is the game in which all the pebbles start off the board, i.e., $\text{dom}(\alpha_0) = \text{dom}(\beta_0) = \text{const}(\tau)$. The reason we include the constants in the domain of every configuration is to make the statement of conditions simpler in what follows. As will be seen, in Ehrenfeucht-Fraïssé games constants behave exactly like pebbles that are fixed at the beginning of the game.

At each move r, $1 \leq r \leq m$, Samson picks up a pair of pebbles and places one of them on an element of one of the two structures. Delilah must then answer by placing the other pebble in the pair on an element of the other structure. Thus, for some $i \in \{1, 2, \ldots, k\}$, pair i of pebbles is placed on $a \in |\mathcal{A}|$ and $b \in |\mathcal{B}|$. Define the next configuration $(\alpha_r, \beta_r) = (\alpha_{r-1}[a/x_i], \beta_{r-1}[b/x_i])$,

$$\alpha_r(x_j) = \begin{cases} \alpha_{r-1}(x_j) & \text{if } i \neq j \\ a & \text{if } i = j \end{cases}, \quad \beta_r(x_j) = \begin{cases} \beta_{r-1}(x_j) & \text{if } i \neq j \\ b & \text{if } i = j \end{cases}.$$

Just after move r, the configuration α_r, β_r determines a relation $\beta_r \circ \alpha_r^{-1} \subseteq |\mathcal{A}| \times |\mathcal{B}|$. We say that *Delilah wins round r of the game* iff the map $\alpha_r(x_j) \mapsto \beta_r(x_j)$, for $x_j \in \text{dom}(\alpha_r)$ determines an isomorphism of the induced substructures[1].

$$\beta_r \circ \alpha_r^{-1} : \langle \text{rng}(\alpha) \rangle^{\mathcal{A}} \cong \langle \text{rng}(\beta) \rangle^{\mathcal{B}} \qquad (6.4)$$

This means in particular that $\beta_r \circ \alpha_r^{-1}$ must be a 1:1 function, so $\alpha_r(x_i) = \alpha_r(x_j)$ iff $\beta_r(x_i) = \beta_r(x_j)$. Furthermore, all constants and relations of the structures must be preserved. For example, if vocabulary τ includes the constant symbol c and the binary relation symbol E, then $\langle c^{\mathcal{A}}, \alpha_r(x_i) \rangle \in E^{\mathcal{A}}$ iff $\langle c^{\mathcal{B}}, \beta_r(x_i) \rangle \in E^{\mathcal{B}}$. This can be more easily written as follows:

$$(\mathcal{A}, \alpha_r) \models E(c, x_i) \quad \Leftrightarrow \quad (\mathcal{B}, \beta_r) \models E(c, x_i).$$

Delilah *wins the game* iff she wins every single round. Delilah must preserve an isomorphism at all times. If a difference between the two structures is ever exposed, then Samson wins.

Since $\mathcal{G}_m^k(\mathcal{A}, \alpha_0, \mathcal{B}, \beta_0)$ is a finite game of perfect information, one of the two players must have a winning strategy. We use the notation $(\mathcal{A}, \alpha_0) \sim_m^k (\mathcal{B}, \beta_0)$ to mean that Delilah has a winning strategy for $\mathcal{G}_m^k(\mathcal{A}, \alpha_0, \mathcal{B}, \beta_0)$. We write $(\mathcal{A}, \alpha_0) \sim^k (\mathcal{B}, \beta_0)$ to mean that for all m, $(\mathcal{A}, \alpha_0) \sim_m^k (\mathcal{B}, \beta_0)$. Similarly $(\mathcal{A}, \alpha_0) \sim_m (\mathcal{B}, \beta_0)$ means that for all k, $(\mathcal{A}, \alpha_0) \sim_m^k (\mathcal{B}, \beta_0)$. □

Delilah wins the game iff after every round $\beta_r \circ \alpha_r^{-1}$ is an isomorphism of the induced substructures. Samson is trying to point out a difference between the two structures, and Delilah is trying to keep them looking the same. An isomorphism preserves all the symbols of τ. It is important to decide whether to include the numeric predicates $\leq$ and BIT in τ. If these relations are available in the language in question and thus as part of the definition of isomorphism, then the game becomes much easier for Samson and much harder for Delilah. For this reason, *in this chapter, we assume unless otherwise noted that ordering and* BIT *are not present.*

Exercise 6.5 Prove that $\sim_m^k$ is an equivalence relation. This is not hard to show directly from the definition of $\mathcal{G}_m^k$. It also explains why we use this notation. Please do not use Theorem 6.10! □

As an example, consider the two-pebble game on the colored graphs G and H shown in Figure 6.6. Here the vocabulary $\tau = \langle E^2, R^1, Y^1, B^1 \rangle$ consists of a binary edge relation and three unary relations which may be thought of as colorings of the vertices.

---

[1]The induced substructure $\langle \text{rng}(\alpha) \rangle^{\mathcal{A}}$ has universe the closure of $\text{rng}(\alpha)$ under all the functions of $\mathcal{A}$. When τ has no function symbols, this simply means that we add all the constants to $\text{rng}(\alpha)$. For $\tau = \langle R_1^{a_1}, \ldots, R_r^{a_r}, c_1, \ldots, c_s \rangle$, $|\langle S \rangle| = S \cup \{c_1^{\mathcal{A}}, \ldots, c_s^{\mathcal{A}}\}$. The meaning of induced substructure is that the relations of $\langle \text{rng}(\alpha) \rangle^{\mathcal{A}}$ are restrictions of the relations of $\mathcal{A}$ to the universe of $\langle \text{rng}(\alpha) \rangle^{\mathcal{A}}$.

94 6. Ehrenfeucht-Fraïssé Games

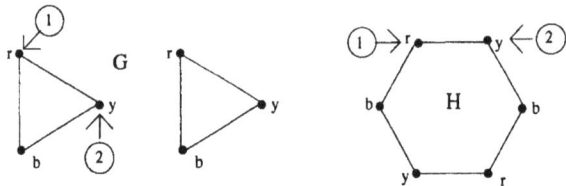

Figure 6.6: A two-pebble game

Assume that the initial configuration has all pebbles off the board, so $\alpha_0 = \beta_0 = \emptyset$. Suppose that Samson's first move is to place pebble 1 on a red vertex in G. Delilah may answer by putting pebble one on either of the red vertices in H. Now suppose Samson puts pebble 2 on an adjacent yellow vertex in H. Delilah has a response because in G, $\alpha_1(x_1)$ also has an adjacent yellow vertex. On the third move, suppose that Samson puts pebble 1 on the blue vertex in H that is not adjacent to $\beta_2(x_2)$. Delilah may answer with the blue vertex in G not adjacent to $\alpha_2(x_2)$. The reader should by now be able to prove by induction that,

Proposition 6.7. *Let G and H be the graphs shown in Figure 6.6. For all m, $G \sim_m^2 H$, i.e., $G \sim^2 H$.*

On the other hand Samson has an easy win for the game $\mathcal{G}_3^3(G, H)$. He can simply choose three points in the same triangle in G on three consecutive moves. Delilah has no response because there is no triangle in H, and thus Samson wins.

Observe that Samson's winning strategy in this three-pebble game is to "play the sentence" Δ which says that a triangle exists. Note that $G \models \Delta$ while $H \models \neg\Delta$.

$$\Delta \equiv (\exists x_1)(\exists x_2)(\exists x_3)(E(x_1, x_2) \wedge E(x_2, x_3) \wedge E(x_3, x_1))$$

Sentence Δ has three variables, corresponding to the number of pebble pairs in the game. Define the *quantifier rank* $(\mathrm{qr}(\varphi))$ of a formula φ to be the depth of nesting of quantifiers in φ. Note that sentence Δ has quantifier rank 3, corresponding to the number of moves in the game.

Example 6.8 Another game example is on the strings $w_1 = 1101$ and $w_2 = 1011$, thought of as structures of vocabulary τ_s, with the ordering relation, but not successor. Samson can win the two-move game on these two strings. He can place the x_1 pebble on the second 1 in w_1. Delilah must answer by placing x_1 on some 1 in w_2. If she answers with the first 1, then Samson can reply by placing x_2 on the first 1 in w_1, and Delilah has no reply. If Delilah instead answers with the second or third 1 in w_2, then Samson replies by placing x_2 on the 0 in w_1. Delilah loses because w_2 has no 0 to the right of x_1. In this case, Samson's winning strategy is to play the following formula that is true of w_1, but not w_2,

$$\varphi \equiv (\exists x_1)(S(x_1) \wedge (\exists x_2)(S(x_2) \wedge x_2 < x_1) \wedge (\exists x_2)(\neg S(x_2) \wedge x_1 < x_2)) \quad \square$$

Definition 6.9 Define language $\mathcal{L}^k$ to be the restriction of language $\mathcal{L}$ in which only variables $x_1, \ldots, x_k$ occur. Define language $\mathcal{L}_m^k$ to be the restriction of $\mathcal{L}^k$ to formulas of quantifier rank at most m. Define $\mathcal{L}_m$ to be the set of formulas of quantifier-rank at most m. Let $\mathcal{A}$ and $\mathcal{B}$ be two structures of some vocabulary τ.

We say that $\mathcal{A}$ and $\mathcal{B}$ are $\mathcal{L}$ equivalent ($\mathcal{A} \equiv_{\mathcal{L}} \mathcal{B}$) iff they agree on all formulas from $\mathcal{L}$,

$$\mathcal{A} \equiv \mathcal{B} \quad \text{iff} \quad \text{for all } \varphi \in \mathcal{L}(\tau), \ \mathcal{A} \models \varphi \Leftrightarrow \mathcal{B} \models \varphi$$

$$\mathcal{A} \equiv_m^k \mathcal{B} \quad \text{iff} \quad \text{for all } \varphi \in \mathcal{L}_m^k(\tau), \ \mathcal{A} \models \varphi \Leftrightarrow \mathcal{B} \models \varphi \ . \qquad \square$$

We can now state and prove the fundamental theorem of Ehrenfeucht-Fraïssé Games. This theorem holds for infinite as well as finite structures.

Theorem 6.10. *Let $\mathcal{A}$ and $\mathcal{B}$ be structures of the same finite, relational vocabulary and let α_0, β_0 be a k-configuration of $\mathcal{A}, \mathcal{B}$. Then the following are equivalent:*

1. $(\mathcal{A}, \alpha_0) \sim_m^k (\mathcal{B}, \beta_0)$
2. $(\mathcal{A}, \alpha_0) \equiv_m^k (\mathcal{B}, \beta_0)$.

Proof We prove the equivalence of (1) and (2) by induction on m. For $m = 0$, Delilah wins the zero move game iff the relation $\beta_0 \circ \alpha_0^{-1}$ is an isomorphism of the induced substructures. This is true iff for every quantifier free formula $\gamma \in \mathcal{L}(\tau)$,

$$(\mathcal{A}, \alpha_0) \models \gamma \quad \Leftrightarrow \quad (\mathcal{B}, \beta_0) \models \gamma \ .$$

Note that α may have as free variables only those variables that occur in $\text{dom}(\alpha_0) = \text{dom}(\beta_0)$. Thus, (1) and (2) are equivalent for $m = 0$.

Assume the theorem is true for m, and suppose that $\mathcal{A}$ and $\mathcal{B}$ disagree on the formula $\varphi \in \mathcal{L}_{m+1}^k$. Note that if φ is $\alpha \wedge \beta$ then $\mathcal{A}$ and $\mathcal{B}$ disagree on one of α and β. Similarly, if φ is $\neg \alpha$, then they disagree on α, so we may assume that φ is $(\exists x_i)\psi$. Suppose that $(\mathcal{A}, \alpha_0) \models \varphi$ and $(\mathcal{B}, \beta_0) \models \neg \varphi$. Samson's first move in $\mathcal{G}_{m+1}^k(\mathcal{A}, \alpha_0, \mathcal{B}, \beta_0)$ is to place pebble i on a witness for ψ in $\mathcal{A}$. Wherever Delilah responds, it will not be a witness for ψ because there is none in $\mathcal{B}$. Thus, after the first move, $(\mathcal{A}, \alpha_1)$ and $(\mathcal{B}, \beta_1)$ disagree on the quantifier depth m formula ψ. By the inductive hypothesis, Samson has a winning strategy for the remaining m-move game. Thus we have shown that (1) implies (2).

Conversely, suppose that $(\mathcal{A}, \alpha_0) \equiv_{m+1}^k (\mathcal{B}, \beta_0)$. Now let Samson make his first move in the game $\mathcal{G}_{m+1}^k(\mathcal{A}, \alpha_0, \mathcal{B}, \beta_0)$. Suppose he places pebble i on an element of $\mathcal{A}$, thus defining α_1. Note that there are only finitely many inequivalent formulas in $\mathcal{L}_m^k$, cf. Exercise 6.11. Let Φ be the conjunction of all these formulas that are satisfied by $(\mathcal{A}, \alpha_1)$. Thus, we know that

$$(\mathcal{A}, \alpha_0) \models (\exists x_i) \Phi \ ,$$

so, by assumption,

$$(\mathcal{B}, \beta_0) \models (\exists x_i) \Phi \ .$$

Delilah's answer is to place the other pebble i on a witness in $\mathcal{B}$ of Φ. Thus, $(\mathcal{A}, \alpha_1)$ and $(\mathcal{B}, \beta_1)$ both satisfy Φ, a complete description of every formula from $\mathcal{L}_m^k$ that $(\mathcal{A}, \alpha_1)$ satisfies. Thus $(\mathcal{A}, \alpha_1) \equiv_m^k (\mathcal{B}, \beta_1)$. It follows by induction that Delilah has a winning strategy for the remaining m moves of the game. $\qquad \square$

In the following exercise, you are asked to prove the lemma that there are only finitely many inequivalent formulas of a given quantifier rank. This was needed for the proof of (2) ⇒ (1) in Theorem 6.10. As the exercise shows, this lemma holds for infinite structures as well, as long as the vocabulary is finite and has no function symbols.

Exercise 6.11 Theorem 6.10 requires the lemma that there are only finitely many inequivalent formulas of quantifier rank r.

1. As usual, let τ be a finite, relational vocabulary. Prove that there are only finitely many inequivalent first-order formulas in $\mathcal{L}_r(\tau)$. [Hint: induction on r.]
2. Let τ be a finite vocabulary which may include function symbols. Let $\Gamma \subset \mathcal{L}(\tau)$ be a first-order theory. Suppose that for every model $\mathcal{A}$ of Γ — including infinite models — and for any finite set $S \subseteq |\mathcal{A}|$, the induced substructure of $\mathcal{A}$ generated by S $\langle S \rangle^{\mathcal{A}}$ is finite. Prove that theory Γ admits only finitely many inequivalent formulas of quantifier rank r. In this case, φ and ψ are equivalent iff $\Gamma \vdash \varphi \leftrightarrow \psi$.
3. Give a counterexample to Theorem 6.10 when $\tau = \langle R_1^1, R_2^1, \ldots \rangle$ consists of infinitely many unary relation symbols.
4. Give a counterexample to Theorem 6.10 when $\tau = \langle R^1, f^1 \rangle$ consists of one unary relation symbol and one unary function symbol. □

The following exercise shows that we never have to consider a move of a game in which Samson pebbles an element that is already pebbled by another pebble or constant.

Exercise 6.12 Prove that in any game $\mathcal{G}_m^k(\mathcal{A}, \alpha_0, \mathcal{B}, \beta_0)$, if Samson has a winning strategy, then he still has a winning strategy if he is never allowed to place a pebble on a constant or an element that already has another pebble sitting on it. □

Theorem 6.10 gives us a way to determine precisely how many variables and how much quantifier rank is needed to express a given query. Here are two examples.

Proposition 6.13. *Let* CLIQUE(k) *be the set of undirected graphs that contain a clique, i.e., a complete subgraph, of size k. In the language without ordering,* CLIQUE(k) *is expressible with k variables but not $k - 1$ variables:*

$$\text{CLIQUE}(k) \in \mathcal{L}^k(\tau_g)(\text{wo}\leq) - \mathcal{L}^{k-1}(\tau_g)(\text{wo}\leq).$$

Proof It is easy to write CLIQUE(k) in $\mathcal{L}^k$:

$$(\exists x_1 x_2 \ldots x_k)(\text{distinct}(x_1, \ldots, x_k) \wedge E(x_1, x_2) \wedge \ldots \wedge E(x_1, x_k) \wedge \ldots E(x_{k-1}, x_k))$$

To prove that k variables are necessary, we prove that $K_k \sim^{k-1} K_{k-1}$, where K_r is the complete graph on r vertices. Delilah has a simple winning strategy for the game $\mathcal{G}_{k-1}(K_k, K_{k-1})$: When Samson places the a pebble on an unpebbled vertex in one of the two graphs, Delilah places the corresponding pebble on any unpebbled vertex in the other graph. Since there are only $k - 1$ pebble pairs, such an unpebbled vertex is always available. Note that this is a winning strategy since edges exist between all points in each graph. Thus, any 1:1 correspondence is an isomorphism. □

As another example, we show that first-order logic without ordering is not strong enough to express any facts about counting, or even parity.

Proposition 6.14. *In the absence of ordering, the boolean query on graphs that is true iff there are an odd number of vertices requires $n+1$ variables, for graphs with n or more vertices. The same is true for the query that there are an odd number of edges.*

Proof Let G_n be the graph on n vertices that has a loop at each vertex but no other edges. We claim that $G_n \sim^n G_{n+1}$. Delilah's strategy is to match each move by Samson on a vertex not already pebbled with a vertex not already pebbled in the other graph. Since each graph has at least n vertices and there are no edges between different vertices, this is a winning strategy for Delilah. It follows that $G_n \equiv^n G_{n+1}$, so the parity of the number of vertices or the number of edges in not expressible in $\mathcal{L}^n$. □

We saw in Chapter 5 that quantifier rank and number of variables are important parameters of parallel complexity. It is useful to have a game that allows us to determine how many variables and how much quantifier rank is needed to describe various queries. As an example, we now use Ehrenfeucht-Fraïssé games to determine the exact number of variables and quantifier rank need to assert the existence of paths in a graph:

Proposition 6.15. *Let the formula* $\text{PATH}_k(x, y) \in \mathcal{L}(\tau_g)$ *mean that there is a path of length at most 2^k from x to y. With or without ordering, quantifier rank k is necessary and sufficient to express* PATH_k. *Furthermore, only three variables are necessary to express* PATH_k. *In symbols,*

$$\text{PATH}_k \quad \in \quad \mathcal{L}_k^3(\tau_g)(\text{wo}\leq) - \mathcal{L}_{k-1}(\tau_g) \ .$$

Proof For the upper bound, we express PATH_k inductively as follows:

$$\text{PATH}_0(x, y) \equiv x = y \vee E(x, y)$$
$$\text{PATH}_{k+1}(x, y) \equiv (\exists z)(\text{PATH}_k(x, z) \wedge \text{PATH}_k(z, y)) \ .$$

Thus, PATH_k is expressible using three variables and quantifier rank k. Only three variables are needed because the right hand side of the inductive definition of $\text{PATH}_{k+1}(x, y)$ may be written in a way that reuses variables:

$$\text{PATH}_k(x, z) \equiv (\exists y)(\text{PATH}_{k-1}(x, y) \wedge \text{PATH}_{k-1}(y, z))$$
$$\text{PATH}_k(z, y) \equiv (\exists x)(\text{PATH}_{k-1}(z, x) \wedge \text{PATH}_{k-1}(x, y))$$

For the lower bound, let $L_n \in \text{STRUC}[\tau_g]$ be a directed line segment of length $n - 1$, so $\|L_n\| = n$. Let the ordering on L_n be from left to right (see Figure 6.16). Letting $n = 2^{k+1} + 1$, we have

$$L_n \models \text{PATH}_{k+1}(0, max); \qquad L_{n+1} \models \neg\text{PATH}_{k+1}(0, max) \ .$$

We now show that $L_n \sim_k L_{n+1}$, and the lower bound follows.

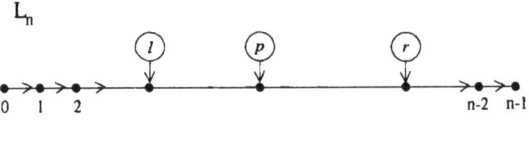

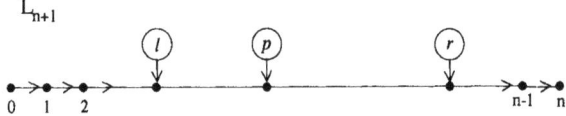

Figure 6.16: With $n = 2^{k+1} + 1$, $L_n \sim_k L_{n+1}$

The idea behind Delilah's winning strategy is that in quantifier rank s — or equivalently, with s moves remaining in the game — distances greater than 2^s are indistinguishable from infinite distances. With this idea in mind, let us define the notion $i =_d j$ to mean that i and j are equal or are both greater than d. Delilah's winning strategy in $\mathcal{G}_k(L_n, L_{n+1})$ is to maintain the following invariant: After the move m of $\mathcal{G}_k(L_n, L_{n+1})$, and for all $p, q \in \text{dom}(\alpha_m)$,

$$\begin{aligned} \text{DIST}(\alpha_m(p), \alpha_m(q)) &=_{2^{k-m}} \text{DIST}(\beta_m(p), \beta_m(q)) \quad \text{and} \\ \alpha_m(p) \leq \alpha_m(q) &\Leftrightarrow \beta_m(p) \leq \beta_m(q) \end{aligned} \quad (6.17)$$

Note that Equation (6.17) implies that Delilah wins the game, because a map that preserves distances of length at most one is an isomorphism. Equation (6.17) holds after move 0 because

$$\text{DIST}(0^{L_n}, max^{L_n}) = 2^{k+1} =_{2^k} 2^{k+1} + 1 = \text{DIST}(0^{L_{n+1}}, max^{L_{n+1}}).$$

Assume inductively that (6.17) holds just after move m. Let Samson begin move $m + 1$ by placing pebble p on some vertex. Let l and r be the closest pebbles to the left and right of p (see Figure 6.16). The inductive assumption tells us that $\text{DIST}(\alpha_m(l), \alpha_m(r)) =_{2^{k-m}} \text{DIST}(\beta_m(l), \beta_m(r))$. Assume without loss of generality that l is the closer of the two pebbles to p or that they are equi-distant. Delilah's response is to place the other pebble p on the point to the right of the other l so that $\text{DIST}(\alpha_{m+1}(l), \alpha_{m+1}(p)) = \text{DIST}(\beta_{m+1}(l), \beta_{m+1}(p))$. It follows, that $\text{DIST}(\alpha_{m+1}(p), \alpha_{m+1}(r)) =_{2^{k-(m+1)}} \text{DIST}(\beta_{m+1}(p), \beta_{m+1}(r))$, so (6.17) holds after move $m + 1$. Thus Delilah wins the game and we have proved that PATH_k is not expressible with quantifier rank less than k, even for ordered structures. □

We remark that the three variables used to express paths in Proposition 6.15 are necessary. In Proposition 6.7, we saw a connected graph H of diameter three and a disconnected graph G such that $G \sim^2 H$. It follows from Theorem 6.10 that CONNECTED is not expressible using 2 variables, no matter what the quantifier rank. Suppose PATH_k were expressible using only two variables for some $k \geq 2$. Then G and H would differ on the $\mathcal{L}^2$ formula $(\forall x_1 x_2) \text{PATH}_k(x_1, x_2)$.

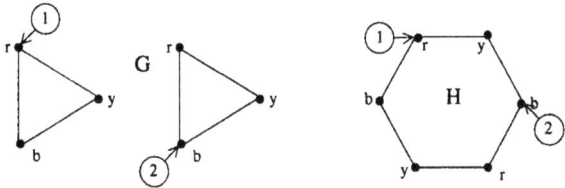

Figure 6.18: The game $\mathcal{G}^2(G, \alpha_0, H, \beta_0)$

In fact, let (α_0, β_0) be the 2-configuration of graphs G, H shown in Figure 6.18. Observe that $(G, \alpha_0) \sim^2 (H, \beta_0)$, but these two structures disagree on the formula $\text{PATH}_1(x_1, x_2)$. It follows that for $k \geq 1$, PATH_k is not expressible in $\mathcal{L}^2$.

6.2 Methodology for First-Order Expressibility

As we now show, Ehrenfeucht-Fraïssé games provide a complete methodology for proving that a query is not first-order. We have already seen that these games are a convenient tool for determining what can be said in first-order logic. This theorem says that if we can show using any method that a query is not first-order expressible, then we can show it using Ehrenfeucht-Fraïssé games.

Theorem 6.19 (Methodology Theorem). *Let $\mathcal{C}$ be any class of finite or infinite structures of some finite, relational vocabulary. Let $S \subseteq \mathcal{C}$ be a boolean query on $\mathcal{C}$. To prove that S is not first-order describable on $\mathcal{C}$ it is necessary and sufficient to show that for all $r \in \mathbf{N}$, there exist structures $\mathcal{A}_r, \mathcal{B}_r \in \mathcal{C}$ such that*

1. $\mathcal{A}_r \in S$ and $\mathcal{B}_r \notin S$, and
2. $\mathcal{A}_r \sim_r \mathcal{B}_r$.

Proof It is easy to see that the methodology is sufficient. The above two conditions imply that $\mathcal{A}_r$ and $\mathcal{B}_r$ agree on all formulas in $\mathcal{L}_r$, but disagree on S. Thus, S is not expressible in $\mathcal{L}_r$ for any r.

Conversely, suppose that S is not first-order expressible over $\mathcal{C}$. Recall from Exercise 6.11 that for any fixed r there are only a bounded number of inequivalent sentences of quantifier rank r.

We say that $\varphi \in \mathcal{L}_r$ is a *complete quantifier-rank r sentence* iff for every other quantifier-rank r sentence ψ of the same vocabulary, $\varphi \vdash \psi$ or $\varphi \vdash \neg\psi$. Let $\varphi_1, \varphi_2, \ldots, \varphi_B$ be a list of all inequivalent, complete quantifier-rank r sentences.

For every quantifier-rank r sentence ψ, each φ_i must assert either ψ or $\neg\psi$. Observe that each structure from $\mathcal{C}$ satisfies a unique φ_i. Suppose there are structures $\mathcal{A}_r \in S$ and $\mathcal{B}_r \in \mathcal{C} - S$ that satisfy the same φ_i. Then $\mathcal{A}_r$ and $\mathcal{B}_r$ satisfy the above conditions. If there is no such pair, then the φ_i's are partitioned by S. In this case, let $Y = \{i \mid (\exists \mathcal{A} \in S)(\mathcal{A} \models \varphi_i)\}$ and let

$$\varphi \equiv \bigvee_{i \in Y} \varphi_i \, .$$

Then φ is a first-order formula of quantifier rank r that expresses S. □

The Methodology Theorem holds whether or not we include ordering in our languages. We now give a few easy applications, proving that various properties are not first-order. Most of these applications do not include ordering.

First, we introduce a general theorem that allows the use of the Methodology Theorem without constructing winning strategies for Delilah by hand.

Definition 6.20 (Gaifman Graph, d-type) Let $\mathcal{A}$ be any structure of vocabulary $\tau = \langle R_1^{a_1}, \ldots, R_r^{a_r}, c_1, \ldots, c_s \rangle$. Define the *Gaifman graph* $G_\mathcal{A} = (|\mathcal{A}|, E_\mathcal{A})$ as follows:

$$E_\mathcal{A} = \left\{ (a,b) \mid (\exists i)(\exists \langle d_1, \ldots, d_{a_i} \rangle \in R_i^\mathcal{A})(a, b \in \{d_1, \ldots, d_{a_i}\}) \right\}.$$

There is an edge between a and b in the Gaifman graph iff a and b occur in the same tuple of some relation of $\mathcal{A}$. As a simple example, if $\mathcal{A} \in \text{STRUC}[\tau_g]$ is a graph, then $G_\mathcal{A} = \mathcal{A}$.

Let $(\mathcal{A}, \alpha_r)$ be the configuration of structure $\mathcal{A}$ after move r of a game. Define the universe of the neighborhood of element a at distance d to be the set of elements of distance at most d from a in the Gaifman graph:

$$|N(a,d)| = \left\{ b \in |\mathcal{A}| \mid \text{DIST}(a,b) \leq d \right\}$$

$N(a, d)$ is almost an induced substructure of $(\mathcal{A}, \alpha_r)$: It inherits the relations from $\mathcal{A}$, but it contains only those constants and pebbled points that are within distance d of a. Define the d-*type* of a to be the isomorphism type of $N(a, d)$. Note that isomorphisms must send each constant $c_j^\mathcal{A}$ to $c_j^\mathcal{B}$ and each pebbled point $\alpha_r(x_i)$ to $\beta_r(x_i)$. (Neighborhood $N(a, d)$ and thus the d-type of a depend on the current configuration $(\mathcal{A}, \alpha_r)$. If the configuration is not clear from the context, then we say the d-type of a with respect to configuration $(\mathcal{A}, \alpha_r)$.) □

The above definitions allow the following,

Theorem 6.21. (Hanf's Theorem) *Let $\mathcal{A}, \mathcal{B} \in \text{STRUC}[\tau]$ and let $r \in \mathbf{N}$. Suppose that for each possible 2^r-type, t, $\mathcal{A}$ and $\mathcal{B}$ have exactly the same number of elements of type t. Then $\mathcal{A} \equiv_r \mathcal{B}$.*

Proof We must show that Delilah wins the game $\mathcal{G}_r(\mathcal{A}, \mathcal{B})$. This is similar to the proof of Proposition 6.15. Delilah's winning strategy is to maintain the following invariant: after move m, $0 \leq m \leq r$,

$$(\mathcal{A}, \alpha_m), (\mathcal{B}, \beta_m) \text{ have same number of each } 2^{r-m}\text{-type.} \tag{6.23}$$

In $\mathcal{G}_r(\mathcal{A}, \mathcal{B})$ there is no bound on the number of pebbles. Therefore we may assume that Samson uses a new pebble at each step. Thus Delilah wins iff she wins at the last round. If Delilah preserves (6.23) then after the last move, the neighborhoods of distance one around each constant or pebbled point are isomorphic to the corresponding neighborhoods in the other structure. It follows that Delilah wins the game.

We have that (6.23) holds for $m = 0$ by assumption. Inductively, assume that it holds after move m. On move $m + 1$, let Samson choose some vertex v. Delilah should respond with any vertex v' of the same 2^{r-m}-type as v.

6.2 Methodology for First-Order Expressibility

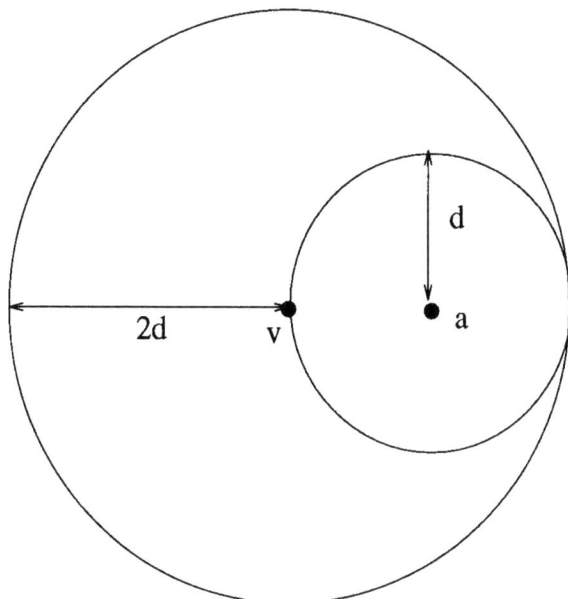

Figure 6.22: Inductive step in proof of Hanf's Theorem: $d = 2^{r-(m+1)}$

We have to show that (6.23) still holds. The inductive assumption immediately implies that,

$(\mathcal{A}, \alpha_m)$, $(\mathcal{B}, \beta_m)$ have same number of each $2^{r-(m+1)}$-type.

Furthermore, the neighborhood $N(a, 2^{r-(m+1)})$ of $(\mathcal{A}, \alpha_m)$ is different from the same neighborhood of $(\mathcal{A}, \alpha_{m+1})$ iff $\mathrm{DIST}(a, v) \leq 2^{r-(m+1)}$. Consider the isomorphism $f : N(v, 2^{r-m}) \to N(v', 2^{r-m})$. It maps every vertex a in $N(v, 2^{r-(m+1)})$ to a corresponding $a' \in N(v', 2^{r-(m+1)})$. Here is the key idea: f maps $N(a, 2^{r-(m+1)})$ isomorphically onto $N(a', 2^{r-(m+1)})$ because these smaller neighborhoods lie inside $\mathrm{dom}(f) = N(v, 2^{r-m})$ (see Figure 6.22). Thus, there is a 1:1 correspondence between the isomorphism types of these neighborhoods close to v and v', so the 1:1 correspondence between the other neighborhoods is undisturbed.

Thus, Delilah's strategy preserves Equation (6.23) and wins the game. □

As a sample application of Theorem 6.21, we prove the following:

Proposition 6.24. *Acyclicity is not first-order expressible.*

Proof Let $\mathcal{A}_r$ be a line segment on 2^{r+3} vertices. Let $\mathcal{B}_r$ be the union of a line segment on 2^{r+2} vertices and a cycle on 2^{r+2} vertices. See Figure 6.25. Observe that $\mathcal{A}_r$ and $\mathcal{B}_r$ both have the same number of each 2^r-type. Therefore, by Theorem 6.21, $\mathcal{A}_r \equiv_r \mathcal{B}_r$. It follows that Acyclicity is not first-order expressible. Note that the same proof works for directed or undirected graphs. □

Exercise 6.26 Using Hanf's Theorem, prove that the following boolean queries are not first-order expressible in the language without ordering.

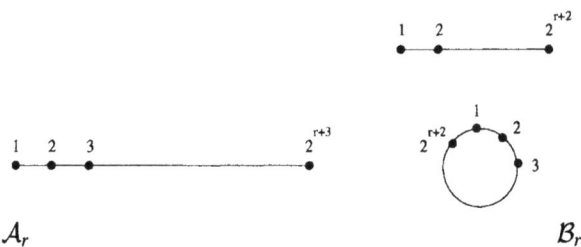

Figure 6.25: Proof of Proposition 6.24.

1. Two-colorability of graphs (cf. Exercise 4.27). [Hint: use $\mathcal{A}_r$ a cycle of size $6d$, and $\mathcal{B}_r$ two cycles of size $3d$ each, with $d = 3^r$.]
2. Consider the following boolean query,

$$\text{CONNECTED} = \{G \mid G \text{ is a connected, undirected graph}\}$$
$$\equiv (\forall xy)(\text{PATH}(x, y) \quad \wedge \quad E(x, y) \rightarrow E(y, x))$$

Prove that

$$\text{CONNECTED} \in \mathcal{L}^3_{2+\lceil\log(n-1)\rceil}(\tau_g)(\text{wo}\leq) - \mathcal{L}_{\lceil\log(n+2)-2\rceil}(\tau_g)(\text{wo}\leq)$$

[Hint: for the upper bound use Proposition 6.15. For the lower bound, use Hanf's theorem with G_r a pair of disjoint cycles of $2^{r+1} - 1$ vertices each and H_r a cycle of $2^{r+2} - 2$ vertices.]
3. Show that REACH is not first-order. [Hint: this is very similar to (2). You just need to place the constants s and t appropriately. Note that s and t may be thought of as the first two pebbled points in the game. Thus, you need $\mathcal{A}_r = G_{r+2}$ and $\mathcal{B}_r = H_{r+2}$ with the appropriately placed s and t.] □

6.3 First-Order Properties Are Local

Hanf's theorem implies that every first-order property is local in the sense that it only concerns neighborhoods of a fixed radius. We have seen that this locality is a useful tool in proving that certain queries are not first-order.

The *degree* of a graph is the maximum number of edges adjacent to any vertex. The *degree* of a structure $\mathcal{A}$ is the degree of its Gaifman graph. We now prove a strengthening of Hanf's theorem for graphs of bounded degree. In this generalization, the number of instances of a given r-type in the two structures need not be equal as long as both numbers are sufficiently large.

Let $\mathcal{A}$ and $\mathcal{B}$ be structures and let n, s be integers. We say that $\mathcal{A}$ and $\mathcal{B}$ are (n, s)-*equivalent* iff for each n-type, σ, $\mathcal{A}$ and $\mathcal{B}$ have the same number of neighborhoods of type σ or they both have more than s neighborhoods of type σ. The following is a generalization of Hanf's theorem for structures of bounded degree.

Theorem 6.27. (Bounded-Degree Hanf Theorem) *Let r and d be fixed. There is an integer s such that for all structures $\mathcal{A}$ and $\mathcal{B}$ of degree at most d, if $\mathcal{A}$ and $\mathcal{B}$ are $(2^r, s)$-equivalent, then $\mathcal{A} \equiv_r \mathcal{B}$.*

Proof This proof is similar to the proof of Theorem 6.21. We must show that Delilah wins the game $\mathcal{G}_r(\mathcal{A}, \mathcal{B})$. Let $s = rd^{2^r} + 1$. Delilah's winning strategy is to maintain the following invariant: after move m, $0 \le m \le r$,

$$(\mathcal{A}, \alpha_m), (\mathcal{B}, \beta_m) \text{ have the same number of each } 2^{r-m}\text{-type.}$$
$$\text{or both have over } (r - m)d^{2^r} + 1 \text{ elements of this type.} \quad (6.28)$$

We have that (6.28) holds for $m = 0$ by assumption. Inductively, assume that it holds after move m. On move $m + 1$, let Samson choose some vertex v. Delilah should respond with any vertex v' of the same 2^{r-m}-type as v.

We have to show that Equation (6.28) still holds. The inductive assumption immediately implies that,

$$(\mathcal{A}, \alpha_m), (\mathcal{B}, \beta_m) \text{ have the same number of each } 2^{r-(m+1)}\text{-type.}$$
$$\text{or both have over } (r - m)d^{2^r} + 1 \text{ elements of this type.}$$

Just as in the proof of Theorem 6.21, the only neighborhoods that change are those within distance $2^{r-(m+1)}$ of v. Furthermore, the same number of neighborhoods change in the same way in $\mathcal{A}$ as in $\mathcal{B}$. The only harm that can be done to Equation (6.28) is that the number of some types can be reduced by the same amount in $\mathcal{A}$ and in $\mathcal{B}$. The number of vertices within distance $\rho = 2^{r-(m+1)}$ of v is at most $d^{\rho+1}/(d-1)$ which is less than d^{2^r}. Thus we have (6.28) holds for $m + 1$ as desired. □

A striking application of Theorem 6.27 is the following theorem of Seese. The definition of linear time in the following is linear time on a unit-cost RAM with $O(\log n)$ bit word size,

Theorem 6.29. *Let $\varphi \in$ FO. Then over bounded degree structures, φ is recognizable in linear time.*

Proof For simplicity, assume that the structures in question are bounded degree graphs and let them be given via adjacency lists.[2] Let r be the quantifier rank of φ and let d be the degree of the graphs in question.

There are a large but bounded number of possible 2^r-types in degree d graphs. The linear time algorithm is to determine the 2^r type of each vertex and count — up to s — how many of each type occurs. This information is what we can call the $2^r, s$ description of G. By Theorem 6.27, the $2^r, s$ description of G determines whether G satisfies φ. We could in principle build — once and for all — a table that lists for each of the finitely many possible $2^r, s$ descriptions, whether or not a graph with this description satisfies φ. From G's description, we can use the table to check in constant additional time whether G satisfies φ. □

---

[2] Adjacency lists are linked lists, one for each vertex, listing all the vertices adjacent to the given vertex, see [AHU74].

6.4 Bounded Variable Languages

A theory Σ satisfies the *k-variable property* iff every first-order formula is equivalent with respect to Σ to a first-order formula that has only k bound variables. Gabbay has shown that the set of models of Σ has the k-variable property for some k iff there exists a finite basis for the set of all temporal-logic connectives over these models [Ga81]. Kozen and Immerman used Ehrenfeucht-Fraïssé games to prove that the theories of linear order and of bounded degree trees have the k-variable property, for appropriate k, (Fact 12.32).

It is interesting and useful to know when a set of structures has the property that all first-order formulas can be expressed using only a bounded number of bound variables. In this section we give one example, showing that the theory of linear order has the 3-variable property.

We begin with a lemma that allows us to give a game-theoretic proof that a theory has the k-variable property. This lemma uses the Compactness Theorem (Theorem 1.35). For this reason, *in this section we consider all structures, not just finite structures*.

Lemma 6.30. *Let $\Sigma \subset \mathcal{L}$ be a first-order theory. Let $\mathcal{L}'$ and $\mathcal{L}''$ be subsets of $\mathcal{L}$ such that $\mathcal{L}'$ is closed under the boolean connectives. Let $k \in \mathbf{N}$. The following conditions are equivalent:*

1. *For all models $\mathcal{A}$ and $\mathcal{B}$ of Σ and all k-configurations α, β of $\mathcal{A}, \mathcal{B}$,*

$$(\mathcal{A}, \alpha) \equiv_{\mathcal{L}'} (\mathcal{B}, \beta) \quad \Rightarrow \quad (\mathcal{A}, \alpha) \equiv_{\mathcal{L}''} (\mathcal{B}, \beta)$$

2. *For all $\varphi \in \mathcal{L}''$ with free variables among $x_1, \ldots, x_k$, there exists $\psi \in \mathcal{L}'$ such that $\Sigma \models \varphi \leftrightarrow \psi$.*

Proof $(2 \to 1)$: If every formula in $\mathcal{L}''$ is equivalent to a formula in $\mathcal{L}'$ and $(\mathcal{A}, \alpha)$ and $(\mathcal{B}, \beta)$ are $\mathcal{L}'$-equivalent, then they are $\mathcal{L}''$-equivalent.

$(1 \to 2)$: If $\Sigma \cup \{\varphi\}$ is inconsistent, then we may take $\psi \equiv$ **false**. Otherwise, let T be the set of all complete $\mathcal{L}'$-types over the variables $x_1, \ldots, x_k$ that is consistent with $\Sigma \cup \{\varphi\}$. Let $\Gamma \in T$ be such a type. Observe that $\Sigma \cup \Gamma \models \varphi$. Otherwise, we could construct models $(\mathcal{A}, \alpha)$ and $(\mathcal{B}, \beta)$ of $\Sigma \cup \Gamma$ that disagree on φ. This is impossible by (1). It follows by the compactness theorem that there is a formula $\psi_\Gamma \in \Gamma$ such that $\Sigma \models \psi_\Gamma \to \varphi$.

Define the following set of formulas,

$$D \;=\; \left\{ \neg \psi_{\Gamma_i} \mid \Gamma_i \in T \right\}.$$

Then $\Sigma \cup D \cup \{\varphi\}$ is inconsistent. By compactness, there must be some finite $F \subseteq T$ such that

$$\Sigma \models \bigwedge_{\Gamma_i \in F} \neg \psi_{\Gamma_i} \to \neg \varphi$$

We can take $\psi = \bigvee_{\Gamma_i \in F} \psi_{\Gamma_i}$. □

Let $\Sigma \subset \mathcal{L}$ be a theory, let $\mathcal{L}' = \mathcal{L}^k$, and let $\mathcal{L}'' = \mathcal{L}$. In this case, Lemma 6.30 implies that condition 1 — which may be proved using Ehrenfeucht–Fraïssé games

— is sufficient to show that every formula in $\mathcal{L}$ that has at most k free variables is equivalent to a formula in $\mathcal{L}^k$. To prove the k-variable property, we must also show that any formula with more than k free variables is equivalent to a formula with at most k bound variables. The following exercise explains how to do this.

Exercise 6.31 Let $\mathcal{L}$ be a first-order relational language with no relation symbols of arity greater than k. Suppose that $\Sigma \subset \mathcal{L}$ is a theory and that $R_1, R_2, \ldots$ are an infinite set of monadic relation symbols from $\mathcal{L}$ that do not occur in Σ. Even though we have infinitely many R_i's, we consider only structures in which only finitely many relations are non-empty. Suppose that for every pair of such structures $\mathcal{A}, \mathcal{B}$ satisfying Σ and every pair of k-configurations α, β, we have

$$(\mathcal{A}, \alpha) \equiv^k (\mathcal{B}, \beta) \quad \Rightarrow \quad (\mathcal{A}, \alpha) \equiv (\mathcal{B}, \beta).$$

Prove that Σ has the k-variable property.

[Hint: this follows essentially from Lemma 6.30. The part you must fill in is how to replace the extra free variables by new monadic relation symbols.] □

The following theorem shows that for structures consisting of a linear order plus a finite number of unary relation symbols, three variables suffice to express all first-order properties. The proof of this theorem produces a winning strategy for Delilah that combines her strategies from several simpler games.

Theorem 6.32. *The set of linear ordered structures satisfies the 3-variable property. These structures may also include any number of monadic relation symbols.*

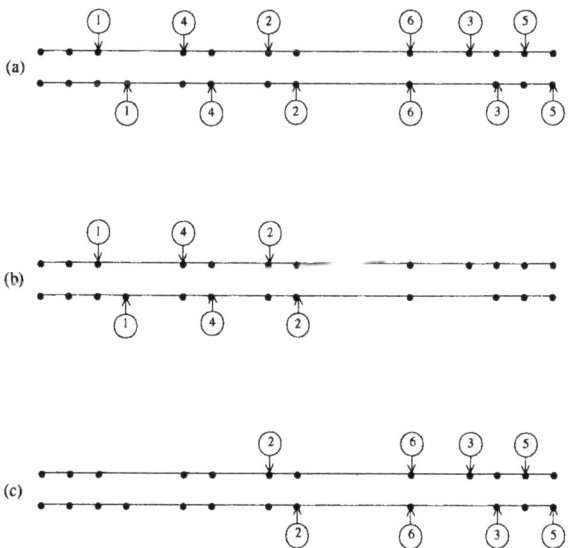

Figure 6.33: Delilah's winning strategy in $\mathcal{G}_{m+1}(\mathcal{A}, \alpha, \mathcal{B}, \beta)$ (a) is built from her winning strategies in $\mathcal{G}_{m+1}(\mathcal{A}, \alpha_\ell, \mathcal{B}, \beta_\ell)$ (b) and $\mathcal{G}_{m+1}(\mathcal{A}, \alpha_r, \mathcal{B}, \beta_r)$ (c).

106   6. Ehrenfeucht-Fraïssé Games

Proof By Exercise 6.31 it suffices to show that for any pair of linear orders $\mathcal{A}, \mathcal{B}$ and any pair of 3-configurations α, β,

$$(\mathcal{A}, \alpha) \equiv^3 (\mathcal{B}, \beta) \quad \Rightarrow \quad (\mathcal{A}, \alpha) \equiv (\mathcal{B}, \beta).$$

We prove the slightly stronger result that for all m,

$$(\mathcal{A}, \alpha) \sim^3_m (\mathcal{B}, \beta) \quad \Rightarrow \quad (\mathcal{A}, \alpha) \sim_m (\mathcal{B}, \beta). \tag{6.34}$$

We prove Equation (6.34) by induction on m. The base case, $m = 0$, is clear because extra pebbles cannot help Samson in the zero move game.

Assume that (6.34) holds for m and suppose that,

$$(\mathcal{A}, \alpha) \sim^3_{m+1} (\mathcal{B}, \beta). \tag{6.35}$$

We now describe a winning strategy for Delilah in the game $\mathcal{G}_{m+1}(\mathcal{A}, \alpha, \mathcal{B}, \beta)$. Suppose that in the initial configuration, $|\text{dom}(\alpha)| < 3$, that is, fewer than three pebbles are on the board. In this case, wherever Samson plays, Delilah can answer using her winning strategy for the game $\mathcal{G}^3_{m+1}(\mathcal{A}, \alpha, \mathcal{B}, \beta)$. Let α_1, β_1 be the resulting configuration. We know that

$$(\mathcal{A}, \alpha_1) \sim^3_m (\mathcal{B}, \beta_1).$$

It follows by the inductive assumption that

$$(\mathcal{A}, \alpha_1) \sim_m (\mathcal{B}, \beta_1),$$

so Delilah wins the remaining m moves of the game.

If $|\alpha| = |\beta| = 3$, then renumber the variables if necessary so that $(\mathcal{A}, \alpha)$ and $(\mathcal{B}, \beta)$ both satisfy $x_1 < x_2 < x_3$. Let α_ℓ, β_ℓ and α_r, β_r be the restrictions of α, β to the domains $\{x_1, x_2\}$ and $\{x_2, x_3\}$ respectively.

By Equation (6.35), Delilah wins the three-variable, $m+1$-move games on these reduced configurations, that is,

$$(\mathcal{A}, \alpha_\ell) \sim^3_{m+1} (\mathcal{B}, \beta_\ell) \quad \text{and} \quad (\mathcal{A}, \alpha_r) \sim^3_{m+1} (\mathcal{B}, \beta_r).$$

Since the domains of these configurations have size less than three, we know by the previous case that,

$$(\mathcal{A}, \alpha_\ell) \sim_{m+1} (\mathcal{B}, \beta_\ell) \quad \text{and} \quad (\mathcal{A}, \alpha_r) \sim_{m+1} (\mathcal{B}, \beta_r).$$

We now combine Delilah's winning strategies for the games $\mathcal{G}_{m+1}(\mathcal{A}, \alpha_\ell, \mathcal{B}, \beta_\ell)$ and $\mathcal{G}_{m+1}(\mathcal{A}, \alpha_r, \mathcal{B}, \beta_r)$ to give her a winning strategy for the game $\mathcal{G}_{m+1}(\mathcal{A}, \alpha, \mathcal{B}, \beta)$. Notice that we are playing a game with an unlimited number of pebbles, so Samson need never reuse a pebble. See Figure 6.33.

Delilah's strategy is as follows: If Samson places a pebble to the left of pebble two, then Delilah answers according to her winning strategy in $\mathcal{G}_{m+1}(\mathcal{A}, \alpha_\ell, \mathcal{B}, \beta_\ell)$. If he places a pebble to the right of pebble two, then she answers according to her winning strategy in $\mathcal{G}_{m+1}(\mathcal{A}, \alpha_r, \mathcal{B}, \beta_r)$. After the $m + 1$ moves, Delilah has won both of the subgames. Thus, the map from the chosen points of $\mathcal{A}$ to the chosen points of $\mathcal{B}$ in the left subgame is an isomorphism, and similarly for the right subgame. Furthermore, all the chosen points in the left subgame are less than x_2

and all the chosen points in the right subgame are greater than x_2. Thus, the map from *all* the pebbled points in $\mathcal{A}$ to the pebbled points in $\mathcal{B}$ is an isomorphism and Delilah has won game $\mathcal{G}_{m+1}(\mathcal{A}, \alpha, \mathcal{B}, \beta)$. □

Exercise 6.36 Show that linear order does not have the two variable property. □

6.5 Zero-One Laws

In this section, we see that with very high probability, structures chosen at random are very simple from a first-order point of view. We begin by writing some sentences called "extension axioms". For any finite vocabulary, with no function or constant symbols, the extension axioms form a complete theory that is true in almost all structures.

The extension axioms can be written for any finite relational vocabulary. We first write them for undirected graphs. Consider the following sentence γ_k, whose meaning is that there are least $k - 1$ distinct vertices and any $k - 1$ tuple of distinct vertices may be extended to a k tuple in any conceivable way:

$$\gamma_k \equiv$$
$$(\exists x_1 \ldots x_{k-1}. \text{distinct}(x_1, \ldots, x_{k-1})) \wedge$$
$$(\forall x_1 \ldots x_{k-1}. \text{distinct}(x_1, \ldots, x_{k-1}))$$
$$\Big((\exists x_k. \text{distinct}(x_1, \ldots, x_k))(E(x_1, x_k) \wedge E(x_2, x_k) \wedge \cdots \wedge E(x_{k-1}, x_k))$$
$$\wedge (\exists x_k. \text{distinct}(x_1, \ldots, x_k))(E(x_1, x_k) \wedge E(x_2, x_k) \wedge \cdots \wedge \neg E(x_{k-1}, x_k))$$
$$\wedge \quad \vdots \quad \vdots \quad \vdots \quad \vdots \quad \vdots \quad \vdots$$
$$\wedge (\exists x_k. \text{distinct}(x_1, \ldots, x_k))(E(x_1, x_k) \wedge E(x_2. x_k) \wedge \cdots$$
$$\wedge E(x_{i-1}, x_k) \wedge \neg E(x_i, x_k) \wedge \cdots \wedge \neg E(x_{k-1}, x_k))$$
$$\wedge \quad \vdots \quad \vdots \quad \vdots \quad \vdots \quad \vdots \quad \vdots$$
$$(\exists x_k. \text{distinct}(x_1, \ldots, x_k))(\neg E(x_1, x_k) \wedge \neg E(x_2, x_k) \wedge \cdots \wedge \neg E(x_{k-1}, x_k))\Big)$$
(6.37)

A simple counting argument shows that almost all graphs satisfy γ_k. Define $\mu_n(\varphi)$ to be the percentage of (ordered) structures of size n that satisfy φ,

$$\mu_n(\varphi) = \frac{|\{G \mid n = \|G\|; G \models \varphi\}|}{|\{G \mid n = \|G\|\}|}$$

Lemma 6.38. *For any fixed $k > 0$,*

$$\lim_{n \to \infty} \mu_n(\gamma_k) = 1 \quad.$$

Proof Let $v_1, \ldots, v_{k-1}$ be a $k - 1$ tuple of distinct vertices from a random graph G of size n, let x be any of the $n + 1 - k$ remaining vertices, and let c be any of the k

108   6. Ehrenfeucht-Fraïssé Games

conjuncts of the sentence γ_k. Conjunct c asserts $k-1$ independent conditions on the existence of edges from x, each of which has probability $1/2$. For this reason, the probability that x does not meet condition c for $v_1, \ldots, v_{k-1}$ is $\alpha = (1 - 2^{-(k-1)})$. Thus, the probability that none of the $(n+1-k)$ x's satisfies condition c is α^{n+1-k}. It follows that the probability that G does not satisfy γ_k is less than

$$k \cdot n^{k-1} \alpha^{n+1-k}.$$

This expression goes quickly to 0 as n goes to infinity.   □

The sentence γ_k says that any next move in the game $\mathcal{G}^k$ can be matched by Delilah. Thus we have the following,

Lemma 6.39. *Let G and H be undirected graphs satisfying γ_k. Then $G \sim^k H$.*

Proof We show by induction on m that $G \sim^k_m H$. In the base case when $m = 0$, there are no chosen points so $G \sim^k_0 H$ holds vacuously.

Suppose that $G \sim^k_m H$ and let Delilah play the $m+1$ move game as follows: For the first m moves Delilah follows her winning strategy for the m move game. Thus she has not lost yet. On the last move, suppose that Samson picks up pair k of pebbles and places one of them on some vertex v of G. We may assume that the previously pebbled points are all distinct (Exercise 6.12). Since $H \models \gamma_k$, there exists a vertex v' of H such that for all $j < k$, there is an edge from v' to $\beta_m(x_j)$ in H iff there is an edge from v to $\alpha_m(x_j)$ in G. Thus, Delilah answers by putting her pebble k on v' and wins the game.   □

We can generalize γ_k as follows: Let $\tau = \langle R_1^{a_1}, \ldots, R_r^{a_r} \rangle$ be a vocabulary with no constant symbols. Let A be the set of all atomic formulas of the form $R_i(y_1, \ldots, y_{a_i})$ such that

$$x_k \in \{y_1, \ldots, y_{a_i}\} \subseteq \{x_1, \ldots, x_k\}$$

Define $\gamma_k(\tau)$ to be the following conjunction, which says that every $(k-1)$-tuple may be extended to a k-tuple in any conceivable way.

$$\gamma_k(\tau) \equiv (\forall x_1 \ldots x_{k-1} . \operatorname{distinct}(x_1, \ldots, x_{k-1}))$$

$$\bigwedge_{S \subseteq A} \left((\exists x_k . \operatorname{distinct}(x_1, \ldots, x_k)) (\bigwedge_{\alpha \in S} \alpha \ \wedge \bigwedge_{\alpha \in A-S} \neg \alpha) \right)$$

It is easy to see that Lemmas 6.38 and 6.39 go through for any such $\gamma_k(\tau)$. The following theorem tells us that any property expressible by a set of sentences from $\mathcal{L}^k(\tau)$ is true in almost all structures, or false in almost all structures. This is sometimes known as the zero-one law for $\mathcal{L}^\omega_{\infty\omega}$.

Theorem 6.40. (Zero-One Law) *Let $S \subseteq \mathcal{L}^k$ be any set of k variable sentences over a finite vocabulary τ with no constant or function symbols. Then the following limit exists and is equal to zero or one.*

$$\lim_{n \to \infty} \mu_n(S).$$

Proof By Lemma 6.39, for every sentence $\varphi \in S$, either $\gamma_k \vdash \varphi$, or $\gamma_k \vdash \neg\varphi$. Thus Lemma 6.38 tells us that the above limit exists and (a) is equal to one if γ_k implies every sentence in S and (b) is equal to 0 if γ_k implies the negation of some sentence in S. □

Corollary 6.41. *Assume that no constant symbols occur. Then a zero-one law holds for the language* FO(wo$\leq$). *Furthermore, a zero-one law holds for all of the following languages:* FO(wo$\leq$)(TC), FO(wo$\leq$)(LFP), *and* FO(wo$\leq$)(PFP). *(The operators* TC *and* PFP *are defined in Chapters 9 and 10, respectively.)*

Proof We see in later chapters that any sentence in one of these languages is equivalent to an infinite disjunction of sentences from $\mathcal{L}^k(\tau)$ for some k and τ. Since γ_k determines the truth of any sentence in $\mathcal{L}^k(\tau)$, it also determines the truth of any infinite disjunction of such sentences. □

Suppose first-order logic has a zero-one law for the class $\mathcal{C}$ of structures. We see in Exercise 12.52 that this means that for each k, $\mathcal{L}_k$ has *bounded expressive power on average* in the following sense: There is a fixed bound b such that almost all elements of $\mathcal{C}$ fall in one of b $\mathcal{L}^k$-equivalence classes. When talking about typical structures, $\mathcal{L}^k$ can express only a bounded number of facts.

The zero-one laws do not hold for ordered structures or for structures with constants. This can be seen from the following equation,

$$\mu_n(E(0,0)) = \frac{1}{2}. \tag{6.42}$$

For ordered structures $\mathcal{A}$ and $\mathcal{B}$, if $\mathcal{A} \equiv^2 \mathcal{B}$, then $\|\mathcal{A}\| = \|\mathcal{B}\|$. Thus, for $k \geq 2$, $\mathcal{L}^k$ is not bounded.

6.6 Ehrenfeucht-Fraïssé Games with Ordering

When the ordering relation is present, Ehrenfeucht-Fraïssé games become much more difficult for Delilah. We present a clear explanation for this. Then we include a few game lower bounds for languages including ordering. It should not be surprising that the lower bounds are more difficult with ordering because here we are really talking about computation. For the same reason, such lower bounds are quite deserving of the effort required. Later, we show some more sophisticated lower bounds with ordering.

The following provides an upper bound on the complexity lower bounds we can prove using games for ordered structures.

Proposition 6.43. *Let G and H be ordered graphs and let $n = \max(\|G\|, \|H\|)$. If $G \sim^3_{\lceil \log(n-1) \rceil + 1} H$, then $G = H$.*

Proof Suppose for the sake of contradiction that $G \sim^3_{\lceil \log(n-1) \rceil + 1} H$ but $G \neq H$. Let $n = \|G\|$ and $m = \|H\|$. Suppose $n > m$. Let PATH$_{<_d}(x, y)$ mean that there

110   6. Ehrenfeucht-Fraïssé Games

is a path of length at most d from x to y, where each step in the path is given by the less than relation. Thus $G \models \text{PATH} <_{n-1} (0, max)$, but $H \models \neg\text{PATH}_{n-1}(0, max)$. From Proposition 6.15 we know that $\text{PATH} <_n \in \mathcal{L}^3_{\lceil\log(n-1)\rceil}$. Thus, $n = \|G\| = \|H\|$.

Since $G \neq H$, there must be a pair of vertices i, j such there is an edge from vertex i to vertex j in one of the graphs but not the other. In the game $\mathcal{G}^3_{\lceil\log(n-1)\rceil+1}(G, H)$, Samson should play the vertices i and j in G in his first two moves. If Delilah answers with vertices i and j from H, then she loses immediately. If she does not answer with these vertices then (G, α_2) and (H, β_2) disagree on a formula of the form $\text{PATH} <_d (x_k, c)$, for $k \in \{1, 2\}$, $c \in \{max, 0\}$, and $d \leq (n - 1)/2$. It follows that Samson wins the remaining $\lceil\log(n - 1)\rceil - 1$ move game.   □

Exercise 6.44 Show that life is even worse when BIT is present: If G and H are ordered graphs that are $\mathcal{L}_{2, \lceil\log_*(n-1)\rceil+2}$ equivalent in the presence of BIT, then $G = H$. (Recall that $\log_* n$ is the smallest k such that log applied k times to n is at most 1.)

[Hint: in the presence of BIT, if Delilah matches vertex number i from one graph, with vertex number $i' \neq i$ from the other graph, how long can she avoid a contradiction?]   □

We saw in Proposition 6.14 that without ordering, we needed $n + 1$ variables to say that a structure has size exactly n or that it has even cardinality. Thus, without ordering, even languages as strong as FO(LFP) cannot express very simple queries.

We now show that with ordering but still without BIT, quantifier rank $\log n$ is necessary to count even mod 2 (cf. Exercise 4.18 where BIT and thus counting mod 2, etc., are shown to be expressible in IND[$\log n$]).

Proposition 6.45.
The sentence EVEN, meaning that the cardinality of the universe is even, is not expressible in quantifier rank $\lceil\log(n - 1)\rceil - 1$ with ordering, but without BIT.

Proof This proposition follows from the proof of Proposition 6.15. The graphs L_n and L_{n+1} shown in Figure 6.16 are $\lceil\log(n - 1)\rceil - 1$-equivalent and disagree on property EVEN.   □

Corollary 6.46. *Boolean query REACH is not expressible in quantifier rank $\lceil\log(n - 1)\rceil - 1$ with ordering, but without BIT.*

Proof Define G_n and G_{n+1} to be graphs that have the same universe and ordering relation as L_n and L_{n+1}, respectively. Let $s = 0$ and $t = max$. Replace the edge predicate by the following relation, meaning that the points are two steps apart in the ordering,

$$E(x, y) \equiv (\exists z)(\text{SUC}(x, z) \wedge \text{SUC}(z, y)).$$

Thus REACH holds for one of G_n, G_{n+1} and not the other. However, G_n and G_{n+1} are still $\lceil\log(n - 1)\rceil - 2$ equivalent because any win by Samson in $\mathcal{G}_r(G_n, G_{n+1})$ can be converted in one more move to a win by Samson in $\mathcal{G}_{r+1}(L_n, L_{n+1})$.   □

Another corollary of Proposition 6.45 is that the language $(aa)^*$, i.e., the set of even-length strings over a single letter alphabet, is not first-order without BIT. By Theorem 1.37, this is equivalent to the fact that $(aa)^*$ is not a star-free regular language.

Exercise 6.47 Prove that in the genealogical database, the ancestor relation is not a first-order query, cf. Examples 1.2, 1.27. This theorem holds even in the presence of the ordering relation. □

Open Problem 6.48 Extend Corollary 6.46 to show that REACH is not expressible in quantifier rank $o[\log n]$ even in the presence of BIT. Class NC_1 is contained in $IND[\log n / \log \log n]$ (Theorem 5.35). Thus, a solution of this problem would prove the very interesting result $NC_1 \neq NL$.

Historical Notes and Suggestions for Further Reading

Games $\mathcal{G}_m(\mathcal{A}, \mathcal{B})$ are called "Ehrenfeucht-Fraïssé games" in honor of their inventors [Ehr61, Fra54]. Barwise invented games that measure the number of variables used as well as the quantifier rank [Bar77]. Immerman reinvented this game, using pebbles [I80]. As seen in later chapters, variations of Ehrenfeucht-Fraïssé games for most logics have been developed. See [KV95] for the games that characterize languages with arbitrary generalized quantifiers. Originally, Samson was referred to as "Player I" and Delilah as "Player II". Joel Spencer coined the names "spoiler" and "duplicator" [ASE92]. From there it was a short step to the more memorable Samson and Delilah. Theorem 6.10 is due to the inventors of the various versions of these games.

We use Ehrenfeucht-Fraïssé games in later chapters to determine that certain queries are not expressible in first-order logic and other, more powerful, languages. The "methodology" for proving lower bounds (Theorem 6.19) was adapted from a lecture by Phokion Kolaitis in [IKL95].

Lemma 6.30 and Theorem 6.32 are due to Kozen and Immerman [IK87]. In addition, exact bounds are proved there on the numbers k such that theories of bounded trees have the k-variable property (Fact 12.32).

Theorems 6.21 and 6.27 are from [FSV95]. Gaifman graphs (Definition 6.20) are named for Haim Gaifman. See [Ga81] for Gaifman's theorem on the locality of first-order properties.

Recently a significant strengthening of Gaifman's locality theorem for first-order logic was proved by Schwentick and Barthelmann [SB98]. They showed that every first-order formula is equivalent to a formula of the form $(\exists x_1 \cdots x_k)(\forall y)\varphi$ where φ is r-local around y. That is, all quantification in φ is restricted to elements of distance at most r from y in the Gaifman graph. Here r depends only on φ.

Theorem 6.29 is from [See95]. See [AHU74] for a discussion of the unit cost RAM.

The Zero-One Law for First-Order Logic is due to Fagin [Fag73, Fag76]. Fagin's original proof used extension axioms (6.37). A good reference for Theorem 6.40 and Corollary 6.41 is [BGK85] by Blass, Gurevich and Kozen. See also [I80].

In our opinion, zero-one laws are inimical to computation. In order to compute we seem to need an ordering on the universe. As we have seen, having an ordering, or even having a single constant, eliminates the possibility of a zero-one law (6.42). Furthermore, the interested reader may look ahead to Fact 12.53, which shows that a language must be very weak indeed — on almost all structures — in order to support a zero-one law. Zero-one laws have very little attention in this book, but there is a large literature on the subject. See [Co88, KV92a, Spe93] for surveys.

7
Second-Order Logic and Fagin's Theorem

Second-order logic consists of first-order logic plus the power to quantify over relations on the universe. We prove Fagin's theorem which says that the queries computable in NP are exactly the second-order existential queries. A corollary due to Stockmeyer says that the second-order queries are exactly those computable in the polynomial-time hierarchy.

7.1 Second-Order Logic

Second-order logic consists of first-order logic plus new relation variables over which we may quantify. For example, the formula $(\forall A^r)\varphi$ means that for all choices of r-ary relation A, φ holds. Let SO be the set of second-order expressible boolean queries.

Any second-order formula may be transformed into an equivalent formula with all second-order quantifiers in front. If all these second-order quantifiers are existential, then we have a second-order existential formula. Let (SO∃) be the set of second-order existential boolean queries. Consider the following example, in which R, Y, and B are unary relation variables. To indicate their arity, we place exponents on relation variables where they are quantified.

$$\Phi_{\text{3-color}} \equiv (\exists R^1)(\exists Y^1)(\exists B^1)(\forall x)\Big[\big(R(x) \vee Y(x) \vee B(x)\big) \wedge (\forall y)\Big(E(x,y) \to \\ \neg\big(R(x) \wedge R(y)\big) \wedge \neg\big(Y(x) \wedge Y(y)\big) \wedge \neg\big(B(x) \wedge B(y)\big)\Big)\Big]$$

Observe that a graph G satisfies $\Phi_{\text{3-color}}$ iff G is 3-colorable. Three colorability of graphs is an NP complete problem (3-COLOR). In Section 7.2, we see that three colorability remains complete via first-order reductions. It will then follow that every query computable in NP is describable in SO∃.

Second-order logic is extremely expressive. For this reason, it is very easy to write second-order specifications of queries. For the same reason, such specifications are not feasible to execute without further refinement (cf. Section 9.6). Recall that the first-order queries are those that can be computed on a CRAM in constant time, using polynomially many processors (Theorem 5.2). We will see that the second-order queries are those that can be computed in constant parallel time, but using exponentially many processors (Corollary 7.27).

Here are a few other examples of SO∃ queries.

Example 7.1 SAT is the set of boolean formulas in conjunctive normal form (CNF) that admit a satisfying assignment (Example 2.18).

The boolean query SAT is expressible in SO∃ as follows:

$$\Phi_{\text{SAT}} \equiv (\exists S)(\forall x)(\exists y)((P(x, y) \wedge S(y)) \vee (N(x, y) \wedge \neg S(y))).$$

Φ_{SAT} asserts that there exists a set S of variables — the variables that should be assigned **true** — that is a satisfying assignment of the input formula. □

Example 7.2 Boolean query CLIQUE is the set of pairs $\langle G, k \rangle$ such that graph G has a complete subgraph of size k (Example 2.10). The vocabulary for CLIQUE is $\tau_{gk} = \langle E^2, k \rangle$.

The SO∃ sentence Φ_{CLIQUE} says that there is a numbering of the vertices such that those vertices numbered less than k form a clique. In order to describe this numbering it is convenient to existentially quantify a function f. This can be replaced by a binary relation in the usual way (Exercise 7.3). Let $\text{Inj}(f)$ mean that f is an injective function,

$$\text{Inj}(f) \equiv (\forall xy)(f(x) = f(y) \rightarrow x = y)$$
$$\Phi_{\text{CLIQUE}} \equiv (\exists f^1.\text{Inj}(f))(\forall xy)((x \neq y \wedge f(x) < k \wedge f(y) < k) \rightarrow E(x, y))$$

□

Exercise 7.3 Show how formula Φ_{CLIQUE} may be rewritten using an existentially quantified relation F of arity two, rather than function f. □

Exercise 7.4 Hamiltonian-Circuit (HC) is the boolean query that is true of an undirected graph iff it has a Hamiltonian circuit, i.e., a path that starts and ends at the same vertex and visits every other vertex exactly once. Write an SO∃ sentence that expresses HC. [Hint: say that there exists a total ordering of the vertices that determines a Hamiltonian circuit.] □

Exercise 7.5 Write an SO∃ sentence that expresses TSP — the traveling salesperson problem. Boolean query TSP has as input an undirected graph G with weights on its edges and an integer L. The TSP query is true iff G admits a Hamiltonian circuit whose total weight is at most L. In order to code TSP instances as logical structures, we must decide on an appropriate range for the integer weights. To be quite general, you should code these integers as binary strings. Let the vocabulary for TSP be $\tau_{\text{tsp}} = \langle W^3, L^1 \rangle$, consisting of a binary string $W(x, y, \cdot)$ for each potential edge (x, y) and a binary string $L(\cdot)$ representing limit L. For pairs (x, y) that are not edges, we can let the edge weight be the maximum value, i.e., the string of all 1's.

[Hint: you can assert the existence of the correct Hamiltonian-Circuit as in Exercise 7.4. To say that the total is at most L, you should assert the existence of a ternary relation R that maintains the running sum.] □

We finish this section by proving the easy direction of Fagin's Theorem.

Proposition 7.6. *The second-order existentially definable boolean queries are all computable in NP. In symbols,* SO∃ ⊆ NP.

Proof Given is a second-order existential sentence $\Phi \equiv (\exists R_1^{r_1}) \ldots (\exists R_k^{r_k}) \psi$. Let τ be the vocabulary of Φ. Our task is to build an NP machine N such that for all $\mathcal{A} \in \text{STRUC}[\tau]$,

$$(\mathcal{A} \models \Phi) \quad \Leftrightarrow \quad (N(\text{bin}(\mathcal{A}))\downarrow) \tag{7.7}$$

Let $\mathcal{A}$ be an input structure to N and let $n = \|\mathcal{A}\|$. What N does is to nondeterministically write down a binary string of length n^{r_1} representing R_1, and similarly for R_2 through R_k. By nondeterministically write down a binary string, we mean that at each step, N nondeterministically chooses whether to write a 0 or a 1. After this polynomial number of steps, we have an expanded structure $\mathcal{A}' = (\mathcal{A}, R_1, R_2, \ldots, R_k)$. N should accept iff $\mathcal{A}' \models \psi$. By Theorem 3.1, we can test if $\mathcal{A}' \models \psi$ in logspace, so certainly in NP. Notice that N accepts $\mathcal{A}$ iff there is some choice of relations R_1 through R_k such that $(\mathcal{A}, R_1, R_2, \ldots, R_k) \models \psi$. Thus, Equivalence 7.7 holds. □

7.2 Proof of Fagin's Theorem

The following theorem characterizes complexity class NP in an elegant and machine independent way. This was originally proved in Ron Fagin's 1973 doctoral thesis. It was the theorem that began the subject of descriptive complexity.

Theorem 7.8. (Fagin's Theorem) NP *is equal to the set of existential, second-order boolean queries,* NP = SO∃. *Furthermore, this equality remains true when the first-order part of the second-order formulas is restricted to be universal.*

116   7. Second-Order Logic and Fagin's Theorem

Proof Let N be a nondeterministic Turing machine that uses time $n^k - 1$ for inputs bin($\mathcal{A}$) with $n = \|\mathcal{A}\|$. We write a second-order sentence,

$$\Phi \;=\; (\exists C_1^{2k} \ldots C_g^{2k} \Delta^k)\varphi$$

that says, "There exists an accepting computation $\overline{C}, \Delta$ of N." More precisely, first-order sentence φ will have the property that $(\mathcal{A}, \overline{C}, \Delta) \models \varphi$ iff $\overline{C}, \Delta$ is an accepting computation of N on input $\mathcal{A}$. That is, Equation 7.7 holds.

We describe how to code N's computation. $\overline{C}$ consists of a matrix $\overline{C}(\bar{s}, \bar{t})$ of n^{2k} tape cells with space $\bar{s}$ and time $\bar{t}$ varying between 0 and $n^k - 1$. We use k-tuples of variables $\bar{t} = t_1, \ldots, t_k$ and $\bar{s} = s_1, \ldots s_k$ each ranging over the universe of $\mathcal{A}$, i.e. from 0 to $n - 1$, to code these values. For each $\bar{s}, \bar{t}$ pair, $\overline{C}(\bar{s}, \bar{t})$ codes the tape symbol σ that appears in cell $\bar{s}$ at time $\bar{t}$ if n's head is not on this cell. If the head is present, then $\overline{C}(\bar{s}, \bar{t})$ codes the pair $\langle q, \sigma \rangle$ consisting of N's state q at time $\bar{t}$ and tape symbol σ. Let $\Gamma = \{\gamma_0, \ldots, \gamma_g\} = (Q \times \Sigma) \cup \Sigma$ be a listing of the possible contents of a computation cell. We will let C_i be a $2k$-ary relation variable for $0 \le i \le g$. The intuitive meaning of $C_i(\bar{s}, \bar{t})$ is that computation cell $\bar{s}$ at time $\bar{t}$ contains symbol γ_i.

At each step, the nondeterministic Turing machine will make one of at most two possible choices.[1] We encode these choices in k-ary relation Δ. Intuitively, $\Delta(\bar{t})$ is true, if step $\bar{t} + 1$ of the computation makes choice "1"; otherwise it makes choice "0". Note that these choices can be determined from $\overline{C}$, but the formula is simplified when we explicitly quantify Δ. See Figure 7.9 for a view of N's computation.

It is now fairly straightforward to write the first-order sentence $\varphi(\overline{C}, \Delta)$ saying that $\overline{C}, \Delta$ codes a valid accepting computation of N. The sentence φ consists of four parts,

$$\varphi \;\equiv\; \alpha \wedge \beta \wedge \eta \wedge \zeta,$$

where α asserts that row 0 of the computation correctly codes input bin($\mathcal{A}$), β says that it is never the case that $C_i(\bar{s}, \bar{t})$ and $C_j(\bar{s}, \bar{t})$ both hold, for $i \ne j$, η says that for all $\bar{t}$, row $\bar{t} + 1$ of $\overline{C}$ follows from row $\bar{t}$ via move $\Delta(\bar{t})$ of N, and ζ says that the last row of the computation includes the accept state. We can write sentence ζ explicitly. We may assume that when N accepts it clears its tape and moves all the way to the left and enters a unique accept state q_f. Let γ_{17} be the member of Γ corresponding to the pair $\langle q_f, 1 \rangle$ of state q_f, looking at the symbol 1. Then $\zeta = C_{17}(\overline{0}, \overline{max})$.

Sentence α must assert that the input is of length $I_\tau(n)$ for some n and that $\mathcal{A}$ has been correctly coded as bin($\mathcal{A}$) (cf. Exercise 2.3). For example, suppose that τ includes relation symbol R_1 of arity one. Assume that cell symbols γ_0, γ_1 are '0','1', respectively. Then α includes the following clauses, meaning that cell $0 \ldots 0 s_k$ contains 1 if $R_1(s_k)$ holds and 0 if it doesn't.

---

[1] A nondeterministic Turing machine can make one of at most a bounded number of choices at any step. By reducing this to a binary choice per step, we slow the machine down by a small constant factor and make the analysis simpler.

	Space						
	0	1	p	$n-1$	n	$n^k - 1$	Δ
Time 0	$\langle q_0, w_0 \rangle$	w_1	$\cdots$	w_{n-1}	⊔	$\cdots$ ⊔	δ_0
1	w_0	$\langle q_1, w_1 \rangle$	$\cdots$	w_{n-1}	⊔	$\cdots$ ⊔	δ_1
	$\vdots$	$\vdots$	$\vdots$		$\vdots$		$\vdots$
t			$a_{-1}a_0a_1$				δ_t
$t+1$			b				δ_{t+1}
	$\vdots$	$\vdots$	$\vdots$		$\vdots$		$\vdots$
$n^k - 1$	$\langle q_f, 1 \rangle$	$\cdots$	$\cdots$		$\cdots$		

Figure 7.9: An NP computation on input $w_0 w_1 \cdots w_{n-1}$; ⊔ denotes blank

$$\cdots \wedge \left(\bar{t} = 0 = s_1 = \ldots = s_{k-1} \wedge s_k \neq 0 \wedge R_1(s_k) \rightarrow C_1(\bar{s}, \bar{t}) \right)$$
$$\wedge \left(\bar{t} = 0 = s_1 = \ldots = s_{k-1} \wedge s_k \neq 0 \wedge \neg R_1(s_k) \rightarrow C_0(\bar{s}, \bar{t}) \right) \wedge \cdots$$

The following sentence η asserts that the contents of tape cell $(\bar{s}, \bar{t} + 1)$ follows from the contents of cells $(\bar{s} - 1, \bar{t})$, $(\bar{s}, \bar{t})$, and $(\bar{s} + 1, \bar{t})$ via the move $\Delta(\bar{t})$ of N. Let $\langle a_{-1}, a_0, a_1, \delta \rangle \xrightarrow{N} b$ mean that the triple of cell contents a_{-1}, a_0, a_1 lead to cell b via move δ of N.

$$\eta_1 \equiv (\forall \bar{t}. \bar{t} \neq \overline{max})(\forall \bar{s}. \bar{0} < \bar{s} < \overline{max}) \bigwedge_{\langle a_{-1}, a_0, a_1, \delta \rangle \xrightarrow{N} b} \left(\neg^\delta \Delta(\bar{t}) \vee \right.$$
$$\left. \neg C_{a_{-1}}(\bar{s} - 1, \bar{t}) \vee \neg C_{a_0}(\bar{s}, \bar{t}) \vee \neg C_{a_1}(\bar{s} + 1, \bar{t}) \vee C_b(\bar{s}, \bar{t} + 1) \right)$$

Here, $\neg^\delta$ is $\neg$ if $\delta = 1$ and is the empty symbol if $\delta = 0$.

Finally, let $\eta \equiv \eta_0 \wedge \eta_1 \wedge \eta_2$ where η_0 and η_2 encode the same information when $\bar{s} = \bar{0}$ and $\overline{max}$ respectively. □

Observe that the first-order part of formula Φ in the proof of Proposition 7.6 is universal and is in conjunctive normal form. Furthermore, if N is a deterministic polynomial-time machine, then we do not need choice relation Δ, so the first-order part of Φ is a Horn formula.[2] We obtain the following corollary, which is part of Grädel's Theorem (Theorem 9.32).

Corollary 7.10. *Every polynomial-time query is expressible as a second-order, existential Horn formula:* $P \subseteq SO\exists\text{-Horn}$.

The proof of Proposition 7.6 shows that nondeterministic time n^k is contained in (SO∃, arity $2k$). Lynch improved this to arity k. His proof uses the numeric predicate PLUS. Fagin's theorem holds even without numeric predicates, since we

---

[2] A Horn formula is a formula in conjunctive normal form with at most one positive literal per clause (Definition 9.26).

can existentially quantify binary relations and assert that they are $\leq$ and BIT. However, without the numeric predicates, we need an existential first-order quantifier to specify time $\bar{t} + 1$, given time $\bar{t}$.

Theorem 7.11. (Lynch's Theorem)

$$\text{For } k \geq 1, \quad \text{NTIME}[n^k] \subseteq \text{SO}\exists(\text{arity } k).$$

Proof This is analogous to Lemma 5.31. We modify the proof of Fagin's theorem so that instead of guessing the entire tape at every step only a bounded number of bits per step is guessed. The following relations need to be guessed.

1. $Q_i(\bar{t})$ meaning that the state at move $\bar{t}$ is q_i,
2. $S_i(\bar{t})$ meaning that the symbol written at move $\bar{t}$ is σ_i,
3. $D(\bar{t})$ meaning that the head moves one space to the right after move $\bar{t}$; otherwise it moves one space to the left.

We must write a first-order formula asserting that $\overline{Q}, \overline{S}, D$ encode a correct accepting computation of N. The only difficulty in doing this is that for each move $\bar{t}$, we must ascertain the symbol $\rho_{\bar{t}}$ that is read by N. $\rho_{\bar{t}}$ is equal to σ_i where $S_i(\bar{t}')$ holds, and $\bar{t}'$ is the last time before $\bar{t}$ that the head was in its present location (or it is the corresponding input symbol if this is the first time the head is at this cell).

To express $\rho_{\bar{t}}$, we need to express the function $\bar{s} = p(\bar{t})$ meaning that at time $\bar{t}$, the head is at position $\bar{s}$. Since we are restricted to relations of arity k, we cannot guess the $k \log n$ bits per time needed to specify the function p. The solution to this problem is to do the next best thing and existentially quantify the current head position once every $\log n$ steps. We do this by quantifying k bits per step in relations $P_i(\bar{t})$, $i = 1, 2, \ldots, k$. When we string $\log n$ of these together, from time $r \log n$ through time $(r + 1) \log n - 1$, we have a total of $k \log n$ bits which encode the head position at time $r \log n$.

The idea is similar to the proof of Bit Sum Lemma 1.18. Recall that numeric predicate BIT allows us to use each first-order variable to store $\log n$ bits. Furthermore, predicate BSUM(x, y), meaning that the number of one's in the binary expansion of x is y, is first-order (Lemma 1.18). This enables us to assert that relations $\overline{P}$ are consistent with the head movements given by D and thus correctly code the head position at $\log n$ step intervals. Finally, using BSUM again, we can ascertain the head position at any time $\bar{t}$. □

The converse of Lynch's Theorem is an open question:

Open Problem 7.12 Prove or disprove: SO∃(arity k) = NTIME[n^k]

The subtlety in Open Problem 7.12 is that the first-order part of an SO∃(arity k) formula may have more than k universal quantifiers. Thus, a first step in answering Open Problem 7.12 may be to answer:

Open Problem 7.13 Is there a fixed k such that FO $\subseteq$ DTIME[n^k]? Is there a fixed k such that FO $\subseteq$ NTIME[n^k]?

Grandjean gave a close relationship between nondeterministic time n^k and the class (SO∃, fun, $k\forall$) of properties expressible by second-order existential sentences including function variables and containing only k universal first-order quantifiers.

Fact 7.14. *For* $k \geq 2$, NTIME$[n^k]$ ⊆ (SO∃, *fun*, $k\forall$) = (SO∃, *fun*, $k\forall$, *arity k*) ⊆ NTIME$[n^k(\log n)^2]$.

By considering the nondeterministic random access machine (NRAM) instead of the Turing machine, Grandjean later gave an exact bound,

Fact 7.15. *For* $k \geq 1$,

$$\text{NRAM-TIME}[n^k] = (\text{SO}\exists, \textit{fun}, k\forall, \textit{arity } k).$$

7.3 NP-Complete Problems

In 1971, Cook proved that SAT (Example 2.18) is NP-complete via polynomial-time Turing reductions [Coo71]. In fact, the problem is NP-complete via significantly weaker reductions:

Theorem 7.16. SAT *is complete for* NP *via first-order reductions.*

Proof This follows from Fagin's theorem. Given any boolean query $B \in$ NP, we know that $B = \text{MOD}[\Phi]$ where $\Phi = (\exists S_1^{a_1} \cdots S_g^{a_g} \Delta^k)(\forall x_1 \cdots x_t)\psi(\bar{x})$, with ψ quantifier-free. We may assume that $\psi(\bar{x}) = \bigwedge_{j=1}^{r} C_j(\bar{x})$ is in conjunctive normal form.

For any input structure $\mathcal{A}$ with $n = \|\mathcal{A}\|$, define the boolean formula $\gamma(\mathcal{A})$ as follows: $\gamma(\mathcal{A})$ has boolean variables: $S_i(e_1, \ldots, e_{a_i})$ and $D(e_1, \ldots, e_k)$, $i = 1, \ldots, g$, $e_1, \ldots, e_{a_i} \in |\mathcal{A}|$. The clauses of $\gamma(\mathcal{A})$ are $C_j(\bar{e})$, $j = 1, \ldots, r$ as $\bar{e}$ ranges over all t-tuples from $|\mathcal{A}|$. In each $C_j(\bar{e})$, there may be some occurrences of numeric or input predicates: $\gamma(\bar{e})$. These should be replaced by **true** or **false** according to whether they are true or false in $\mathcal{A}$.

It is clear from the construction that

$$\mathcal{A} \in B \quad \Leftrightarrow \quad \mathcal{A} \models \Phi \quad \Leftrightarrow \quad \gamma(\mathcal{A}) \in \text{SAT}.$$

Furthermore, the mapping from $\mathcal{A}$ to $\gamma(\mathcal{A})$ is a $t+1$-ary first-order query. □

Now that we know that SAT is NP-complete via first-order reductions, we can reduce SAT to other SO∃ boolean queries. This is possible iff these other problems are also NP-complete via first-order reductions (Exercise 2.15).

Proposition 7.17. *Let* 3-SAT *be the subset of* SAT *in which each clause has at most three literals. Then* 3-SAT *is NP-complete via first-order reductions.*

Proof We show that SAT $\leq_{\text{fo}}$ 3-SAT. Here is an example of the idea behind the reduction. Let $C = (\ell_1 \vee \ell_2 \vee \cdots \vee \ell_7)$ be a clause with more than three literals. Observe that $C \in$ SAT iff $C' \in$ 3-SAT, where C' is the following clause in which new variables $d_1, \ldots, d_4$ are introduced.

120   7. Second-Order Logic and Fagin's Theorem

$$C' \equiv (\ell_1 \vee \ell_2 \vee d_1) \wedge (\overline{d_1} \vee \ell_3 \vee d_2) \wedge (\overline{d_2} \vee \ell_4 \vee d_3) \wedge$$
$$(\overline{d_3} \vee \ell_5 \vee d_4) \wedge (\overline{d_4} \vee \ell_6 \vee \ell_7)$$

The first-order reduction from SAT to 3-SAT proceeds as follows. Let $\mathcal{A} \in$ STRUC[$\langle P^2, N^2 \rangle$] be an instance of SAT with $n = \|\mathcal{A}\|$. Each clause c of $\mathcal{A}$ is replaced by $2n$ clauses as follows:

$$c' \equiv ([x_1]^c \vee d_1) \wedge (\overline{d_1} \vee [x_2]^c \vee d_2) \wedge (\overline{d_2} \vee [x_3]^c \vee d_3) \wedge \cdots \wedge$$
$$(\overline{d_n} \vee [\overline{x_1}]^c \vee d_{n+1})(\overline{d_{n+1}} \vee [\overline{x_2}]^c \vee d_{n+2}) \wedge \cdots \wedge (\overline{d_{2n-1}} \vee [\overline{x_n}]^c)$$

Here $[\ell]^c$ means the literal ℓ if this occurs in c and **false** otherwise. It is not hard to see that c' is satisfiable iff c is satisfiable and that c' is definable in a first-order way from c. □

Proposition 7.18. 3-COLOR *is NP-complete via first-order reductions.*

Proof We will show that 3-SAT $\leq_{\text{fo}}$ 3-COLOR. We are given an instance $\mathcal{A}$ of 3-SAT and we must produce a graph $f(\mathcal{A})$ that is three colorable iff $\mathcal{A} \in$ 3-SAT. Let $n = \|\mathcal{A}\|$, so $\mathcal{A}$ is a boolean formula with at most n variables and n clauses.

The construction of $f(\mathcal{A})$ is shown in Figure 7.19. Notice the triangle, with vertices labeled T, F, R. Any three-coloring of the graph must color these three vertices distinct colors. We may assume without loss of generality that the colors used to color T, F, R are true, false, and red, respectively.

Graph $f(\mathcal{A})$ also contains a ladder each rung of which is a variable x_i and its negation $\overline{x_i}$. Each of these is connected to R, meaning that any valid three-coloring colors one of $x_i, \overline{x_i}$ true and the other false.

Finally, for each clause $C_i = \ell_1 \vee \ell_2 \vee \ell_3$, $f(\mathcal{A})$ contains the gadget G_i consisting of six vertices. G_i has three inputs a_i, b_i, c_i, connected to literals ℓ_1, ℓ_2, ℓ_3, respectively, and it has one output, f_i. See Figure 7.19 where gadget G_1 corresponds to clause $C_1 = \overline{x_1} \vee x_2 \vee \overline{x_3}$.

Observe that the triangle a_1, b_1, d_1 serves as an "or"-gate in that d_1 may be colored true iff at least one of its inputs $\overline{x_1}, x_2$ is colored true. Similarly, output f_1

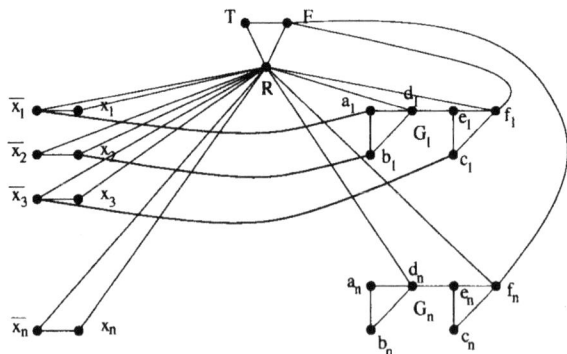

Figure 7.19: 3-SAT $\leq_{\text{fo}}$ 3-COLOR; G_1 encodes clause $C_1 = (\overline{x_1} \vee x_2 \vee \overline{x_3})$

may be colored true iff at least one of d_1 and the third input, $\overline{x_3}$ is colored true. Since f_i is connected to both F and R, f_i can only be colored true. It follows that a three coloring of the literals can be extended to color G_i iff the corresponding truth assignment makes C_i true. Thus, $f(\mathcal{A}) \in$ 3-COLOR iff $\mathcal{A} \in$ 3-SAT.

The details of first-order reduction f are easy to fill in. $f(\mathcal{A})$ consists of one triangle, a ladder with n rungs, and n copies of the gadget. The only dependency on the input $\mathcal{A}$ — as opposed to its size — is that there is an edge from literal ℓ to input j of gadget G_i iff ℓ is the jth literal occurring in C_i. □

7.4 The Polynomial-Time Hierarchy

We defined the polynomial-time hierarchy (PH) to be the set of boolean queries accepted in polynomial time by alternating Turing machines making a bounded number of alternations between existential and universal states (Equation (2.35)). The original definition of the polynomial-time hierarchy was via nondeterministic polynomial-time Turing reductions (Definition 2.9).

Definition 7.20 (**Polynomial-Time Hierarchy via Oracles**) Let $\Sigma_0^p =$ P be level 0 of the polynomial-time hierarchy. Inductively, let

$$\Sigma_{i+1}^p = \left\{ L(M^A) \mid M \text{ is an NP oracle TM}, A \in \Sigma_i^p \right\}$$

Equivalently, Σ_{i+1}^p is the set of boolean queries that are nondeterministic polynomial-time Turing reducible to a set from Σ_i^p,

$$\Sigma_{i+1}^p = \left\{ B \mid B \leq_{np}^t A, \text{ for some } A \in \Sigma_i^p \right\}$$

Define Π_i^p to be co-Σ_i^p, $\Pi_i^p = \left\{ \overline{A} \mid A \in \Sigma_i^p \right\}$. Finally, PH $= \bigcup_{k=1}^{\infty} \Sigma_k^p$. □

The relationship between second-order boolean queries and the levels of the polynomial hierarchy are given by the following:

Theorem 7.21. *Let $S \subseteq$ STRUC$[\tau]$ be a boolean query, and let $k \geq 1$. The following are equivalent,*

1. $S = MOD[\Phi]$, *for some $\Phi \in \Sigma_k^{SO}$. (Here Σ_k^{SO} is the set of all second-order sentences with second-order quantifier prefix $(\exists \overline{R_1})(\forall \overline{R_2}) \ldots (Q_k \overline{R_k})$.)*
2. $S = \left\{ x \mid (\exists y_1.|y_1| \leq |x|^c)(\forall y_2.|y_2| \leq |x|^c) \cdots (Q_k y_k.|y_k| \leq |x|^c) R(x, \overline{y}) \right\}$ *where R is a deterministic polynomial-time predicate on $k+1$ tuples of binary strings and c is a constant.*
3. $S \in$ ATIME-ALT$[n^{O[1]}, k]$.
4. $S \in \Sigma_k^p$.

Corollary 7.22. *A boolean query is in the polynomial-time hierarchy iff it is second-order expressible,*

$$\text{PH} = \text{SO}.$$

From Theorem 4.10 and Corollary 7.22, we obtain the following descriptive characterization of the P? = NP question: P is equal to NP iff every second-order query — over finite, ordered structures — is expressible as a first-order inductive definition.

Corollary 7.23. *The following conditions are equivalent:*

1. P = NP.
2. *Over finite, ordered structures*, FO(LFP) = SO.

Proof If FO(LFP) = SO, then P $\subseteq$ NP $\subseteq$ PH = P. Conversely, if P = NP, then PH = NP, so FO(LFP) = SO. □

Exercise 7.24 Prove Theorem 7.21. [Hint: By induction on k. The subtle part is relating Σ_k^p to the other conditions. For this, note that an NP machine with an oracle $A \in \Sigma_{k-1}^p$ can guess all the answers to its oracle queries. Then, at the end of its computation, it can check that these answers were all correct. This is a polynomial number of Σ_{k-1}^p and Π_{k-1}^p questions.] □

As seen in the following, the polynomial-time hierarchy is robust enough to finesse the difficulty that occurs in Open Problem 7.12,

Exercise 7.25 Prove that for any k,

$$\text{SO}(\text{arity } k) = \text{PH-TIME}[n^k] = \text{ATIME-ALT}[n^k, O(1)] \qquad \square$$

Exercise 7.26 Fagin's Theorem (Theorem 7.8) is a generalization of the Spectrum Theorem. Define the *spectrum* of a first-order sentence φ to be the set of cardinalities of the finite models of φ,

$$\text{spec}(\varphi) = \{n \mid n = |\mathcal{A}| \text{ for some } \mathcal{A} \in \text{MOD}[\varphi]\}.$$

As an example let φ_{field} be the conjunction of the field axioms, so $\text{spec}(\varphi_{\text{field}})$ is the set of prime powers. An interesting question is whether the set of spectra of first-order sentences is closed under complementation, i.e., if S is a spectrum then is $\overline{S} = \mathbf{Z}^+ - S$ one also? As we now see, this is equivalent to an important open question in complexity theory. The Spectrum Theorem says that a set $S \subseteq \mathbf{Z}^+$ is the spectrum of a first-order sentence iff $S \in \text{NTIME}[2^{O[n]}]$. Fagin originally called the finite models of SO∃ sentences "generalized spectra".

1. Write a first-order sentence whose spectrum is the set of even positive integers
2. Modify part 1 to get a first-order sentence whose spectrum is the set of odd positive integers.
3. Prove the Spectrum Theorem.
   [Hint: Show how it follows from Theorem 7.8. Note that a problem $S \subseteq \mathbf{Z}^+$ is assumed to be a set of binary strings coding natural numbers. Thus $S \in \text{NTIME}[2^{O[n]}]$ iff S coded in unary is in NP.]
4. Show using the Spectrum Theorem that $\overline{\text{spec}(\varphi_{\text{field}})}$ is a spectrum. □

As a corollary to the proof of Theorem 5.2, we obtain the following characterization of PH as a parallel complexity class. Up to this point, we had been assuming for notational simplicity that a CRAM has at most polynomially many processors. However, the class CRAM-PROC[$t(n)$, $p(n)$] still makes sense for numbers of processors $p(n)$ that are not polynomially bounded.

Corollary 7.27. PH *is equal to the set of boolean queries recognizable by a CRAM using exponentially many processors and constant time,*

$$\text{PH} \;=\; \bigcup_{k=1}^{\infty} \text{CRAM-PROC}[1, 2^{n^k}]$$

Proof The inclusion SO $\subseteq$ CRAM-PROC[$1, 2^{n^{O[1]}}$] follows just as in the proof of Lemma 5.4. A processor number is now large enough to give values to all the relational variables as well as to all the first-order variables. Thus, as in Lemma 5.4, the CRAM can evaluate each first or second-order quantifier in three steps.

The inclusion CRAM-PROC[$1, 2^{n^{O[1]}}$] $\subseteq$ SO follows just as in the proof of Lemma 5.3. The only difference is that we use second-order variables to specify the processor number. □

In fact, Corollary 7.27 can be extended to,

Corollary 7.28. *For all constructible* $t(n)$,

$$\text{SO}[t(n)] \;=\; \text{CRAM-PROC}[t(n), 2^{n^{O[1]}}] \,.$$

Observe that Corollary 7.27 suggests that PH is a rather strange complexity class. No one would ever buy exponentially many processors and then use them only for constant time. See Corollary 10.30 for an interesting characterization of the much more robust complexity class PSPACE as exponentially many processors running in polynomial time.

Historical Notes and Suggestions for Further Reading

Theorem 7.8 (Fagin's theorem) was proved in Fagin's thesis, [Fag73, Fag74]. The idea of using choice relation Δ is due to Papadimitriou [Pap94]. The Spectrum Theorem discussed in Exercise 7.26 is due to Jones and Selman [JS74]. See [B82] for a history of the spectrum problem.

Theorem 7.16 was first proved by Lovász and Gács [LG77]. Dahlhaus proved that SAT is NP-complete via quantifier-free, first-order reductions [Da84].

The polynomial-hierarchy (PH) was defined by Stockmeyer [Sto77]. Corollary 7.22 appears there as well. Item 2 of Theorem 7.21 is due to Wrathall [Wra76].

Some of the simple ways to write NP-complete problems as SO∃ formulas, like CLIQUE (Example 7.2) are due to Jose Antonio Medina [MI94].

Theorem 7.11 is due to Lynch, [Lyn82]. Facts 7.14 and 7.15 are due to Grandjean; see [Gra84, Gra85, Gra89] for their proofs. An interesting place to start investigating Open Problem 7.13 is to consider the deterministic time complexity of problem CLIQUE(k) — the set of graphs containing a k-clique — for a fixed k. The best known algorithm is due to Boppana and Halldórsson [BH92].

Exercise 7.25 is from [I83]. Corollary 7.27 is from [I89a].

8
Second-Order Lower Bounds

Although second-order logic is very powerful, when we restrict second-order quantifiers to monadic relations, we can prove some strong lower bounds using Ehrenfeucht-Fraïssé games.

8.1 Second-Order Games

The second-order version of Ehrenfeucht-Fraïssé games gives players the power not only to pebble elements of the universe but to choose arbitrary new relations over the universe. We consider mainly the monadic case, because the choice of c monadic relations amounts to coloring the elements of the universe with one of 2^c colors. This is more manageable than playing relations of greater arity.

In this section we define the second-order existential, monadic game. We also define a variant called the Ajtai-Fagin game. We prove a complete methodology theorem for this game: A boolean query is inexpressible in second-order existential, monadic logic iff Delilah wins the corresponding Ajtai-Fagin game. We include a simple and elegant application.

The remaining two sections of this chapter include more sophisticated applications. They can be safely skipped by anyone who does not yet love to play Ehrenfeucht-Fraïssé games.

Definition 8.1 (SO∃(monadic) **Games**) Let $\mathcal{A}, \mathcal{B}$ be structures of the same vocabulary. For $c, m \in \mathbf{N}$, define the *second-order(monadic) c-color, m-move game* on $\mathcal{A}, \mathcal{B}$ as follows. The two players start with the coloring phase in which Samson chooses c monadic relations $C_1^{\mathcal{A}}, C_2^{\mathcal{A}}, \ldots, C_c^{\mathcal{A}}$ on $|\mathcal{A}|$. Delilah answers with c

monadic relations $C_1^B, C_2^B, \ldots, C_c^B$ on $|\mathcal{B}|$. Observe that the coloring phase is not symmetric, in that Samson must play on $\mathcal{A}$.

Next, the two players play the m-move Ehrenfeucht-Fraïssé game on the two expanded structures, i.e., they play

$$\mathcal{G}_m((\mathcal{A}, C_1^A, C_2^A, \ldots, C_c^A), (\mathcal{B}, C_1^B, C_2^B, \ldots, C_c^B)) \qquad \square$$

Not surprisingly we have,

Theorem 8.2. *Let $\mathcal{A}$ and $\mathcal{B}$ be two not necessarily finite structures of the same finite, relational vocabulary and let $c, m \in \mathbf{N}$. Then the following two conditions are equivalent,*

1. *For any formula $\Phi \equiv (\exists C_1^1 \cdots C_c^1)(\varphi)$ with φ first-order of quantifier rank m,*
   $\mathcal{A} \models \Phi \Rightarrow \mathcal{B} \models \Phi$
2. *Delilah has a winning strategy for the second-order(monadic) c-color, m-move game on $\mathcal{A}, \mathcal{B}$.*

Proof This theorem follows easily from Theorem 6.10 and the fact that there are only finitely many inequivalent formulas in $\mathcal{L}_m(\tau \cup \{C_1^1, \ldots, C_c^1\})$ (Exercise 6.11). Assume 1 and let $C_1^A, C_2^A, \ldots, C_c^A$ be Samson's move in the coloring phase. Let φ be the conjunction of all quantifier-rank m sentences that are true of $(\mathcal{A}, C_1^A, C_2^A, \ldots, C_c^A)$. By 1 it follows that

$$\mathcal{B} \models (\exists C_1^1 \cdots C_c^1)\varphi$$

Delilah thus can play $C_1^B, C_2^B, \ldots, C_c^B$ that are witnesses of φ. It follows that

$$(\mathcal{A}, C_1^A, C_2^A, \ldots, C_c^A) \equiv_m (\mathcal{B}, C_1^B, C_2^B, \ldots, C_c^B)$$

so Delilah wins the first-order part of the game.

Conversely, suppose 1 is false and that $\mathcal{A} \models \Phi$ but $\mathcal{B} \models \neg \Phi$. Samson plays the coloring phase by choosing $C_1^A, C_2^A, \ldots, C_c^A$ witnessing the truth of Φ. However Delilah responds, the two structures $(\mathcal{A}, C_1^A, C_2^A, \ldots, C_c^A)$ and $(\mathcal{B}, C_1^B, C_2^B, \ldots, C_c^B)$ disagree on the quantifier-rank m formula φ, so Samson wins the first-order part of the game. $\qquad \square$

As in Theorem 6.19, the SO∃(monadic) Ehrenfeucht-Fraïssé game gives a complete methodology for determining whether a boolean query is expressible in SO∃(monadic). Much of the reason for this completeness goes back to Exercise 6.11, which showed that there are only finitely many inequivalent formulas in a given finite relational language restricted to a given quantifier rank and with a given number of free variables. The next definition provides a useful way to think about SO∃(monadic) games.

Definition 8.3 *Let $\mathcal{L}$ be a language. We say that φ is a* complete formula *of $\mathcal{L}$ if it is consistent and maximal in the sense that if $\psi \in \mathcal{L}$ is another consistent formula that implies φ, then φ and ψ are equivalent.* Consider the second-order(monadic) c-color, m-move game on structures of the finite, relational vocabulary τ. Let

$C = C(c, m, \tau)$ be the finite number of such inequivalent, complete formulas in $\mathcal{L}_m(\tau \cup \{C_1^1, \ldots, C_c^1\})$ that have one free variable. □

Suppose we play the second-order(monadic) c-color, m-move game on structures of the finite, relational vocabulary τ. The result of Samson's coloring a structure with c new monadic relations is that he partitions the universe into a larger but still finite number $C = C(c, m, \tau)$ of equivalence classes. The equivalence relation is as follows. Let $a, a' \in |\mathcal{A}|$, where $\mathcal{A} \in \text{STRUC}[\tau \cup \{C_1^1, \ldots, C_c^1\}]$. Then a, a' are equivalent iff $(\mathcal{A}, a) \sim_m (\mathcal{A}, a')$.

Since the SO∃(monadic) Ehrenfeucht-Fraïssé game is still fairly difficult for Delilah to play, Ajtai and Fagin invented an equivalent game. The games are equivalent in that Delilah has a winning strategy in one iff she has a winning strategy in the other. However, Delilah's winning strategies in the Ajtai-Fagin game can be much simpler.

Definition 8.4 (Ajtai-Fagin Game) Let $I \subseteq \text{STRUC}[\tau]$ be a boolean query. Define the Ajtai-Fagin game on I as follows:

1. Samson chooses natural numbers c and m.
2. Delilah chooses a structure $\mathcal{A} \in \text{STRUC}[\tau]$ such that $\mathcal{A} \in I$.
3. Samson chooses c monadic relations $C_1^{\mathcal{A}}, C_2^{\mathcal{A}}, \ldots, C_c^{\mathcal{A}}$ on $|\mathcal{A}|$.
4. Delilah chooses a structure $\mathcal{B} \in \text{STRUC}[\tau]$ such that $\mathcal{B} \notin I$. Delilah also chooses c monadic relations $C_1^{\mathcal{B}}, C_2^{\mathcal{B}}, \ldots, C_c^{\mathcal{B}}$ on $|\mathcal{B}|$.
5. Finally, the two players play

$$\mathcal{G}_m((\mathcal{A}, C_1^{\mathcal{A}}, C_2^{\mathcal{A}}, \ldots, C_c^{\mathcal{A}}), (\mathcal{B}, C_1^{\mathcal{B}}, C_2^{\mathcal{B}}, \ldots, C_c^{\mathcal{B}}))$$

□

The Ajtai-Fagin game makes life easier for Delilah because she does not have to choose structure $\mathcal{B}$ until after Samson has already colored $\mathcal{A}$. However, the Ajtai-Fagin game still gives a complete methodology for proving that a boolean query is not second-order existential, monadic:

Theorem 8.5. (Ajtai-Fagin Methodology Theorem) *Let $I \subseteq \text{STRUC}[\tau]$ be a boolean query. Then the following are equivalent,*

1. *Delilah has a winning strategy for the Ajtai-Fagin game on I.*
2. *$I \notin \text{SO∃}(\text{monadic})$*

Proof Suppose $I = \text{MOD}[\Phi]$, where

$$\Phi \equiv (\exists C_1^1 \cdots C_c^1)(\varphi); \quad \varphi \text{ of quantifier rank } m \tag{8.6}$$

Then Samson wins the Ajtai-Fagin game on I as follows: His first move is to choose c, m. Let $\mathcal{A} \in I$ be chosen by Delilah. Samson chooses colorings $C_1^{\mathcal{A}}, C_2^{\mathcal{A}}, \ldots, C_c^{\mathcal{A}}$ such that

$$(\mathcal{A}, C_1^{\mathcal{A}}, C_2^{\mathcal{A}}, \ldots, C_c^{\mathcal{A}}) \models \varphi$$

Delilah then chooses a structure $\mathcal{B} \notin I$, so $\mathcal{B} \models \neg \Phi$. Thus, whatever coloring $C_1^\mathcal{B}, C_2^\mathcal{B}, \ldots, C_c^\mathcal{B}$ is chosen by Delilah, we know that $(\mathcal{B}, C_1^\mathcal{B}, C_2^\mathcal{B}, \ldots, C_c^\mathcal{B}) \models \neg \varphi$. It follows that

Samson wins the game $G_m((\mathcal{A}, C_1^\mathcal{A}, C_2^\mathcal{A}, \ldots, C_c^\mathcal{A}), (\mathcal{B}, C_1^\mathcal{B}, C_2^\mathcal{B}, \ldots, C_c^\mathcal{B}))$.

Conversely, suppose $I \notin \mathrm{SO\exists}(\mathrm{monadic})$. We describe a winning strategy for Delilah. Let Samson choose the numbers c, m. Let S_m be a maximal set of inequivalent sentences of the form of Equation (8.6), where the first-order part φ is a complete sentence of quantifier rank m. S_m is finite by Exercise 6.11. The sentences that cause us trouble are those that are not satisfied by any structure not in I,

$$T \equiv \left\{ \Phi \in S_m \mid \forall \mathcal{B} \in (\mathrm{STRUC}[\tau] - I), \mathcal{B} \models \neg \Phi \right\}.$$

Let Ψ be the disjunction of all sentences in T. Note that Ψ is of the form of (8.6) because the disjunction can be pushed through the second-order existential quantifiers. By assumption, there must be a structure $\mathcal{A} \in I$ such that $\mathcal{A} \models \neg \Psi$. Otherwise, Ψ would express I. Delilah should play this $\mathcal{A}$. Now, let Samson choose colors $C_1^\mathcal{A}, C_2^\mathcal{A}, \ldots, C_c^\mathcal{A}$, and let φ_0 be the complete quantifier-rank m sentence satisfied by $(\mathcal{A}, C_1^\mathcal{A}, C_2^\mathcal{A}, \ldots, C_c^\mathcal{A})$. Let $\Phi_0 = (\exists \overline{C})\varphi_0$. Then $\mathcal{A} \models \Phi_0$, so $\Phi_0 \notin T$. Therefore, by the definition of T, there exists $\mathcal{B} \in (\mathrm{STRUC}[\tau] - I)$ such that $\mathcal{B} \models \Phi_0$. Delilah should play this $\mathcal{B}$ together with a coloring that witnesses Φ_0. It follows that

Delilah wins the game $G_m((\mathcal{A}, C_1^\mathcal{A}, C_2^\mathcal{A}, \ldots, C_c^\mathcal{A}), (\mathcal{B}, C_1^\mathcal{B}, C_2^\mathcal{B}, \ldots, C_c^\mathcal{B}))$. □

Our first application of the Ajtai-Fagin Methodology is,

Theorem 8.7. *The connectivity problem for undirected graphs is not expressible in monadic second-order existential logic without numeric relations:*

$$\mathrm{CONNECTED} \notin \mathrm{SO\exists}(\mathrm{monadic})(\mathrm{wo}\leq).$$

Proof We show that Delilah has a winning strategy for the Ajtai-Fagin game on CONNECTED. Suppose that Samson chooses constants c, m. Delilah responds with a sufficiently large cycle, $\mathcal{A}$. Sufficiently large means that $\|\mathcal{A}\| \geq h(2^{ch} + 1)$, where $h = 2^{m+1} + 1$.

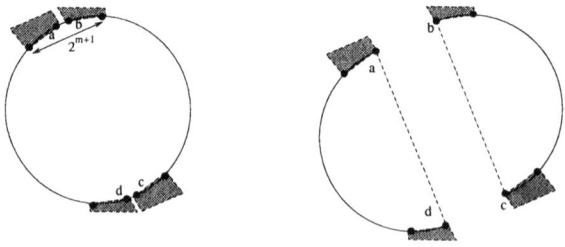

Figure 8.8: An Ajtai-Fagin game showing CONNECTED $\notin \mathrm{SO\exists}(\mathrm{monadic})$

Let $C_1^{\mathcal{A}}, C_2^{\mathcal{A}}, \ldots, C_c^{\mathcal{A}}$ be the coloring on $\mathcal{A}$ played by Samson. The neighborhood $N(a, 2^m)$ of any vertex a contains h vertices. The number of possible colorings of such a neighborhood is thus 2^{ch}. Since $\|\mathcal{A}\|$ is chosen to be at least $h(2^{ch} + 1)$ it must contain at least two disjoint neighborhoods $N(a, 2^m)$ and $N(c, 2^m)$ containing identical colorings in clockwise order around the cycle. See Figure 8.8.

Let b be the next vertex after a and d the next vertex after c in clockwise order around $\mathcal{A}$. To construct $\mathcal{B}$, Delilah removes edges (a, b) and (c, d) and replaces them by edges (a, d) and (b, c). Thus, $\mathcal{B}$ consists of two disjoint cycles. Delilah colors $\mathcal{B}$ exactly as Samson has colored $\mathcal{A}$.

It follows that the structures $(\mathcal{A}, C_1^{\mathcal{A}}, C_2^{\mathcal{A}}, \ldots, C_c^{\mathcal{A}})$ and $(\mathcal{B}, C_1^{\mathcal{B}}, C_2^{\mathcal{B}}, \ldots, C_c^{\mathcal{B}})$ have the same number of each 2^m type. Thus, by Hanf's Theorem (Theorem 6.21),

Delilah wins the game $G_m((\mathcal{A}, C_1^{\mathcal{A}}, C_2^{\mathcal{A}}, \ldots, C_c^{\mathcal{A}}), (\mathcal{B}, C_1^{\mathcal{B}}, C_2^{\mathcal{B}}, \ldots, C_c^{\mathcal{B}}))$. □

The negation of connectivity is easily expressible in SO∃(monadic)(wo≤). The following sentence does it by asserting the existence of a set S that is not empty and does not contain all vertices, but contains all the neighbors of its elements. Thus, S contains a proper connected component of the graph, which is therefore not connected.

$$\overline{\text{CONNECTED}} \equiv (\exists S^1)[(\exists xy)(S(x) \wedge \neg S(y)) \wedge$$
$$(\forall xy)((S(x) \wedge E(x, y)) \rightarrow S(y))]$$

Corollary 8.9. *SO∃(monadic)(wo≤) is not closed under complementation.*

Corollary 8.9 holds with arbitrary numeric predicates. Let ALL-EVEN-DEGREE be true of undirected graphs all of whoses vertices have an even number of edges. Ajtai showed that ALL-EVEN-DEGREE is not in SO∃, in the presence of arbitrary numeric relations [Ajt83]. However, in the presence of an ordering relation, ALL-EVEN-DEGREE is expressible in second-order, monadic, universal logic. This is done as follows: We say that for all two-colorings of the graph, and for all vertices v, if the coloring of the neighbors of v alternates between the two colors, then v's first neighbor has a different color than its last neighbor. Note also, that it is an SO∃ property that a particular vertex has even degree. It follows that

Corollary 8.10. *The language SO∃(monadic) is not closed under complementation, or first-order quantification. In particular, ALL-EVEN-DEGREE is expressible in SO∀(monadic) and in the form $(\forall x)$SO∃(monadic), using ordering as the only numeric predicate. However, ALL-EVEN-DEGREE is not expressible in second-order existential, monadic logic in the presence of arbitrary numeric predicates.*

8.2 SO∃(monadic) Lower Bound on Reachability

In 1986, Paris Kanellakis observed to Ron Fagin that undirected graph reachability (REACH$_u$) is expressible in SO∃(monadic) and asked whether the same is true for

8. Second-Order Lower Bounds

directed reachability (REACH). Ajtai and Fagin showed that it is not. Their proof was later simplified by Arora and Fagin, and we present the latter proof in this section. We begin with Kanellakis' observation:

Proposition 8.11. *The undirected reachability query is expressible in second-order, existential, monadic logic, without numeric relations. In symbols,*

$$\text{REACH}_u \in \text{SO}\exists(\text{monadic})(\text{wo}\leq) \ .$$

Proof To express the existence of an undirected path from s to t we assert the existence of a set of vertices S with the following properties:

1. s and t are members of S.
2. s and t each have unique neighbors in S.
3. All other members of S have exactly two neighbors in S.

Clearly, these three conditions are first-order expressible using S. Furthermore, any such set S must include a path from s to t. Conversely, the vertices along a shortest path from s to t constitute such a set S. □

We saw in Exercise 1.29 that connectivity is first-order reducible — in fact quantifier-free reducible — to REACH_u. We see in Theorem 8.22 that connectivity is not expressible in monadic, second-order existential logic, even in the presence of ordering. Whether this holds in the presence of BIT is an open question.
The following corollary says that monadic second-order existential logic is not very robust.

Corollary 8.12. *The language* $\text{SO}\exists(\text{monadic})(\text{wo BIT})$ *is not closed under quantifier-free reductions.*

In the remainder of this section, we prove the following theorem:

Theorem 8.13. $\text{REACH} \notin \text{SO}\exists(\mathit{monadic})(\text{wo}\leq)$.

Proof The reader may have noticed that the proof of Proposition 8.11 does not work for directed graphs. The reason is that a graph $G \in \text{REACH}$ may have the property that every set of vertices forming a path from s to t admits a "back edge", i.e., an edge from a vertex close to t, to a vertex farther away from t.

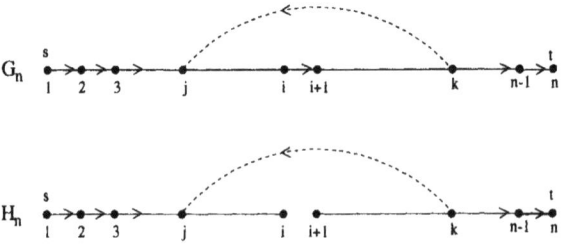

Figure 8.14: G_n and H_n are identical except that edge $(i, i + 1)$ is not in H_n. Each back edge (k, j), with $j < k$, is included with probability $n^{\sigma-1}$.

To prove Theorem 8.13, we now show that Delilah wins the Ajtai-Fagin game (Definition 8.4) on REACH. Let Samson begin by playing c and m. Recall that this indicates Samson's intention to define c new monadic relations $C_1, \ldots, C_c$ and then play the m-move first-order game. Let $C = C(c, m, \tau)$ be the number of inequivalent, quantifier-rank m formulas in the language of graphs extended by relations $C_1, \ldots, C_c$ (Definition 8.3).

Consider the set of random graphs G_n, consisting of a directed path:

$$s = g_1, \quad g_2, \quad \ldots, \quad g_{n-1}, \quad g_n = t,$$

together with some random back edges, (g_i, g_j), for $j < i$. These edges will be chosen independently at random with probability $p(n) = n^{\sigma-1}$. The small constant σ is chosen by Delilah, depending on constants m and c chosen by Samson.

See Figure 8.14 for a drawing of G_n, together with the graph H_n that Delilah chooses later. H_n is the same as G_n and colored the same as G_n, except that one forward edge from h_i to h_{i+1} is missing. Note that $G_n \in$ REACH and $H_n \notin$ REACH.

Delilah now plays one of the random graphs G_n meeting the four conditions of the following probabilistic lemma, whose proof we defer. Assume for simplicity in the following that $n > 100m^2C^2$ and $\epsilon = 0.01$. All discussions of distances, paths, and cycles in the following concern undirected paths, i.e., paths in the Gaifman graph (Definition 6.20).

Lemma 8.15. *Let C and m be natural numbers, and let $\epsilon > 0$. Let $\sigma > 0$ be sufficiently small and let n be sufficiently large. Let G_n be chosen at random with each back edge occurring with probability $n^{\sigma-1}$. With high probability, the following conditions hold:*

1. *G_n has fewer than n^ϵ undirected cycles of length at most 2^m.*
2. *For every vertex $v \in G_n$, the number of vertices of distance at most 2^m from v is less than n^ϵ.*
3. *No matter how the vertices of G_n are colored, using C colors, a fraction of $(1 - \epsilon)$ of the vertices g_i of G_n have at least m back edges to vertices colored the same color as g_{i+1}.*
4. *No matter how the vertices of G_n are colored, using C colors, a fraction of $(1 - \epsilon)$ of the vertices g_{i+1} of G_n have at least m back edges from vertices colored the same color as g_i.*

Next, Samson colors G_n with c color relations, forming G_n^c. This induces a coloring of each vertex of G_n^c by its quantifier-rank $m2^m$ type. Thus, each vertex has one of C colors. Call this expansion G_n^C.

By our choice of G_n and the values of n and ϵ, we know that some vertex v satisfies the following conditions:

1. The distance of g_i to any cycle of length $\leq 2^m$ is greater than $2^m + 1$.
2. In G_n^C, g_i has at least m back edges to vertices of the same color as g_{i+1}. Similarly, there are at least m back edges to g_{i+1} from vertices of the same color as g_i.

3. There are $2m$ vertices $a_1, \ldots, a_m$, and $b_1, \ldots, b_m$ in G_n^C such that each pair of these vertices are distance greater than $2^m + 1$ from each other and from g_i. The color of the a_i's is the same as g_i and the color of the b_i's is the same as g_{i+1}.

Let H_n^c be G_n^c with the edge from g_i to g_{i+1} removed. Delilah plays H_n^c.

Claim 8.16. $G_n^c \sim_m H_n^c$

Proof Each vertex v in G_n^C has been colored by the m-color of v, i.e., by a complete quantifier-rank m description of v. We now show that the coloring of each vertex v of H_n^C is still the unchanged quantifier-rank m description of v.

This follows from the above three conditions. Especially important is that g_i is not within distance $2^m + 1$ of a cycle of length at most 2^m. It follows that for the m-move game, the set of vertices near g_i, g_{i+1} and h_i, h_{i+1} form a forest[1]. Furthermore, the m-colors of g_i and g_{i+1} guarantee this property of not being near a small cycle. Thus the a_i's and b_i's from condition (3) above also have this property: their 2^m neighborhoods are disjoint trees.

Delilah's winning strategy for the m-move game on G_n^c and H_n^c is now clear. At the first move, she answers any play v of Samson's with a vertex v' of the same color. From then on, Delilah answers almost any move of Samson's according to her winning strategy in the game on (G_n, v) and (G_n, v'). The exception is if this strategy would call for a move near g_i or g_{i+1} when a vertex near the other of these points has already been played. If the move is near g_i, Delilah should answer by substituting for g_i a "new"[2] vertex w where w is one of the a_i's, if Delilah is answering a vertex that is not near g_{i+1}, and w is one of the vertices of the same color as g_i having a back edge to g_{i+1}, otherwise. By assumption, m such vertices are available. □

Delilah has won the Ajtai-Fagin game on the boolean query REACH. This proves Theorem 8.13. □

It remains to prove Lemma 8.15. To do so, we use the law of large numbers:

Fact 8.17. (The Weak Law of Large Numbers) *Suppose there are n independent trials, and each has probability p of success. Let S_n be the number of successful trials and let $M = pn$ be the expected number of successful trials. Then, for any $\rho > 0$,*

$$\lim_{n \to \infty} \text{Prob}[|S_n - M| \geq \rho M] = 0.$$

Proof Conditions (1) and (2) of Lemma 8.15 are easily met by letting $\sigma \leq \epsilon 4^{-m}$, and then letting n grow. In particular, the expected number of cycles of length at most 2^m is less than $n^{2^m}(n^{-1}(n^\sigma + 2))^{2^m}$, which is less than $n^{\epsilon/2}$ for large n.

---

[1] A forest is an acyclic graph, i.e., a set of trees.
[2] "New" means of distance greater than 2^r from any chosen vertex, when there are r remaining moves.

Similarly, the expected number of neighbors of distance at most 2^m of any vertex v is less than,

$$(n^\sigma + 2)^{2^m} < n^{\epsilon/2}.$$

To prove condition (3), let $\alpha = \frac{\epsilon}{2C}$. Let A be a set of αn vertices from G_n and let v be a vertex to the right of all the vertices in A. The expected number of back edges from v to A is αn^σ. Let S_v be the number of such back edges in a randomly chosen G_n. Let

$$\delta = \text{Prob}[S_v < \frac{\alpha}{2} n^\sigma].$$

By Fact 8.17, we can choose n so large that δ is as small as we like. Choose n so that $\delta^\alpha < 1/8$.

Let B be a set of αn vertices all to the right of all the vertices in A. Let $E(A, B)$ be the event that $S_v < \frac{\alpha}{2} n^\sigma$ for all v in B. Thus,

$$\text{Prob}[E(A, B)] \leq \delta^{\alpha n}.$$

Clearly there are fewer than 2^{2n} choices of such sets A and B. Thus the probability that any of them satisfy event $E(A, B)$ is at most,

$$2^{2n} \delta^{\alpha n} \leq 2^{2n} \left(\frac{1}{8}\right)^n = 2^{-n}.$$

Thus, with high probability, there are no such sets A and B. Let G_n be a graph with random back edges having no sets A and B satisfying $E(A, B)$.

Suppose that each vertex in G_n is colored one of C colors.

Let S be a set of vertices g_i of G_n such that g_{i+1} has color C_ℓ and g_i has fewer than m back edges to vertices of color C_ℓ. Suppose for the sake of contradiction that $|S| > (n\epsilon/C)$.

Let B be the rightmost half of the vertices of S. Let A be all the vertices of color C_ℓ to the left of all the vertices of B. Thus,

$$|A|, |B| \geq \frac{n\epsilon}{2C} = \alpha n.$$

It would follow that $E(A, B)$ holds. But this is impossible since G_n was chosen with no such sets A and B. It follows that $|S| \leq (n\epsilon/C)$. Thus, the number of vertices g_i that do not have at least m back edges to vertices of the same color as g_{i+1} is at most $n\epsilon$. This proves condition (3). Condition (4) is similar and is left to the reader. □

8.3 Lower Bounds Including Ordering

In this section, we strengthen Theorem 8.7 by proving that, even in the presence of ordering, graph connectivity is not expressible in monadic, second-order existential logic. The argument is subtle in that every two vertices appear together in a tuple

8. Second-Order Lower Bounds

in the ordering relation. Thus, every Gaifman graph (Definition 6.20) has diameter one. It follows that a proof using Hanf's Theorem is not possible.

The main interest in this result is that it introduces a new way for Delilah to win a game in the presence of ordering. Tight complexity lower bounds on nondeterministic time can be proved in the presence of ordering and addition (Theorem 7.11).

Schwentick proved the lower bound with ordering by a more careful game-theoretic argument. In this construction we fix the constants c and m chosen by Samson in the first move of the Ajtai-Fagin game. Delilah answers by choosing a sufficiently large n and playing graph $\mathcal{A}_n$, which we now describe.

Let $S_n = \{\sigma_1, \sigma_2, \ldots, \sigma_{n!}\}$ be the set of all permutations on n objects. Let $s = \pi_1, \ldots, \pi_r$ be a sequence of elements of S_n. Define the ordered graph $P_s^n = (V_s^n, E_s^n)$ as follows:

$$V_s^n = \{1, 2, \ldots, r+1\} \times \{1, 2, \ldots, n\}$$
$$E_s^n = \{(\langle i, j \rangle, \langle i+1, \pi_i(j) \rangle) \mid 1 \leq i \leq r, 1 \leq j \leq n\}$$

Thus, P_s^n consists of n disjoint paths of length r. The ordering on P_s^n is the lexicographic ordering. See Figure 8.18 for a drawing of P_s^4, for $s = (1234), (12), (23), e, (1234)$, where e is the identity permutation.

For any permutation $\sigma_i \in S_n$, let Q_i be the sequence consisting of 2^m copies of the identity permutation, followed by σ_i, and then followed by another 2^m identities:

$$Q_i = \underbrace{e, \ldots, e}_{2^m}, \sigma_i, \underbrace{e, \ldots, e}_{2^m}.$$

Let σ_{i_0} be the inverse of the product of all $n!$ permutations in S_n,

$$\sigma_{i_0} = \left(\prod_{i=1}^{n!} \sigma_i\right)^{-1}.$$

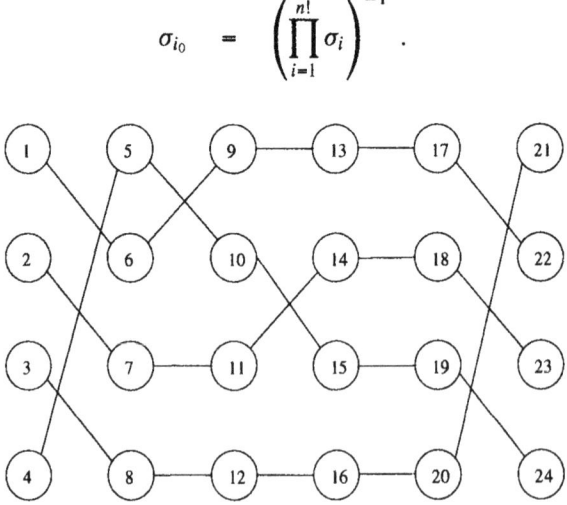

Figure 8.18: The graph P_s^4, where $s = (1234), (12), (23), e, (1234)$

Define the following sequence of permutations,

$$T = Q_{i_0}, Q_1, Q_2, \ldots, Q_{n!}.$$

The graph P_T^n consists of n disjoint paths of length $\ell = (2^{m+1} + 1)(n! + 1)$. Since the product of sequence T is the identity permutation, these paths connect vertices i and $n\ell + i$, for $i = 1, 2, \ldots, n$.

Finally, let sequence Z consist of N copies of T followed by a single copy of Q_{i_1}, where σ_{i_1} is the n-cycle $(12 \cdots n)$. The length of this sequence is $L = N\ell + (2^{m+1} + 1)$ and the product of sequence Z is just the n-cycle $(12 \cdots n)$. Thus P_Z^n consists of n paths of length L, connecting vertex i on the left to vertex $i + 1 \bmod n$ on the right, for $i = 1, 2, \ldots, n$.

Let $\mathcal{A}_n$ be the graph P_Z^n together with the n edges $\{(i, nL + i) \mid 1 \leq i \leq n\}$ connecting the first and last vertices in each row. $\mathcal{A}_n$ consists of a single long cycle and is thus connected.

Let us now play the Ajtai-Fagin game on boolean query CONNECTED. At the first move, Samson chooses constants c and m. Delilah plays the graph $\mathcal{A}_n$. The numbers n and N will be specified later, to be sufficiently large.

Let Samson now choose a coloring of $\mathcal{A}_n$ using c new color relations, $C_1, \ldots, C_c$. Let $\mathcal{A}_n^c$ be the structure $\mathcal{A}_n$ together with these new color relations. Each vertex $v \in |\mathcal{A}_n^c|$ has one of at most $C = C(c, m, \langle E^2, \leq^2 \rangle)$ complete descriptions in the language $\mathcal{L}_m(C_1, \ldots, C_c)$ (Definition 8.3). Consider $\mathcal{A}_n^C$ as colored with these complete descriptions.

Thus, each vertex has one of C possible colorings. The number of possible colorings of a copy of P_T^n in $\mathcal{A}_n^C$ is $C^{n\ell}$. If we choose $N > n!C^{n\ell}$, then,

There are at least $n!$ identically colored copies of P_T^n in $\mathcal{A}_n^C$ \hfill (8.19)

Let T^c be such a copy of P_T^n in $\mathcal{A}_n^c$. For each part P_{Q_i} of T^c, there are $C^{n(2^{m+1}+1)}$ possible colorings of all the vertices in P_{Q_i}. Note that $C^{n(2^{m+1}+1)} \leq B^n$ for some constant B. For sufficiently large n, $n!$ is much greater than B^n. It follows that there is some set of permutations $A \subseteq S_n$ of size at least $n!/B^n$ such that for all $\sigma_i, \sigma_j \in A$, the colorings of P_{Q_i} and P_{Q_j} in T^c are identical.

The following lemma says that Delilah can interchange any such P_{Q_i}'s and P_{Q_j}'s without detection.

Lemma 8.20. *Let $\mathcal{B}_n^c$ result from $\mathcal{A}_n^c$ by replacing any number of parts P_{Q_i} in a copy of T^c by the part P_{Q_j}, for pairs $\sigma_i, \sigma_j \in A$. Then $\mathcal{A}_n^c \sim_m \mathcal{B}_n^c$.*

Proof We show that Delilah wins the m-move game on $\mathcal{A}_n^c$ and $\mathcal{B}_n^c$. The only difference between the two structures is in the middle two columns of any P_{Q_i} that has been changed to P_{Q_j}.

With r moves to go, we say that a newly pebbled vertex is "near" another chosen vertex if the distance between their respective columns is at most 2^r.

Let Samson put the first pebble on any vertex in either structure. Delilah should answer with the vertex of the same number in the other structure. Let the chosen vertices be $a = \alpha_1(x_1)$ and $b = \beta_1(x_1)$.

Suppose that a and b are inside parts P_{Q_i} and P_{Q_j}, for $i \neq j$. Since $\sigma_i, \sigma_j \in A$, a and b have the same complete description in $\mathcal{L}_{m-1}$. Thus Delilah has a winning strategy in the game $\mathcal{G}_{m-1}(\mathcal{A}_n^c, a, \mathcal{A}_n^c, b')$ where b' is the piece corresponding to b in part P_{Q_j} of $\mathcal{A}_n^c$. From now on, for moves near already chosen points in this part, Delilah should answer according to her winning strategy in $\mathcal{G}_{m-1}(\mathcal{A}_n^c, a, \mathcal{A}_n^c, b')$.

Suppose on the other hand that the chosen vertices are inside unchanged and therefore identical parts of $\mathcal{A}_n^c$ and $\mathcal{B}_n^c$. In this case, from now on, for moves near this part, Delilah will keep moving according to the isomorphism between these two parts.

In successive moves, if the newly pebbled point is near an already chosen point, then Delilah should answer according to her winning strategy in the subgame of the already chosen point. If the newly pebbled point is not near any such subgame, then Delilah answers with the vertex of the same number in the other structure, and this pair establishes a new subgame.

Delilah thus wins all the subgames. Furthermore, if $\alpha_m(x_u)$ and $\alpha_m(x_v)$ belong to different subgames, then these subgames were each started with points of the same number in the two structures. Thus,

$$\alpha_m(x_u) < \alpha_m(x_v) \quad \Leftrightarrow \quad \beta_m(x_u) < \beta_m(x_v).$$

It follows that Delilah wins the whole game. $\square$

Lemma 8.20 tells us that when we transplant a part of the structure P_{Q_j} in place of the different, but identically colored P_{Q_i}, then the colors — that is, the complete descriptions in $\mathcal{L}_{m-1}$ of the vertices — remain the same!

This transplanting changes the product of the corresponding permutations, but it is not detectable in language $\mathcal{L}_m$. The reason we defined the sequences Q_i to have a length 2^m buffer on each side was so that Delilah's winning strategy for the game $\mathcal{G}_{m-1}(\mathcal{A}_n^c, a, \mathcal{A}_n^c, b')$ can be used for the subgames.

If we can change the product enough so that it is no longer an n-cycle, then $\mathcal{B}_n^c$ will not be connected and we have our theorem.

We use the following fact.

Fact 8.21. *For sufficiently large n, suppose that for H a subgroup of S_n such that for all $h \in H$, the product $(12 \cdots n)h$ is an n-cycle. Then,*

$$|H| \leq n!(6/\log n)^n.$$

Fix $\sigma_i \in A$. For each $\sigma_j \in A$, let $T_j = Q_{i_0}, Q_1, Q_2, \ldots, Q_{i-1}, Q_j, Q_{i+1} Q_{n!}$ be the sequence T with Q_i replaced by Q_j. Let ρ_j be the product of the sequence T_j and define H to be the subgroup of S_n generated by all the ρ_j's. Obviously H is at least as big as A and thus of size at least $n!/B^n$. It follows from Fact 8.21 that there exists $h \in H$ such that $(12 \cdots n)h$ is not a cycle. By the definition of H, we can write h as the product

$$h = \rho_{j_1} \cdot \rho_{j_2} \cdots \rho_{j_t},$$

and we know that $t \leq n!$.

Define $\mathcal{B}_n^c$ to replace P_{Q_i} by $P_{Q_{j_k}}$ in t successive copies of T^c in $\mathcal{A}_n^c$. It follows from Lemma 8.20 that $\mathcal{A}_n^c \sim_m \mathcal{B}_n^c$. However, $\mathcal{A}_n$ is connected and $\mathcal{B}_n$ is not. Therefore we have proved:

Theorem 8.22. *Connectivity is not expressible in monadic, second-order existential logic with ordering as the only numeric predicate.*

Historical Notes and Suggestions for Further Reading

The Ajtai-Fagin game was introduced in [AF90]. Theorem 8.7 is due to Fagin [Fag73, Fag75]. The simplified proof we present here is due to Fagin, Stockmeyer, and Vardi [FSV95]. Recently, Ajtai, Fagin and Stockmeyer have studied the closure of monadic SO∃ under first-order quantification. This gives a more robust version of what they call "monadic NP" [AFS97].

Theorem 8.13 is due to Ajtai and Fagin [AF90]. In our treatment, we follow that reference, and also a simplified and strengthened treatment of it by Arora and Fagin [AF97]. Both of those papers prove a stronger theorem in that they show that the result goes through with numeric relations of low degree such as successor. Fact 8.17 is standard, see [Fel50].

Theorem 8.22 is due to Schwentick [Sch94]. Fact 8.21 is due to Coppersmith [Cop94]. It was previously shown by de Rougemont that connectivity is not expressible in monadic, second-order logic when the only numeric predicate is successor [deR84]. See Fagin's article [Fag97] for a survey and discussion of monadic second-order games.

A great deal is known about monadic second-order logic. Some authors consider the stronger language in which subsets of edges as well as subsets of vertices can be quantified. An important theorem of Courcelle says that for graphs of bounded tree width, all monadic second-order properties — even in this stronger sense — are linear-time computable! See [Cou90] for a discussion of Courcelle's theorem and [Cou97] for a survey of second-order monadic expressibility.

Matz and Thomas have proved using automata-theoretic methods that there is a strict hierarchy for the expressive power of monadic second-order logic over graphs [MT97]. Schweikardt presents an Ehrenfeucht-Fraïssé game argument for this result over grid graphs [Sch97].

9

Complementation and Transitive Closure

Over infinite structures, there is a strict hierarchy of languages that is obtained by alternating uses of the least-fixed point operator and negation. For finite structures, we show that the hierarchy collapses to its first-level.

The transitive-closure operator (TC) *allows a simple kind of inductive definition. We prove that the transitive closure operator captures the power of nondeterministic logarithmic space. Over ordered structures, every formula in* FO(TC) *collapses to a single positive application of transitive closure. It follows that for all* $s(n) \geq \log n$, NSPACE$[s(n)]$ *is closed under complementation.*

9.1 Normal Form Theorem for FO(LFP)

The language FO(LFP) suggests a potential hierarchy of queries beginning with FO, then LFP[FO] — the single application of the least fixed point operator to a first-order formula. Next we may apply quantifiers and boolean operations, then another application of fixed point, and so on. For the infinite structure consisting of the natural numbers with its ordering relation, the first fixed-point level is equal to the set of second-order, universal queries,

Fact 9.1. *Over the infinite structure* $(\mathbf{N}, \leq)$, LFP[FO] = SO$\forall$.

It follows that over infinite structures, the fixed-point hierarchy is infinite. On this basis, Chandra and Harel conjectured that the fixed-point hierarchy for finite structures would also be infinite. We have already seen that for finite ordered structures, the fixed-point hierarchy collapses to its first fixed-point level, LFP[FO] (Corollary 4.11). As we show in this section, this remains true for unordered

140   9. Complementation and Transitive Closure

structures as well, assuming that there are at least two constants (Proviso 1.15). Without this assumption, the normal form would not be as nice because it would require a quantification outside LFP.

Finite structures are fundamentally different from infinite structures in this case. The crucial difference is that for finite structures, every least fixed point is completed at a fixed stage. We are able to detect this completion. We then define the negation of a fixed point as the set of tuples that enter only after this last stage, that is, that never enter the fixed point.

We begin the proof of Theorem 9.6 by showing all the easy cases.

Lemma 9.2. *For any class of finite structures, LFP[FO] is closed under quantification, conjunction, disjunction, and applications of LFP.*

Proof We will show by induction that any formula in FO(LFP) that has no negations of fixed points can be written in the form $(\text{LFP}\varphi)(\bar{0})$, where φ is first-order. There are four cases.

1. $\Phi_1 \equiv (\text{LFP}_{S^k, x_1 \ldots x_k} \varphi)(u_1, \ldots, u_k)$. Define,

$$\gamma_1(\bar{x}, b) \equiv (b = 0 \wedge R(\bar{u}, 1)) \vee (b = 1 \wedge \varphi(R(\star\star, 1)/S))$$

Thus, $\Phi_1 \equiv (\text{LFP}_{R^{k+1}, x_1 \ldots x_k b} \gamma_1)(\bar{0})$

Here and below we write "$\varphi(R(\star, \bar{v})$" to mean the formula φ with all occurrences of $S(u_1, \ldots, u_k)$ replaced by $R(u_1, \ldots, u_k, \bar{0}, \bar{v})$. The tuple $\bar{0}$ consists of dummy values in case the arity of R is greater than k plus the arity of $\bar{v}$.

2. $\Phi_2 \equiv (\text{LFP}_{S^k, x_1 \ldots x_k} \varphi)(\bar{0})$ op $(\text{LFP}_{T^r, x_1 \ldots x_r} \psi)(\bar{0})$, where op $\in \{\wedge, \vee\}$. Let $a = \max(k, r)$, and let,

$$\gamma_2(x_1, \ldots, x_a, b) \equiv b = 0 \wedge (R(\bar{0}, 1) \text{ op } R(\bar{0}, 2)) \vee$$
$$b = 1 \wedge \varphi(R(\star, 1)/S) \vee$$
$$b = 2 \wedge \psi(R(\star, 2)/S) .$$

What γ_2 does is simultaneously compute the two fixed points ($b = 1, 2$) and apply "op" to them ($b = 0$). Thus,

$$\Phi_2 \equiv (\text{LFP}_{R^{a+1}, x_1 \ldots x_a b} \gamma_2)(\bar{0}) .$$

3. $(Qz)\Phi_3$, where $\Phi_3 \equiv (\text{LFP}_{S^k, x_1 \ldots x_k} \varphi)(z, \bar{0})$ and $Q \in \{\exists, \forall\}$. Let

$$\gamma_3(x_1, \ldots, x_k, b) \equiv b = 0 \wedge (Qz)(R(z, \bar{0}, 1)) \vee$$
$$b = 1 \wedge \varphi(R(\star, 1)/S) .$$

Here γ_3 computes the fixed point in parallel for all values of z ($b = 1$) and then quantifies the fixed point ($b = 0$). Thus,

$$\Phi_3 \equiv (\text{LFP}_{R^{k+1}, x_1 \ldots x_k b} \gamma_3)(\bar{0})$$

4. $\Phi_4 \equiv (\text{LFP}_{S^k, x_1 \ldots x_k} \varphi(S, \bar{x}, (\text{LFP}_{T^r, y_1 \ldots y_r} \psi(S, \bar{y}, T)(\bar{0}))))$. Let,

$$\gamma_4(\bar{x}, \bar{y}, b) \equiv b = 0 \wedge \varphi(R(\star, \bar{0}, 0), \bar{x}, R(\bar{0}, \star, 1)) \vee$$
$$b = 1 \wedge \psi(R(\star, \bar{0}, 0), \bar{y}, R(\bar{0}, \star, 1))$$

9.1 Normal Form Theorem for FO(LFP)

In the formal definition of Φ_4, we would normally first compute the fixed point T_1 of ψ, with $S = \emptyset$. Then, we would compute a step of φ with $T = T_1$, getting S_1. Next we would compute the fixed point T_2 of ψ with $S = S_1$, and so on. In the fixed point of γ_4, we concurrently build the least fixed points of φ ($b = 0$) and ψ ($b = 1$). Since everything is monotone, even though the fixed point of γ_4 may be reached earlier, nothing that should not will enter the fixed point.

More explicitly, it can be proved inductively that if a tuple $\bar{z}$ is an element of S_t, then it enters the relation $R(\star, \bar{0}, 0)$ no later than round tn^r. Conversely, if a tuple $\bar{z}$ enters $R(\star, \bar{0}, 0)$ at round t', then it is also a member of $S_{t'}$. Thus we have,

$$\Phi_4 \equiv (\text{LFP}_{R^{k+r+1}, x_1 \ldots x_k y_1 \ldots y_r b} \gamma_4)(\bar{0}) . \qquad \square$$

The main difference between taking fixed points over finite rather than infinite structures is that for finite structures, the fixed point finishes after a finite number of stages. Moschovakis showed that the stages of a fixed point can be identified as the fixed point is being computed. Using his stage comparison theorem, we will be able to find a tuple $\bar{m}$ that enters the fixed point at its last state. We will then be able express the fact that a tuple $\bar{t}$ never enters the fixed point, by saying that the stage at which $\bar{t}$ enters is greater than the state at which $\bar{m}$ enters.

Let $\varphi(x_1 \ldots x_r, R)$ be an R-monotone formula and let $\mathcal{A}$ be a finite structure. Each tuple $\langle a_1 \ldots a_r \rangle \in I_\varphi$ comes in at some stage of the induction. Let $|\bar{a}|_\varphi$, the *rank* of $\bar{a}$ with respect to φ, be the step at which $\bar{a}$ enters I_φ:

$$|\bar{a}|_\varphi = \begin{cases} r & \text{if } \bar{a} \in I_\varphi^r - I_\varphi^{r-1} \\ \infty & \text{if } \bar{a} \notin I_\varphi \end{cases}$$

Define the relation $\bar{x} \leq_\varphi \bar{y}$ to mean $\bar{x} \in I_\varphi$ and $|\bar{x}|_\varphi \leq |\bar{y}|_\varphi$. Similarly, $\bar{x} <_\varphi \bar{y}$ means $\bar{x} \in I_\varphi$ and $|\bar{x}|_\varphi < |\bar{y}|_\varphi$

Theorem 9.3. *(Stage Comparison Theorem) Given monotone formula $\varphi(\bar{x}, R)$, relations $\leq_\varphi$ and $<_\varphi$ can be written as positive fixed points.*

Proof We view formula $\varphi(\bar{x}, R)$ as $\varphi'(\bar{x}, R, \neg R)$ which is positive in R and $\neg R$.

We define relations $\leq_\varphi$ and $<_\varphi$ by simultaneous positive inductions. We use the following abbreviations:

$$L_{\bar{z}} = \{\bar{u} \mid \bar{u} \leq_\varphi \bar{z}\}, \qquad G_{\bar{z}} = \{\bar{w} \mid \bar{z} <_\varphi \bar{w}\}$$

We will see that for every $\bar{z}$ already in the fixed point, $L_{\bar{z}}$ and $G_{\bar{z}}$ are complements. The following are the positive inductive definitions of $\leq_\varphi$ and $<_\varphi$:

$$\bar{x} \leq_\varphi \bar{y} \equiv \varphi'(\bar{x}, \textbf{false}, \textbf{true}) \vee$$
$$(\exists \bar{z})(\bar{z} <_\varphi \bar{y} \wedge \varphi'(\bar{x}, L_{\bar{z}}, G_{\bar{z}}))$$

$$\bar{x} <_\varphi \bar{y} \equiv (\varphi'(\bar{x}, \textbf{false}, \textbf{true}) \wedge \neg \varphi'(\bar{y}, \textbf{false}, \textbf{true})) \vee$$
$$(\exists \bar{z})(\bar{z} <_\varphi \bar{y} \wedge \varphi'(\bar{x}, L_{\bar{z}}, G_{\bar{z}}) \wedge \neg \varphi'(\bar{y}, \neg G_{\bar{z}}, \neg L_{\bar{z}}))$$

142   9. Complementation and Transitive Closure

Let α and β be the positive first-order formulas in the above definitions of $\leq_\varphi$ and $<_\varphi$. For $\langle \bar{z}, \bar{z} \rangle \in I_\alpha^r$, let

$$L_{\bar{z}}^r = \{\bar{u} \mid \langle \bar{u}, \bar{z} \rangle \in I_\alpha^r\}, \qquad G_{\bar{z}}^r = \{\bar{w} \mid \langle \bar{z}, \bar{w} \rangle \in I_\beta^r\}$$

It is not hard to show by induction that for all r,

$$I_\alpha^r = \{\langle \bar{x}, \bar{y} \rangle \mid |\bar{x}|_\varphi \leq r \text{ and } |\bar{x}|_\varphi \leq |\bar{y}|_\varphi\}$$

$$I_\beta^r = \{\langle \bar{x}, \bar{y} \rangle \mid |\bar{x}|_\varphi \leq r \text{ and } |\bar{x}|_\varphi < |\bar{y}|_\varphi\}$$

$$L_{\bar{z}}^r = \overline{G_{\bar{z}}^r}, \quad \text{for } \langle \bar{z}, \bar{z} \rangle \in I_\alpha^r$$

Combining the definitions of $\leq_\varphi$ and $<_\varphi$ into a single definition (Exercise 4.9), the theorem is proved. □

A useful corollary of Theorem 9.3 is that a monotone — but not necessarily positive — inductive definition may be rewritten as a positive one: $(\text{LFP}\varphi)(\bar{a}) \equiv (\text{LFP}\alpha)(\bar{a}, \bar{a})$.

Corollary 9.4. *Let $\varphi(R, \bar{x})$ be monotone, but not necessarily positive in R. Then the least fixed point of φ is expressible as the least fixed point of a positive formula.*

In order to negate fixed points we express the fact that some tuple $\bar{m}$ has maximum possible rank. The following formula says that $\bar{m}$ has maximum rank by saying that it is in the fixed point and no tuple enters the fixed point exactly one step after $\bar{m}$.

$$\text{MAX}(\bar{m}) \equiv \bar{m} \leq_\varphi \bar{m} \;\land\; (\forall \bar{x})(\bar{x} \leq_\varphi \bar{m} \lor \neg \varphi'(\bar{x}, \neg G_{\bar{m}}, \neg L_{\bar{m}})) \;.$$

It now follows that for any monotone φ, we can express the negation of the fixed point of φ as a positive least fixed point:

$$\neg(\text{LFP}_{R, x_1 \ldots x_k} \varphi)(\bar{a}) \equiv (\exists \bar{m})(\text{MAX}(\bar{m}) \land \bar{m} <_\varphi \bar{a}) \;. \tag{9.5}$$

Combining (9.5), Theorem 9.3, and Lemma 9.2, we have proved,

Theorem 9.6. *For any class of finite structures, the fixed point hierarchy collapses at its first fixed point level. In symbols, FO(LFP) = LFP[FO].*

Gurevich and Shelah defined the inflationary fixed point operator IFP. Let R be a new k-ary relation symbol that occurs not necessarily monotonically in $\varphi(R, x_1, \ldots x_k)$. Define the *inflationary fixed point* operator IFP as follows,

$$\text{IFP}(\varphi(R, \bar{x})) \equiv \text{LFP}(\varphi(R, \bar{x}) \lor R(\bar{x}))$$

IFP may be applied to any inductive definition — there is no syntactic restriction. If φ is monotone, then $\text{IFP}(\varphi) = \text{LFP}(\varphi)$.

Let $\psi(R, \bar{x}) = \varphi(R, \bar{x}) \lor R(\bar{x})$. Whether or not φ is monotone, the following sequence is monotonically increasing and its union is $\text{IFP}(\varphi)$,

$$\emptyset \subseteq \psi(\emptyset) \subseteq \psi^2(\emptyset) \subseteq \psi^3(\emptyset) \subseteq \cdots \tag{9.7}$$

Even though ψ may not be monotone, the monotonicity of the sequence in (9.7) suffices for the proof of Theorem 9.3 to go through. Thus, the stage comparison formulas $\psi_\leq$ and $\psi_<$ are expressible as least fixed points of positive formulas. An immediate corollary is that FO(LFP) and FO(IFP) have the same expressive power. We will find, however, that IFP is usually more convenient than LFP since, when using IFP, we do not have to worry about keeping our definitions positive.

Corollary 9.8. FO(IFP) = FO(LFP).

9.2 Transitive Closure Operators

The transitive closure is a particularly important case of an inductive definition. Let $\varphi(x_1, \ldots, x_k, x'_1 \ldots x'_k)$ be a formula of some vocabulary τ with $2k$ free variables. The formula φ describes a query I_φ from STRUC[τ] to graphs. For a structure $\mathcal{A} \in \text{STRUC}[\tau]$,

$$I_\varphi(\mathcal{A}) = \langle |\mathcal{A}|^k, E \rangle; \quad E = \left\{ \langle a_1, \ldots, a_k, a'_1, \ldots, a'_k \rangle \mid \mathcal{A} \models \varphi(\bar{a}, \bar{a}') \right\}.$$

We write $(\text{TC}_{x_1 \ldots x_k x'_1 \ldots x'_k} \varphi)$ to denote the reflexive, transitive closure of binary relation $\varphi(\bar{x}, \bar{x}')$. Let FO(TC) be the closure of first-order logic with arbitrary occurrences of TC. We know from Proposition 4.17 and Theorem 4.10 that,

$$\text{FO(TC)} \subseteq \text{FO}[\log n] \subseteq \text{FO}[n^{O(1)}] = \text{FO(LFP)}. \tag{9.9}$$

Let FO(pos TC) be the restriction of FO(TC) in which TC never occurs within a negation. Then,

Theorem 9.10. FO(pos TC) = NL.

Proof ($\subseteq$): The set of relations computable in NSPACE[$\log n$] is closed under first-order quantifiers, ($\forall x$) and ($\exists x$), because with space $\log n$ we can cycle through all the values of x. Thus it suffices to show that if $\varphi(\bar{x}, \bar{x}')$ is computable in NSPACE[$\log n$], then so is $(\text{TC}_{\bar{x}\bar{x}'} \varphi)$. We can test if structure $\mathcal{A}$ satisfies $(\text{TC}_{\bar{x}\bar{x}'} \varphi)(\bar{a}, \bar{a}')$ as follows: If $\bar{a} = \bar{a}'$ then accept. Else, guess $\bar{b}$ and check that $\mathcal{A} \models \varphi(\bar{a}, \bar{b})$. Next throw away $\bar{a}$ and guess $\bar{c}$ such that $\mathcal{A} \models \varphi(\bar{b}, \bar{c})$. Repeat this process until we guess $\bar{z}$ such that $\mathcal{A} \models \varphi(\bar{y}, \bar{z})$ and $\bar{z} = \bar{a}'$, in which case we accept. The space needed is $3k \log n$ plus the space to check if $\varphi(\bar{x}, \bar{x}')$ holds, where k is the arity of $\bar{x}$.

($\supseteq$): Recall that problem REACH is complete for NL via first-order reductions (Theorem 3.16). REACH is expressible in FO(pos TC) as follows,

$$\text{REACH} \equiv (\text{TC}_{xy}(E(x, y)))(s, t)$$

Since FO(pos TC) is closed under first-order reductions, it follows that NL $\subseteq$ FO(pos TC). □

We next define a deterministic version of transitive closure DTC. Given a first order relation, $\varphi(\bar{x}, \bar{y})$, define its deterministic reduct,

$$\varphi_d(\bar{x}, \bar{y}) \equiv \varphi(\bar{x}, \bar{y}) \wedge [(\forall \bar{z}) \neg \varphi(\bar{x}, \bar{z}) \vee (\bar{y} = \bar{z})]$$

That is, $\varphi_d(\bar{x}, \bar{y})$ is true just if $\bar{y}$ is the unique descendent of $\bar{x}$. Now define:

$$(\text{DTC}\, \varphi) \equiv (\text{TC}\, \varphi_d).$$

Theorem 9.11. FO(DTC) = L.

Proof This proof is similar to the proof of Theorem 9.10. To see that L contains FO(DTC), suppose $\varphi(x_1, \ldots, x_k, y_1, \ldots, y_k) \in$ L. Observe that Algorithm 3.22 determines in logspace whether or not $(\text{DTC}\, \varphi)(\bar{s}, \bar{t})$ holds, where instead of checking whether there is an edge from $\bar{b}$ to $\bar{a}$, we check that $\varphi(\bar{b}, \bar{a})$ holds.

Conversely, FO(DTC) contains L. We know that REACH_d is complete for L via first-order reductions (Theorem 3.23) and that FO(DTC) is closed under first-order reductions. Thus, it suffices to show that REACH_d is expressible in FO(DTC),

$$\text{REACH}_d \equiv (\text{DTC}_{xy}(E(x, y)))(s, t).\qquad \square$$

9.3 Normal Form for FO(TC)

In this section we show that every query in FO(TC) is expressible in a very simple form, namely as $(\text{TC}\, \alpha)(\bar{0}, \overline{max})$, where α is quantifier-free. As we see later, this implies that REACH is complete for NL via quantifier-free reductions. This assumes the presence of the successor relation, SUC. We first prove this result for FO(pos TC) and then deal separately with negation.

Lemma 9.12. *In the presence of the successor relation, every formula $\varphi \in$ FO(pos TC) is equivalent to a single application of transitive closure to a quantifier-free formula,*

$$\varphi \equiv (\text{TC}\, \alpha)(\bar{0}, \overline{max})$$

Proof By induction on the complexity of φ. There are five cases. The most interesting of these are cases 3, 4, and 5, which are shown in Figure 9.13.

1. Formula φ is atomic or is the negation of an atomic formula. In this case let u, v be variables not occurring in φ. Then $\varphi \Leftrightarrow (\text{TC}_{uv}\varphi)(0, max)$.
2. $\varphi \equiv (\text{TC}_{\bar{x}\bar{y}}\psi)(\bar{q}, \bar{r})$. We wish to replace $\bar{q}, \bar{r}$ with $\bar{0}, \overline{max}$. Put

$$\rho(s_1, t_1, \bar{x}, s_2, t_2, \bar{y}) \equiv$$
$$[s_1 = 0 \wedge t_1 = 0 \wedge \bar{x} = \bar{0} \wedge s_2 = 0 \wedge t_2 = max \wedge \bar{y} = \bar{q}]$$
$$\vee [s_1 = 0 \wedge t_1 = max \wedge s_2 = 0 \wedge t_2 = max \wedge \psi(\bar{x}, \bar{y})]$$
$$\vee [s_1 = 0 \wedge t_1 = max \wedge \bar{x} = \bar{r} \wedge s_2 = t_2 = max \wedge \bar{y} = \overline{max}]$$

Variables t, s split the ρ-path in three stages: ($st = 00$): Set $\bar{x}$ to $\bar{q}$ and go to next stage. ($st = 0max$): Take a ψ step and stay in this stage. When $\bar{r}$ is reached, go to next stage. ($st = max\, max$): Set $\bar{x} = \overline{max}$ and stop. Thus as desired:

$$\varphi \Leftrightarrow (\text{TC}_{s_1 t_1 \bar{x} s_2 t_2 \bar{y}} \rho)(\bar{0}, \overline{max}).$$

9.3 Normal Form for FO(TC) 145

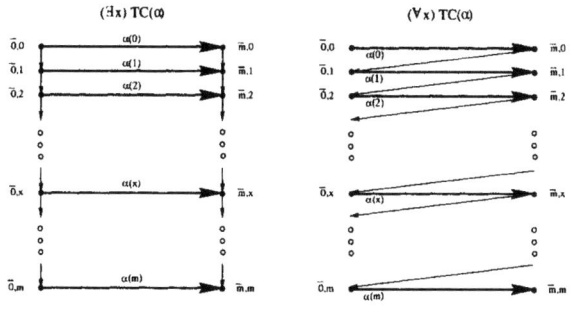

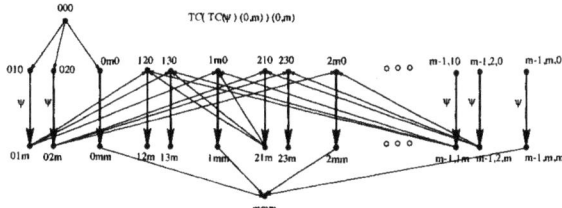

Figure 9.13: Cases 3, 4, and 5 of Lemma 9.12

3. $\varphi \equiv (\exists x)(\text{TC}_{\bar{u}\bar{v}}\alpha(x))(\bar{0}, \overline{max})$. Here the notation means that the transitive closure is taken over the relation $\alpha(\bar{u}, \bar{v})$ and variable x occurs free in α. Put

$$\chi(\bar{u}, x_1, \bar{v}, x_2) \equiv \; [\bar{u} = \bar{0} \wedge \text{SUC}(x_1, x_2)]$$
$$\vee \; [\alpha(\bar{u}, \bar{v}; x_1) \wedge x_1 = x_2]$$
$$\vee \; [\bar{u} = \overline{max} \wedge \bar{v} = \overline{max} \wedge \text{SUC}(x_1, x_2)]$$

χ allows a guess of x on the first step, so $\varphi \Leftrightarrow (\text{TC}_{\bar{u}x_1\bar{v}x_2}\chi)(\bar{0}, \overline{max})$.

4. $\varphi \equiv (\forall x)(\text{TC}_{\bar{u}\bar{v}}\alpha(x))(\bar{0}, \overline{max})$. In this case, we simulate $(\forall x)$ by searching through all x's in order using SUC. Put

$$\nu(\bar{u}, x_1, \bar{v}, x_2) \equiv \; [\bar{u} \neq \overline{max} \wedge \alpha(\bar{u}, \bar{v}; x_1) \wedge x_1 = x_2]$$
$$\vee \; [\bar{u} = \overline{max} \wedge \bar{v} = \bar{0} \wedge \text{SUC}(x_1, x_2)]$$

Thus, $\varphi \Leftrightarrow (\text{TC}_{\bar{u}x_1\bar{v}x_2}\nu)(\bar{0}, \overline{max})$.

5. $\varphi \equiv \left(\text{TC}_{\bar{u}\bar{v}}[\text{TC}_{\bar{x}\bar{y}}\psi](\bar{0}, \overline{max})\right)(\bar{0}, \overline{max})$. In this case, formula ψ has free variables $\bar{x}, \bar{y}, \bar{u}, \bar{v}$. The inner transitive closure is on $\bar{x}, \bar{y}$, treating the other variables as parameters. The outer transitive closure is on $\bar{u}, \bar{v}$. We combine these two TC's into a single transitive closure on δ defined as follows:

$$\delta(\bar{u}_1, \bar{v}_1, \bar{x}, \bar{u}_2, \bar{v}_2, \bar{y}) = \; [\bar{x} = \bar{y} = \bar{0} \wedge \bar{u}_1 = \bar{v}_1 = \bar{u}_2 = \bar{0}]$$
$$\vee \; [\bar{x} \neq \overline{max} \wedge \bar{u}_1 \neq \bar{v}_1 \wedge \bar{u}_1 = \bar{u}_2 \wedge \bar{v}_1 = \bar{v}_2 \wedge \psi(\bar{x}, \bar{y}; \bar{u}_1, \bar{v}_1)]$$
$$\vee \; [\bar{x} = \overline{max} \wedge \bar{v}_1 \neq \overline{max} \wedge \bar{y} = \bar{0} \wedge \bar{u}_2 = \bar{v}_1]$$
$$\vee \; \vee \; [\bar{x} = \overline{max} \wedge \bar{v}_1 = \overline{max} \wedge \bar{y} = \overline{max} \wedge \bar{u}_2 = \bar{v}_2 = \overline{max}].$$

We claim $\varphi \Leftrightarrow (\text{TC}_{\bar{u}_1\bar{v}_1\bar{x}\bar{u}_2\bar{v}_2\bar{y}}\delta)(\bar{0}, \overline{max})$. This holds because a δ path consists exactly of a series of $\psi(\cdot, \cdot; u, v)$ paths from $\bar{0}$ to $\overline{max}$ with u, v fixed. At the end of any such path we know that $(\text{TC}_{\bar{x}\bar{y}}\psi(\bar{u}, \bar{v}))(\bar{0}, \overline{max})$ holds. The δ path may now appropriately step from $(\overline{max}, \bar{u}, \bar{v})$ to $(\bar{0}, \bar{v}, \bar{w})$, i.e., reach v and begin trying to move from v to w.

The cases of disjunction and conjunction follow easily from cases 3 and 4 respectively. □

Exercise 9.14 Prove that Lemma 9.12 fails in the absence of the successor relation, even if we have ordering. [Hint: Lemma 9.12 implies that every formula in FO(pos TC) is equivalent on structures of at most a given size, to a first-order existential formula. Starting with a universal formula, such as $\varphi \equiv (\forall x)(R(x))$, show that it is not equivalent to an existential formula, in the absence of SUC. This can be done by showing that the truth of φ is not preserved under superstructures, unlike the case for all existential formulas (Exercise 1.13).] □

The normal form for FO(pos TC) goes through for FO(DTC) as well:

Lemma 9.15. *Every formula $\varphi \in$ FO(DTC) is equivalent to a single application of deterministic transitive closure to a quantifier-free formula,*

$$\varphi \equiv (\text{DTC}\,\alpha)(\bar{0}, \overline{max}) \,.$$

Proof We modify the construction in the proof of Lemma 9.12 so that a deterministic path is never turned into a nondeterministic path. The most interesting case is the existential quantifier: $\varphi \equiv (\exists x)(\text{DTC}_{\bar{u},\bar{u}'}\alpha(x))(\bar{0}, \overline{max})$. Here, instead of letting the path finder guess the correct x we force the path to try all x's and go to $\overline{max}$ when a correct one is found. We use the fact that there is a path in an n^k vertex graph iff there is such a path of length at most $n^k - 1$. Let $k = \text{arity}(\bar{z}) = \text{arity}(\bar{w}) = \text{arity}(\bar{u}) = \text{arity}(\bar{s})$. In the following, counter $\bar{z}$ is used to cut off a cycling α-path, $\bar{w}$ is used to find the α-successor of u if one exists, and $\bar{s}$ is used to store this α-successor while we check that there are no others. We abuse notation and write $\text{SUC}(\bar{z}, \bar{z}')$ to mean that $\bar{z}'$ is the successor of $\bar{z}$ in the lexicographical ordering induced by successor relation SUC. Let

$$\chi'(\bar{u}, \bar{z}, \bar{w}, \bar{s}, x, \bar{u}', \bar{z}', \bar{w}', \bar{s}', x') \equiv \delta_1 \vee \delta_2 \vee \delta_3 \vee \delta_4 \vee \delta_5 \vee \delta_6 \vee \delta_7,$$

where the meaning of the mutually exclusive δ_i's are as follows:

1. $(\bar{u} = \overline{max})$: success, so set all primed variables to $\overline{max}$ and halt.
2. $(\bar{z} = \overline{max})$: failure on x because the counter has overflowed, so set $x' = x + 1$.
3. $(\bar{w} = \overline{max}) \wedge \neg\alpha(\bar{u}, \bar{w}; x) \wedge \neg\alpha(\bar{u}, \bar{s}; x)$: failure on x because there is no α-edge leaving $\bar{u}$, so set $x' = x + 1$.
4. $\alpha(\bar{u}, \bar{s}; x) \wedge \alpha(\bar{u}, \bar{w}; x) \wedge \bar{s} \neq \bar{w}$: failure on x because $\bar{u}$ has more than one α-successor, so set $x' = x + 1$.
5. $(\bar{w} = \overline{max}) \wedge \neg\alpha(\bar{u}, \bar{w}; x) \wedge \neg\alpha(\bar{u}, \bar{s}; x)$: failure on x because there is no α-path leaving $\bar{u}$, so set $x' = x + 1$.

6. $\overline{w} = \overline{max} \wedge (\alpha(\bar{u}, \bar{s}; x) \oplus \alpha(\bar{u}, \overline{w}; x))$: there is a unique α-successor of $\bar{u}$. Increment $\bar{z}$ and set $\bar{u}'$ to its successor.
7. $\neg \alpha(\bar{u}, \overline{w}; x)$: increment $\overline{w}$ and keep looking for an α-edge leaving $\bar{u}$.

For completeness we include the first-order definitions of the δ_i's:

$\delta_1 \equiv \bar{u} = \bar{u}' = \bar{z}' = \overline{w}' = \overline{max} \wedge x' = max$

$\delta_2 \equiv \bar{u} \neq \overline{max} \wedge \text{SUC}(x, x') \wedge \bar{z} = \overline{max} \wedge \bar{u}' = \bar{z}' = \overline{w}' = \bar{0}$

$\delta_3 \equiv \bar{u} \neq \overline{max} \wedge \text{SUC}(x, x') \wedge \bar{z} \neq \overline{max} \wedge \overline{w} = \overline{max} \wedge \neg \alpha(\bar{u}, \overline{w}; x)$
$\quad \wedge \neg \alpha(\bar{u}, \bar{s}; x) \wedge \bar{u}' = \bar{z}' = \overline{w}' = \bar{0}$

$\delta_4 \equiv \bar{u} \neq \overline{max} \wedge \text{SUC}(x, x') \wedge \bar{z} \neq \overline{max} \wedge \alpha(\bar{u}, \bar{s}; x) \wedge \alpha(\bar{u}, \overline{w}; x)$
$\quad \wedge \bar{s} \neq \overline{w} \wedge \bar{u}' = \bar{z}' = \overline{w}' = \bar{0}$

$\delta_5 \equiv \bar{u} \neq \overline{max} \wedge \text{SUC}(x, x') \wedge \bar{z} \neq \overline{max} \wedge (\overline{w} = \overline{max}) \wedge \neg \alpha(\bar{u}, \overline{w}; x)$
$\quad \wedge \neg \alpha(\bar{u}, \bar{s}; x) \wedge \bar{u}' = \bar{z}' = \overline{w}' = \bar{0}$

$\delta_6 \equiv \bar{u} \neq \overline{max} \wedge x' = x \wedge \text{SUC}(\bar{z}, \bar{z}') \wedge \overline{w} = \overline{max} \wedge \alpha(\bar{u}, \bar{u}'; x)$
$\quad \wedge (\alpha(\bar{u}, \bar{s}; x) \oplus \alpha(\bar{u}, \overline{w}; x)) \wedge \overline{w}' = \bar{0}$

$\delta_7 \equiv \bar{u} \neq \overline{max} \wedge \bar{z} \neq \overline{max} \wedge \neg \alpha(\bar{u}, \overline{w}; x) \wedge \text{SUC}(\overline{w}, \overline{w}')$
$\quad \wedge \bar{u}' = \bar{u} \wedge \bar{z}' = \bar{z} \wedge x' = x$

It follows that φ is equivalent to $(\text{DTC } \chi)(\bar{0}, \overline{max})$ as desired.
The remaining case is negation:

$$\varphi \equiv \neg (\text{DTC}_{x_1 \ldots x_k y_1 \ldots y_k} \psi)(\bar{0}, \overline{max}) .$$

We can handle this case in a similar way to the above case: We add k-tuples of variables: $\bar{z}, \bar{z}'$ to serve as a counter, $\overline{w}, \overline{w}'$ to run through possible ψ-successors, and $\bar{s}, \bar{s}'$ to store the candidate ψ-successor while checking that it is unique.

We start at $\bar{0}$, find a unique ψ-successor of $\bar{x} = \bar{0}$, and increment the counter and repeat. If we ever would get to $\bar{y} = \overline{max}$, then instead we return to $\bar{0}$, i.e., reject. If ever the counter overflows ($\bar{z} = \overline{max}$) or there are zero or more than one ψ-successors of $\bar{x}$, then we go to $\overline{max}$, i.e., accept. □

Next, we show that BIT is expressible in FO(wo BIT)(DTC). It follows that Theorems 9.11, 9.20, and 4.10 go through for the languages with successor but not BIT. Similarly, Theorems 9.6, 9.20, and Lemma 9.15 do not require BIT.

Proposition 9.16. *Relation* BIT *is definable in* FO(wo BIT)(DTC), *and thus also in* FO(wo BIT)(TC) *and* FO(wo BIT)(LFP).

Proof We first show that the relations PLUS is definable, using DTC and SUC. We say that there is an α-edge from $\langle x, y \rangle$ to $\langle u, v \rangle$ iff $u = x - 1$ and $v = y + 1$:

$$\alpha(x, y, u, v) \equiv \text{SUC}(u, x) \wedge \text{SUC}(y, v) .$$

Using transitive closure we then get:

$$\text{PLUS}(x, y, z) \equiv (\text{DTC} \, \alpha)(x, y, 0, z) .$$

Now define β as follows:

$$\beta(w_1, j_1, w_2, j_2) \equiv \bigl(\exists z (\text{PLUS}(w_2, w_2, z) \land (w_1 = z \lor \text{SUC}(z, w_1)))$$
$$\land \quad \text{SUC}(j_2, j_1)\bigr) \,.$$

Note that $\beta(w, j, w', j+1)$ holds iff $w' = \lfloor w/2 \rfloor$. Let $\text{ODD}(z)$ abbreviate $\exists x \exists y (\text{PLUS}(x, x, y) \land \text{SUC}(y, z))$. It follows that

$$\text{BIT}(w, j) \equiv (\exists z)\bigl(\text{ODD}(z) \land (\text{DTC}\,\beta)(w, j, z, 0)\bigr) \,. \qquad \square$$

From Lemma 9.15 and Proposition 9.16 we obtain the following corollary.

Corollary 9.17. *In the presence of the successor relations, problems* REACH_d, *REACH and* REACH_a *are complete for* P, NL *and* l, *respectively, via quantifier-free reductions.*

Proof Lemma 9.15 shows how to write any formula in l as a quantifier-free reduction to REACH_d, and Lemma 9.12 does the same thing for NL and REACH. We can define an alternating transitive closure operator ATC that similarly formalizes alternating reachability. A similar proof gives the same quantifier-free normal form for FO(ATC) (Exercise 11.12). $\qquad \square$

9.4 Logspace is Primitive Recursive

Gurevich gave a different characterization of logspace via functions [Gur83]. He defined an algebra of partial global functions f where for each ordered finite structure $\mathcal{A}$ of the appropriate vocabulary, $f^{\mathcal{A}} : |\mathcal{A}|^r \to |\mathcal{A}|^s$. Let us fix a vocabulary τ, which may include some function symbols. Define the *initial functions* to be the following,

1. **Constant functions**: 0 and max are 0-ary constant functions.
2. **Successor function**: for each $r > 0$, $\text{SUC}(x_1, \ldots, x_r) = \bar{x} + 1$ is the successor of $\bar{x}$ in lexicographic order and is undefined if $\bar{x} = \overline{max}$.
3. **Projection functions**: for $\ell > 0$ and $1 \leq i_1 < i_2 \cdots < i_r \leq \ell$,

$$\pi^{\ell}_{i_1,\ldots,i_r}(x_1, \ldots, x_{\ell}) = (x_{i_1}, x_{i_2}, \ldots, x_{i_r})$$

4. **Input symbols**: for each function or constant symbol in τ we have the corresponding function. For each relation symbol, we have the corresponding characteristic function.

The initial functions are then closed under the following operations:

1. **Composition**: if $h_1, \ldots h_r$ are functions from s-tuples to a_i tuples and g is a function on $a_1 + a_2 + \cdots + a_r$-tuples, then the composition of g and $h_1, \ldots, h_r$ is defined by,

$$g \circ (h_1, \ldots, h_r)(x_1, \ldots x_s) = g(h_1(\bar{x}), h_2(\bar{x}), \ldots, h_r(\bar{x})) \,.$$

2. Primitive recursion: if g and h are functions of appropriate arity, then the following scheme defines f by primitive recursion from g and h,

$$f(\bar{x}, \bar{0}) = g(\bar{x})$$
$$f(\bar{x}, \text{SUC}(\bar{t})) = h(\bar{x}, \bar{t}, f(\bar{x}, \bar{t})) .$$

Define the *primitive recursive functions on finite structures* to be the closure of initial functions under composition and primitive recursion. Gurevich proved that,

Theorem 9.18. *The primitive recursive functions on finite structures are the partial functions computable in logspace.*

Exercise 9.19 Prove Theorem 9.18. [Hint: for the upperbound, show that REACH is primitive recursive using Algorithm 3.22. Next show that the primitive recursive functions are closed under quantifier-free reductions.] □

9.5 NSPACE[$s(n)$] = co-NSPACE[$s(n)$]

Corollaries 9.23 and 9.24, which follow from the following theorem, had been open for about twenty-five years. Virtually everyone had conjectured their negation. Nondeterministic space corresponds to an existential search, and it seemed natural — based on our intuition concerning existential quantification — that this class would not be closed under complementation. Surprisingly, in this case, existential search and universal search are equivalent.

Theorem 9.20. *For any class of finite, ordered structures,*

$$\text{FO(pos TC)} \;=\; \text{FO(TC)} .$$

Proof By Lemma 9.12 it suffices to show that the relation $\neg(\text{TC}_{uu'} E(u, u'))(0, max)$ meaning that there is no path from 0 to *max*, is expressible in FO(pos TC).

To do this, we count the number of reachable vertices. Fix a graph $G \in \text{STRUC}[\tau_g]$. As usual, we consider the elements of G as numbers as well as vertices. In one setting, as distances, we think of these numbers as ranging from 0 to $n - 1$. In another setting, as counts of the number of reachable vertices, we have numbers ranging from 1 to n. Writing these two sets of numbers as numbers rather than as vertices makes our notation simpler to understand.

Define n_d to be the number of vertices in G that are reachable from 0 in a path of length at most d. Given number n_d, we show how to compute number n_{d+1}. As a first step, we show that n_d allows us to say in FO(pos TC) that there is *not* a path of length at most d from 0 to a given vertex.

Claim 9.21. *The following formulas are expressible in* FO(pos TC):

1. DIST(x, d), meaning that there is a path of length at most d from 0 to x
2. NDIST$(x, d; m)$, which—when $m = n_d$—means that there is no path of length at most d from 0 to x.

150   9. Complementation and Transitive Closure

Proof There is no trouble writing DIST(x, d) positively,

$$\text{DIST}(x, d) \equiv \text{TC}(\alpha)(0, 0, x, d), \quad \text{where}$$

$$\alpha(a, i, b, j) \equiv (E(a, b) \vee a = b) \wedge \text{SUC}(i, j)$$

We write the formula, NDIST$(x, d; m) \in$ FO(pos TC) to mean the following:

$$\text{NDIST}(x, d; m) \equiv (\text{There are at least } m \text{ vertices } v)(v \neq x \wedge \text{DIST}(x, d)).$$

It will then follow that when $m = n_d$, NDIST$(x, d; m)$ is equivalent to $\neg$DIST(x, d).
Define edge relation β on pairs of vertices as follows,

$$\beta(v, c, v', c') \equiv \quad \equiv \quad 0 \neq x \wedge \text{SUC}(v, v')$$
$$\wedge \quad (c = c' \vee (\text{SUC}(c, c') \wedge \text{DIST}(v', d) \wedge v' \neq x)).$$

Suppose that c is the number of vertices — not including x — that are at most v and reachable from 0 in at most d steps. Then we can take a β-step from $\langle v, c \rangle$ to $\langle v + 1, c \rangle$ guessing that $v + 1$ is not reachable from 0 in d steps; or, we may take a β-step to $\langle v + 1, c + 1 \rangle$ **if we prove** that $v + 1$ is not equal to x and is reachable from 0 in d steps.

Thus, there is a path from $\langle 0, 1 \rangle$ to $\langle v, c \rangle$ iff there are at least c vertices not equal to x and at most v such that DIST(v, d):

$$\text{TC}(\beta)(0, 1, v, c) \quad \Leftrightarrow \quad b \leq |\{w \mid w \leq v \wedge \text{DIST}(w, d)\}|.$$

NDIST can now be defined as follows,

$$\text{NDIST}(x, d; m) \quad \equiv \quad \text{TC}(\beta)(0, 1, max, m). \qquad \square$$

Using Claim 9.21, we now define the relation $\delta(d, m, d', m')$ so that if $m = n_d$, then $m' = n_{d+1}$. We simply cycle through all the vertices, counting how many of them are reachable in $d + 1$ steps:

$$\delta(d, m, d', m') \equiv \text{SUC}(d, d') \wedge \text{TC}(\gamma)(0, 1, max, m')$$

$$\gamma(v, c, v', c') \equiv \text{SUC}(v, v') \wedge \Big([\text{SUC}(c, c') \wedge \text{DIST}(v', d + 1)] \vee$$
$$[c = c' \wedge (\forall z)(\text{NDIST}(z, d; m) \vee (z \neq v' \wedge \neg E(z, v')))]\Big)$$

It follows that formula TC$(\delta)(0, 1, n - 1, m)$ holds iff $m = n_{n-1}$ is the number of vertices in G that are reachable from 0. Using this m, we can express the nonexistence of a path positively as claimed,

$$\neg \text{TC}(E)(0, x) \quad \equiv \quad (\exists m)(\text{TC}(\delta)(0, 1, n - 1, m) \wedge \text{NDIST}(x, n - 1; m)). \quad \square$$

As a corollary, we obtain a nicer characterization of NL.

Corollary 9.22. NL = FO(TC). *Furthermore every formula $\varphi \in$ FO(TC) is equivalent to a single application of transitive closure to a quantifier-free formula,* $\varphi \equiv (\text{TC}\,\alpha)(\overline{0}, \overline{max})$.

One can generalize Theorem 9.20 to the following two corollaries. Corollary 9.24 settles a question dating back to 1964.

Corollary 9.23. *For any $s(n) \geq \log n$, $NSPACE[s(n)] = co\text{-}NSPACE[s(n)]$.*

Proof Recall from Theorems 9.20 and 9.10 that NL = FO(pos TC) = FO(TC). If we take any NL property and negate it, then it is still in FO(TC) and thus NL. It follows that NL = co-NL.

Suppose $s(n) \geq \log n$ and we are given an $NSPACE[s(n)]$ machine N and its input w of length $n = |w|$. The computation graph of N on input w has $m = 2^{O(s(n))}$ nodes. The question whether N rejects w is the non-reachability problem on this computation graph. By Theorem 9.20, it is thus solvable in $NSPACE[\log(m)]$, i.e. $NSPACE[s(n)]$. (By the way, even if we do not know what $s(n)$ is, we can do the same construction by starting with $s = 1$ and incrementing s each time a reachable configuration in the computation graph of size $s + 1$ is found.) □

Corollary 9.24. *The class of context sensitive languages is closed under complementation.*

Proof Kuroda showed in 1964 that CSL = $NSPACE[n]$ [Kur64]. □

Exercise 9.25 In the proof of Theorem 9.20, show that formula $\delta(d, m, d+1, m')$ is true iff one of the following conditions hold,

1. $m = n_d$ and $m' = n_{d+1}$, or
2. $m < n_d$ and $m' \leq n_{d+1}$.

□

9.6 Restrictions of SO

Grädel and others have given elegant characterizations of P and NL via restrictions of second-order logic.

Definition 9.26 (Horn and Krom Formulas) Let Φ be a second-order formula in prenex form,

$$\Phi \equiv (Q_1 P_1^{a_1}) \cdots (Q_k P_k^{a_k})(\forall \bar{x})\alpha$$

Let the first-order part of Φ be universal, and let the quantifier-free part — α — be in conjunctive normal form, i.e., a conjunction of clauses, each of which is a disjunction. We say that Φ is a second-order *Horn formula* iff the quantifier-free part has at most one positive occurrence of a quantified predicate, P_i, per clause. Φ is a second-order *Krom formula* iff the quantifier-free part has at most two occurrences of a quantified predicate per clause. Let SO-Horn and SO-Krom be the set of boolean queries describable by second-order Horn and second-order Krom formulas, respectively. □

We see later in Theorem 9.32 that SO-Horn = P and SO-Krom = NL. In order to prove this, we begin with the Horn and Krom restrictions of SAT,

Proposition 9.27. *Let* HORN-SAT *and* 2-SAT *be the restrictions of the boolean satisfiability problem to Horn and Krom formulas respectively. Then,*

1. HORN-SAT *is complete for* P *via quantifier-free reductions.*
2. 2-SAT *is complete for* NL *via quantifier-free reductions.*

Proof 1. One way to see that HORN-SAT is in P is to express it in FO(LFP). Inductively, define a variable to be true if it occurs positively in a clause all of whose other variables are true. A Horn formula is satisfiable iff this inductively defined assignment satisfies the formula. Here is the inductive definition,

$$\varphi(R, x) \equiv (\exists c)(P(c, x) \wedge (\forall y. N(c, y)) R(y))$$

Let $T \equiv (\text{LFP}\varphi)$. Then,

$$\text{HORN-SAT} \equiv (\forall c)(\exists x)(P(c, x) \wedge T(x) \vee N(c, x) \wedge \neg T(x))$$

We know that REACH_a is complete for P via quantifier-free reductions (Theorem 3.26 and Corollary 9.17). It follows that the complementary problem $\overline{\text{REACH}_a}$ is complete for co-P = P. To show that HORN-SAT is complete for P it therefore suffices to show that $\overline{\text{REACH}_a} \leq_{\text{qf}}$ HORN-SAT. The idea of the reduction is simple. Let G be an alternating graph. Since REACH_a remains complete when graphs are restricted to outdegree two, we assume that the outdegree of G is two. Formula $I(G)$ consists of the following clauses:

- t
- $(e \vee \neg f_1 \vee \neg f_2)$ where e is a universal node and has edges to $f_1 \neq f_2$.
- $(e \vee \neg f_1)$ where there is an edge from e to f_1 and e is existential.
- $\neg s$.

I is quantifier-free definable and $I(G) \in$ HORN-SAT iff $G \in \overline{\text{REACH}_a}$.

2. 2-SAT is in NL because a clause with two literals, $\ell_1 \vee \ell_2$, can be understood as two edges in a graph: $\overline{\ell_1} \to \ell_2, \overline{\ell_2} \to \ell_1$. Let 2-CNF be the set of CNF formulas that have at most two literal per clause. Thus, 2-SAT = SAT$\cap$2-CNF.

It is not hard to see that a 2-CNF formula φ is satisfiable iff there is no variable x for which there is a path in the corresponding graph from x to $\bar{x}$ and from $\bar{x}$ to x.[1] We now write this in FO(TC). In the following, $(x, 0)$ encodes literal x and $(x, 1)$ encodes $\bar{x}$. Formula δ encodes the edges from literal to literal, and PATH is the transitive closure of this edge relation.

$$\text{Occur}(c, x, b) \equiv (b = 0 \wedge P(c, x)) \vee (b = 1 \wedge N(c, x))$$
$$\delta(x, b, x', b') \equiv (\exists c)(\text{Occur}(c, x, 1 - b) \wedge \text{Occur}(c, x', b') \wedge x \neq x')$$
$$\text{PATH}(u, d, u', d') \equiv (\text{TC}_{xbx'b'}\delta)(u, d, u', d')$$

---

[1] One way to see this is that the condition is equivalent to there not being a resolution proof of $\neg \varphi$.

9.6 Restrictions of SO

$$\text{2-SAT} \equiv (\forall x)\neg(\text{PATH}(x, 0, x, 1) \wedge \text{PATH}(x, 1, x, 0))$$

We know from Theorem 3.16 and Corollary 9.17 that $\overline{\text{REACH}}$ is complete for NL via quantifier-free reductions. The completeness of 2-SAT will follow when we show that $\overline{\text{REACH}} \leq_{qf} \text{2-SAT}$. Given a graph G, the boolean formula $I(G)$ will have the following clauses,

- s.
- $\neg a \vee b$, when (a, b) is an edge of G.
- $\neg t$.

I is quantifier-free definable and $I(G) \in \text{2-SAT}$ iff $G \in \overline{\text{REACH}}$. It follows that HORN-SAT and 2-SAT are complete via quantifier-free reductions. $\square$

The proof of Proposition 9.27 also shows how to express the negations of boolean queries REACH$_a$ and REACH as SO-Horn and SO-Krom formulas.

$$\overline{\text{REACH}_a} \equiv (\exists T^1)(\forall e f_1 f_2)(T(t) \wedge \neg T(s) \wedge$$
$$(T(e) \vee \neg T(f_1) \vee A(e) \vee \neg E(e, f_1)) \wedge$$
$$(T(e) \vee \neg T(f_1) \vee \neg T(f_2) \vee \neg A(e) \vee f_1 = f_2 \vee$$
$$\vee \neg E(e, f_1) \vee \neg E(e, f_2) \vee \neg f_1 \vee \neg f_2))$$

$$\overline{\text{REACH}} \equiv (\exists T^1)(\forall ab)(T(s) \wedge \neg T(t) \wedge (T(b) \vee \neg T(a) \vee \neg E(a, b)))$$

Surprisingly, SO-Horn and SO-Krom collapse to their second-order existential parts,

Lemma 9.28. *The following equations hold for all sets of structures — finite or infinite,*

1. SO-Horn $=$ SO$\exists$-Horn
2. SO-Krom $=$ SO$\exists$-Krom

Proof It suffices to show that Horn or Krom formulas of the form,

$$\Psi \equiv (\forall P)(\exists Q_1 \cdots Q_r)(\forall \bar{z})\alpha$$

are equivalent to SO$\exists$-Horn and SO$\exists$-Krom formulas respectively. In the Horn case we first observe the following

Claim 9.29. *If the Horn formula Ψ holds for every P that is false on at most one tuple, then Ψ holds (for every P).*

From Claim 9.29, it follows that that we can replace P by either the true relation, or the relation that is true everywhere but on a fixed tuple $\bar{u}$,

$$\Psi \equiv (\exists Q_1 \cdots Q_r)(\forall \bar{u})(\forall \bar{z})\alpha(\mathbf{true}/P) \wedge \alpha(\neg \bar{u}/P)$$

In the Krom case, let us introduce the notation $A \xrightarrow{\Psi} B$ to mean that there is a path in the graph determined by Ψ from literal A to literal B and all intermediate literals are existential, i.e., Q-literals. We have the following generalization of the satisfaction condition described in the proof of part 2 of Proposition 9.27,

154   9. Complementation and Transitive Closure

Claim 9.30. *An ∀∃-Krom formula Ψ is false iff at least one of the following holds,*

1. *There are ∀ literals X, X' such that $X \xrightarrow{\Psi} X'$.*
2. *There is ∃ literal Y such that $Y \xrightarrow{\Psi} \neg Y$ and $\neg Y \xrightarrow{\Psi} Y$.*

It follows from Claim 9.30 that Ψ is equivalent to the SO∃-Krom formula in which P is replaced by relations that are false at at most two tuples.   □

Exercise 9.31 Prove Claims 9.29 and 9.30. [Hint for the first: for each relation $P^{\bar{y}}$ that is true except at $\bar{y}$, let $\overline{Q^{\bar{y}}}$ be the corresponding witness that makes Ψ true. If P is now a relation false at more than one point, let,

$$\overline{Q} \equiv \bigcap_{\bar{y} \notin P} \overline{Q^{\bar{y}}}$$

be the witness for P.]   □

Now, combining Proposition 9.27 and Lemma 9.28 we obtain the following characterizations of polynomial time and logspace,

Theorem 9.32. *The following equations hold for finite structures that include a successor relation,*

1. SO-Horn = P
2. SO-Krom = NL

Proof That SO-Horn $\subseteq$ P and SO-Krom $\subseteq$ NL follows from Proposition 9.27 and Lemma 9.28. We have also seen that languages SO-Horn and SO-Krom express problems that are complete for P and NL respectively via quantifier-free reductions. It therefore suffices to show that SO-Horn and SO-Krom are closed under quantifier-free reductions. We leave this as an exercise.   □

Exercise 9.33 Complete the proof of Theorem 9.32 by showing that the languages SO-Horn and SO-Krom are closed under quantifier-free reductions.   □

Historical Notes and Suggestions for Further Reading

Fact 9.1 is due to Kleene and Spector, see [Mos74]. The consequence that the fixed point hierarchy is strict over infinite structures then follows from the fact that the analytic hierarchy is strict. This is Suslin's Theorem. See [Mos80] for details.

Theorem 9.3 is originally due to Moschovakis [Mos74]. The formulation given here — showing that a monotone inductive definition leads to positive definitions of $\leq_\varphi$ and $<_\varphi$ — comes from Ebbinghaus and Flum [EF95]. Theorem 9.6 is from [I82]. Corollary 9.8 is due to Gurevich and Shelah [GS85].

Theorems 9.10 and 9.11 are from [I83]. Theorem 9.20 is from [I88]. Corollaries 9.23 and 9.24 were independently proved by Immerman and Szelepcsényi [I88],

[Sze88]. The presentation here borrows from [EF95]. Grädel and McColm proved in [GM96] that Theorem 9.20 does not hold in the absence of ordering.

Exercise 9.25 was suggested by David Mix Barrington.

Versions of Proposition 9.27 regarding HORN-SAT and 2-SAT appear in [JLL76] and [JL77]. Theorem 9.32 and Lemma 9.28 are from Grädel [Grä91l], see also Leivant [Lei89] for another characterization of complexity classes via restrictions of second-order formulas.

The complementation of fixed points (Theorem 9.6) increases the arity of the relations involved. We suspect that this is necessary. Bradfield has shown that the hierarchy of fixed points and negations is strict for the modal mu-calculus [Bra96]. The modal mu-calculus allows only fixed point of arity one, see Chapter 14.

10
Polynomial Space

Polynomial Space is the largest complexity class studied in this book. It consists of what we could compute with a feasible amount of hardware, but with no time limit. PSPACE has several apparently quite different descriptive characterizations.

PSPACE is a large and robust complexity class. With polynomially many bits of memory, we can search any implicitly-defined graph of exponential size. This leads to complete problems such as reachability on exponentially-large graphs (Proposition 10.7). Similarly, we can search the game tree of any board game whose configurations are describable with polynomially-many bits. This leads to complete problems concerning winning strategies (Exercise 10.3).

There are several descriptive characterizations of PSPACE (Corollary 10.29, Corollary 10.30). Two of these are: the set of boolean queries expressible by first-order quantifier blocks iterated exponentially and the set of boolean queries expressible by second-order quantifier blocks iterated polynomially:

$$\text{PSPACE} \;=\; \text{FO}[2^{n^{O(1)}}] \;=\; \text{SO}[n^{O(1)}]. \tag{10.1}$$

There is a perfectly tight relationship between space and the number of variables (Theorem 10.16).

10.1 Complete Problems for PSPACE

We know from Theorem 2.25 that PSPACE is equal to alternating polynomial time ($\text{ATIME}[n^{O(1)}]$). This characterization leads to our first complete problem for PSPACE. Recall QSAT, the quantified satisfiability problem (Definition 2.30).

Proposition 10.2. *The quantified boolean satisfaction problem (QSAT) is complete for PSPACE via first-order reductions.*

Proof We saw in Proposition 2.31 that QSAT is in ATIME[n] and thus in PSPACE.

To show completeness, let M be an alternating machine that makes n^k moves for inputs of size n. We know that M can be put in a normal form in which it writes down its n^k alternating choices $\bar{c} = c_1 c_2 \ldots c_{n^k}$ and then deterministically evaluates its input, using choice vector $\bar{c}$. Let the corresponding deterministic time n^k machine be D. We have that for all inputs $\mathcal{A}$,

$$M(\text{bin}(\mathcal{A}))\!\downarrow \quad \Leftrightarrow \quad (\exists c_1)(\forall c_2)\cdots(Q_{n^k} c_{n^k})(D(\bar{c}, \text{bin}(\mathcal{A}))\!\downarrow)$$

Thinking of D as an NP machine, we know by Theorem 7.16 that there is a first-order reduction from the language accepted by D to SAT. Let f be the first-order query such that for all $\bar{c}$ and $\mathcal{A}$,

$$D(\bar{c}, \text{bin}(\mathcal{A}))\!\downarrow \quad \Leftrightarrow \quad f(\mathcal{A}) \in \text{SAT}$$

Let the new boolean variables in $f(\mathcal{A})$ be $d_1 \ldots d_{t(n)}$. Then, finally, we have,

$$M(\text{bin}(\mathcal{A}))\!\downarrow \quad \Leftrightarrow \quad \text{"}(\exists c_1)(\forall c_2)\cdots(Q_{n^k} c_{n^k})(\exists d_1 \ldots d_{t(n)}) f(\mathcal{A})\text{"} \in \text{QSAT} . \quad \square$$

Exercise 10.3 Show that the standard universal complete problem for PSPACE, U_{PSPACE}, is in fact complete for PSPACE via first-order reductions.

$$U_{\text{PSPACE}} = \{M\#w\#^r \mid M \text{ accepts } w \text{ using at most } r \text{ tape cells}\} . \quad \square$$

Exercise 10.4 GEOGRAPHY is a game played by players E and A, played on a directed graph with start node s. At the first move, E chooses a vertex v_1 with an edge from s; then A chooses v_2 having an edge from v_1, and so on. No vertex may be chosen more than once. The last player able to move wins. (This is a generalization of the game geography, a popular pastime for kids during long car rides. In that game, E chooses a place name, e.g., Ithaca, and A must answer with a name that starts with the letter that the previous name ended with. In this case A could say Amherst. Then E must choose a place name beginning with "T" and so on. Whoever cannot think of an appropriate unused name loses.)

Show that GEOGRAPHY is PSPACE complete via first-order reductions. [Hint: reduce QSAT to GEOGRAPHY. Map the QSAT problem $(\exists a)(\forall b)(\exists c) \cdots$ to a graph that begins as in Figure 10.5.] $\quad \square$

Another complete problem for PSPACE is a generalization of REACH to graphs with a polynomial-size description, but exponentially many nodes. Define a *k-local*

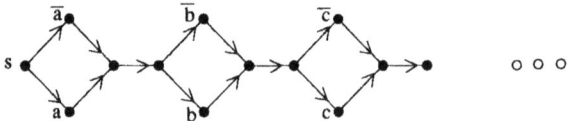

Figure 10.5: Image of $(\exists a)(\forall b)(\exists c)\cdots$ in reduction QSAT $\leq_{\text{fo}}$ GEOGRAPHY

graph to be a graph on vertex set $\{0, 1\}^n$ such that for each vertex u there is a unique next vertex v, and bit i of v is determined uniformly by bits $i-k, i-k+1, \ldots, i+k$ of u. Note that a k-local graph can be presented as a table of size 2^{2k+1}. In order to keep the tables size at most n, and thus codeable by a single unary input relation, R, we insist that $k \leq \lfloor (\log n)/2 - 1 \rfloor$.

Let $\mathcal{A}$ be a structure of vocabulary $\tau_\ell = \langle R^1, S^1, T^1 \rangle$ consisting of three unary relations: R encodes the transition relation, S is the source node and T is the terminal node. Define boolean query REACH$_{d\ell}$ to be the set of structures $\mathcal{A} \in$ STRUC$[\tau_\ell]$ such that there is an $R^\mathcal{A}$ path from $S^\mathcal{A}$ to $T^\mathcal{A}$.

Let unary relation variable A denote our present position. To define the next move relation, for each i, read the $\lfloor \log n \rfloor$ bits $w = A[[i-k \ldots i+k]]$; then $R(w)$ yields bit i of A'. Note that in the following, $\log n$ can be computed because it is the largest r such that BIT(max, r) holds, and the additions are mod n, so we think of the n bits of A as being in a loop. In the following, $\alpha(i, w, A)$ means that the binary representation of w encodes bits $A[[i-k..i+k]]$ and δ is the edge relation on unary relations that results.

$$\alpha(i, w, A) \equiv (\exists k.k = \lfloor (\log n)/2 - 1 \rfloor)(\forall j.j \leq \lfloor \log n \rfloor)(A(i-k+j) \leftrightarrow \text{BIT}(w, j))$$

$$\delta(A, A') \equiv (\forall i)(\exists w)(\alpha(i, w, A) \wedge A'(i) \leftrightarrow R(w))$$

With the above definitions, we can express REACH$_{d\ell}$ in the language SO(monadic, DTC) — second-order logic restricted to monadic relation variables, plus the deterministic transitive closure operator.

$$\text{REACH}_{d\ell} \equiv (\text{DTC}_{AA'}\delta)(S, T) . \qquad (10.6)$$

Proposition 10.7. *Problem* REACH$_{d\ell}$ *is complete for* PSPACE *via first-order reductions. In fact,* REACH$_{d\ell}$ *remains complete when k is restricted to a fixed constant.*

Proof REACH$_{d\ell}$ is in PSPACE, and in fact in DSPACE$[n]$. The algorithm is the same as Algorithm 3.22 which showed that REACH$_d \in$ L. We need n bits to record the current position and another n bits to keep a counter to avoid looping.

Let M_0 be a linear-space Turing-machine that accepts U_{PSPACE} (Exercise 10.3). Since M_0 has a fixed state set, we may encode its instantaneous description in a way that is k-local for appropriate k — twice the log of the number of states of M_0 suffices. Then a first-order reduction from U_{PSPACE} to REACH$_{d\ell}$ maps input string w to starting vertex S, lets T be M_0's fixed accept ID, and lets R be the encoding of M_0's fixed transition relation. $\square$

As seen in the next exercise, it is quite natural to express REACH$_{d\ell}$ in FO$[2^{n^{O(1)}}]$ and in SO$[n^{O(1)}]$ cf. Equation 10.1. The language FO$[2^{n^{O(1)}}]$ allows exponential quantifier depth but only $O(\log n)$ bits of variables. By contrast, SO$[n^{O(1)}]$ allows polynomial quantifier depth and polynomially many variable bits. See Open Problem 11.5, where the tradeoff exhibited by this problem for quantifier-depth versus number of variable bits is discussed.

Exercise 10.8 1. Show that REACH$_{de}$ can be expressed in FO[2^n] using only two first-order variables.

2. Show that REACH$_{de}$ can be expressed in SO[n] using only three relation variables, all of which are monadic. □

10.2 Partial Fixed Points

Recall from Theorem 5.2 that for all polynomially bounded and parallel time constructible $t(n)$,

$$\text{CRAM}[t(n)] = \text{IND}[t(n)] = \text{FO}[t(n)].$$

This theorem breaks down for IND[$t(n)$] when $t(n)$ is super-polynomial, because monotone inductive definitions are polynomially bounded (Theorem 4.3). The reason that monotone inductions are polynomially bounded is that a relation of arity k that is being defined has n^k possible tuples. Monotonicity implies that after a tuple is added to a relation it can never be removed. Thus, such a computation cannot make more than n^k steps.

We now generalize inductive definitions to *iterative definitions* in which the requirement of monotonicity is removed. Any tuple in a k-ary relation may be added or removed, depending on the whole relation at the previous time step. Such an algorithm can thus usefully take no more than 2^{n^k} steps.

Definition 10.9 Define ITER[$t(n)$, arity k] to be the set of properties definable by iterating the simultaneous first-order definitions of a set of c relations of arity k $t(n)$ times, for some constant c. Let ITER[$t(n)$] be the set of boolean queries definable by $O(t(n))$ iterations, regardless of the arity,

$$\text{ITER}[t(n)] = \bigcup_{k=1}^{\infty} \text{ITER}[t(n), \text{arity } k].$$

Of course, after $t(n) = 2^{cn^k}$ iterations, these relations will either reach a fixed point or be in a cycle. Thus, define

$$\text{ITER}[\text{arity } k] = \bigcup_{c=1}^{\infty} \text{ITER}[2^{cn^k}, \text{arity } k]. \quad \square$$

Observe that ITER[$t(n)$] is exactly the generalization of IND[$t(n)$] that removes the monotonicity requirement:

Exercise 10.10 Show that for all polynomially-bounded $t(n)$,

$$\text{IND}[t(n)] = \text{ITER}[t(n)].$$

[Hint: You can make an iterative definition of a relation R inductive by adding a time component. You can simultaneously define the new values $R'(\bar{x}, \bar{t} + 1)$ and $\overline{R}'(\bar{x}, \bar{t} + 1)$ in terms of $R'(\cdot, \bar{t})$ and $\overline{R}'(\cdot, \bar{t})$.] □

10.2 Partial Fixed Points   161

Example 10.11 The following iterative definition proves REACH$_{d\ell}$ ∈ ITER[arity 1]. In this definition, we simultaneously define two booleans start and accept, and the relation P indicating our current position. Recall from Equation (10.6) that $\alpha(i, w, P)$ means that w encodes the relevant bits of P: $P[[i - k \ldots i + k]]$.

$$\text{start}' := \textbf{false}$$

$$\text{accept}' := \text{accept} \lor (\forall x)(T(x) \leftrightarrow P(x))$$

$$P'(i) := (\text{start} \land S(i)) \lor (\neg\text{start} \land \text{accept} \land P(i))$$
$$\lor (\neg\text{start} \land \neg\text{accept} \land (\exists w)(\alpha(i, w, P) \land R(w)))$$

□

Definition 10.12 (Partial Fixed Point) The iterative definitions in Example 10.11 were designed to stop once T is reached. In general, however, such iterative definitions do not have to reach a fixed point; they could cycle forever. Thus, the appropriate generalization of the least fixed point operator is called *partial fixed point* (PFP). Given an iterative — not necessarily monotone — definition $\varphi(R^k, x_1, \ldots, x_k)$ define its *partial fixed point* as follows. For any structure $\mathcal{A}$,

$$(\text{PFP}_{R, x_1 \ldots x_k} \varphi)^{\mathcal{A}} = \begin{cases} (\varphi^{\mathcal{A}})^r(\emptyset) & \text{if } (\varphi^{\mathcal{A}})^r(\emptyset) = (\varphi^{\mathcal{A}})^{r+1}(\emptyset) \\ \emptyset & \text{if there is no such } r \end{cases}$$

The above definition defines all uses of partial fixed point, even if they define a loop. In practice, we never need to write a looping iterative definition. Define FO(PFP) to be the closure of first-order logic under applications of the partial fixed point operator. □

Now we show that the iteration and PFP operators appropriately generalize Theorem 5.2 to super-polynomial time.

Theorem 10.13. *For any parallel-time constructible $t(n)$,*

$$\text{CRAM}[t(n)] = \text{ITER}[t(n)] = \text{FO}[t(n)]$$

In particular, PSPACE $=$ FO(PFP) $=$ FO$[2^{n^{O(1)}}]$.

Proof This proof follows along the lines of the proof of Theorem 5.2, with surprisingly few changes. The containment CRAM$[t(n)] \subseteq$ ITER$[t(n)]$ has almost the same proof as that CRAM$[t(n)] \subseteq$ IND$[t(n)]$ for polynomially bounded $t(n)$ (Lemma 5.3). The only change is that before we kept track of the time t. This allowed us to avoid keeping track of memory, thus obtaining a corollary that for $t(n)$ polynomially bounded, having more than polynomial memory is not useful (Corollary 5.8). Now, we keep track of each of the polynomial memory locations in our iterative definition, just as we keep track of all the registers of all the processors. The contents of a memory location at the next is the same as it was at the previous step unless it was assigned at the previous step. With this change, we no longer need to maintain the time variable, and the proof goes through without further change.

The containment $\text{IND}[t(n)] \subseteq \text{FO}[t(n)]$ depended on the fact that the inductive definition was monotone and that Lemma 4.20 applies. Recall the statement of this lemma:

An R-positive first-order formula φ can be written as:

$$\varphi(R, x_1, \ldots, x_k) \equiv (Q_1 z_1.M_1)\ldots(Q_s z_s.M_s)(\exists x_1 \ldots x_k.M_{s+1})R(x_1, \ldots, x_k).$$

where the M_i's are quantifier-free formulas in which R does not occur.

If the inductive definition φ closes in at most $t(n)$ iterations, it follows from Lemma 4.20 that,

$$\text{LFP}(\varphi) \equiv [(Q_1 z_1.M_1)\ldots(Q_s z_s.M_s)(\exists x_1 \ldots x_k.M_{s+1})]^{t(n)}\textbf{false}.$$

Notice that the quantifier-free formula after the iterated quantifier block can be taken to be "**false**". This corresponds to the fact that the inductive definition can be evaluated starting with the emptyset and repeatedly applying φ.

Now, let $\varphi(R, x_1, \ldots, x_k)$ be a not-necessarily-monotone, first-order formula. Let b be a new boolean variable and let R' be an arity $k+1$ relation symbol, where $R'(\bar{x}, 0)$ will correspond to $R(\bar{x})$ and $R'(\bar{x}, 1)$ will correspond to $\neg R(\bar{x})$. In the following, we generalize the iterative definition of φ to φ' by replacing all occurrences of $R(\bar{y})$ in φ by $R'(\bar{y}, 0)$ and all occurrences of $\neg R(\bar{y})$ by $R'(\bar{y}, 1)$. We do the opposite for $\neg \varphi$,

$$\varphi'(R', \bar{x}, b) \equiv (b = 0 \wedge \varphi) \vee (b = 1 \wedge \neg \varphi).$$

Thus, φ' is positive and Lemma 4.20 applies. Let QB be the resulting quantifier-block, so that,

$$\varphi'(R', \bar{x}, b) \equiv [\text{QB}]R'(\bar{x}, b).$$

If iterative definition φ closes in at most $t(n)$ iterations, it follows that,

$$\text{PFP}(\varphi) \equiv [\text{QB}]^{t(n)}(b = 1).$$

Here the quantifier-free formula is $b = 1$, corresponding to the initial step of the iterative definition in which for all $\bar{x}$, $R'(\bar{x}, 0)$ is false and $R'(\bar{x}, 1)$ is true. Thus, $\text{ITER}[t(n)] \subseteq \text{FO}[t(n)]$, as desired.

Finally, the containment $\text{FO}[t(n)] \subseteq \text{CRAM}[t(n)]$ goes through unchanged. This completes the proof of Theorem 10.13. □

10.3 DSPACE[n^k] = VAR[$k+1$]

In Corollary 5.10, we saw that the number of variables in a first-order inductive definition is closely related to the number of processors in a CRAM. The bounds, however, were not exact because the connection pattern of a CRAM is different from the connection pattern of a "first-order machine", so there is some inefficiency in simulating one by the other.

In this section, we explore the relationship between number of variables, arity of iterated definitions, and deterministic space. We see in Theorem 10.16 that the relationship is tight. We begin with the following definition of FO-VAR$[t(n), v]$. Note that we allow a bounded number of boolean variables in addition to the v first-order variables. Quantified boolean variable, $(\forall b)$, $(\exists b)$, are similar to the boolean connectives "and" and "or". The inclusion of boolean variables makes the definitions of the following classes more robust.

Definition 10.14 A query is in class FO-VAR$[t(n), v]$ iff it is in FO$[t(n)]$ and the relevant quantifier-block contains only v first-order variables — although it may contain some constant number of additional boolean variables.

As in the definition of ITER[arity v], the truth assignment of all the variables will cycle or stabilize after at most $t(n) = 2^{cn^{v-1}}$ iterations. Thus,

$$\text{VAR}[v+1] = \bigcup_{c=1}^{\infty} \text{FO-VAR}[2^{cn^v}, v+1] \, . \qquad \square$$

The above definition considers only a bounded number of variables. In Definition 11.1, we generalize this definition to any number of variables. In Theorem 10.16, we carefully count the number of domain variables. Thanks to the boolean variables, the number of domain variables need not be increased to take care of conjunctions and disjunctions.

Lemma 10.15. *Suppose that we have two quantifier blocks with identical quantifiers in identical order (ignoring any boolean quantifiers):*

$$QB_1 = [(Q_1 v_1.M_1)\ldots(Q_s v_s.M_s)] \, ,$$
$$QB_2 = [(Q_1 v_1.N_1)\ldots(Q_s v_s.N_s)] \, .$$

Then the conjunction and disjunction of these quantifier blocks may be written in the same form.

Proof The conjunction, for example, can be written with an extra, universally quantified boolean variable:

$$[QB_1]\varphi \wedge [QB_2]\varphi \equiv [(\forall b)(Q_1 v_1.R_1)\ldots(Q_s v_s.R_s)]\varphi$$

where,

$$R_i = (b=0 \wedge M_i) \vee (b=1 \wedge N_i) \, . \qquad \square$$

Theorem 10.16. *For $k = 1, 2, \ldots,$*

$$\text{DSPACE}[n^k] = \text{VAR}[k+1] = \text{ITER}[\text{arity } k] \, .$$

Proof The proof is accomplished using three lemmas proving the following containments:

$$\text{DSPACE}[n^k] \subseteq \text{VAR}[k+1] \subseteq \text{ITER}[\text{arity } k] \subseteq \text{DSPACE}[n^k] \, .$$

164   10. Polynomial Space

Lemma 10.17. $DSPACE[n^k] \subseteq VAR[k+1]$.

Proof Let M be a DSPACE[n^k] Turing machine. M's work space consists of n^k tape cells each of which holds a symbol from some finite alphabet, Σ.

The contents of M's tape at time $t+1$ is a deterministic, local transformation of the contents at time t. Namely, the contents of cell p at time $t+1$ is a function of the contents of cells $p-1$, p, $p+1$ at time t.

We write a logical formula $C_t(\bar{x}, \bar{b})$ meaning that after step t of M's computation, the cell at position $\bar{x}$ is $\bar{b}$. Here $\bar{x} = x_1, \ldots, x_k$ is a k-tuple of variables ranging over the set $\{0, \ldots, n-1\}$ and $\bar{b}$ is a tuple of boolean variables coding an element of Σ.

The following is an iterative definition of C_t:

$$C_{t+1}(\bar{x}, \bar{b}) = \bigvee_{\langle \bar{a}_{-1}, \bar{a}_0, \bar{a}_1 \rangle \to \bar{b}} \Big(C_t(\bar{x}-1, \bar{a}_{-1}) \wedge C_t(\bar{x}, \bar{a}_0) \wedge C_t(\bar{x}+1, \bar{a}_1) \Big). \tag{10.18}$$

Here, the disjunction is over the finite set of quadruples $(\bar{a}_{-1}, \bar{a}_0, \bar{a}_1, \bar{b})$ such that the first three symbols lead to the fourth symbol in one move of M. Note that this set of quadruples is exactly a representation of M's state table.

We have already seen how to write C_0 with $k+1$ domain variables, for example, in Theorem 7.8. Furthermore, M accepts its input iff it eventually reaches its accept state. Let $\overline{\text{true}}$ code the appropriate accept symbol. Thus M accepts its input iff eventually $C_t(\bar{0}, \overline{\text{true}})$ holds.

The lemma will be proved once we show the following:

Claim 10.19. *There is a quantifier block QB containing $k+1$ domain variables such that Equation (10.18) may be rewritten as:* $C_{t+1}(\bar{x}, \bar{b}) = [QB] C_t(\bar{x}, \bar{b})$.

The proof of Claim 10.19 is purely symbol manipulation. We first write quantifier blocks QB_+ and QB_- whose job it is to replace $\bar{x}$ by $\bar{x}+1$ and $\bar{x}-1$ respectively, i.e., for any formula φ, we have,

$$\varphi(\bar{x}+1) \equiv [QB_+]\varphi(\bar{x}),$$
$$\varphi(\bar{x}-1) \equiv [QB_-]\varphi(\bar{x}).$$

These quantifier blocks can be written with $k+1$ domain variables. The idea is to add one to $\bar{x}$ by replacing x_k with its successor, or, if $x_k = max$, by replacing x_k by 0 and x_{k-1} by its successor, or, etc. We existentially quantify a tuple of boolean variables, $\bar{c}$, to guess for which i, $1 \leq i \leq k$, x_i will be incremented. For $j > i$, it must be that $x_j = max$ and $x'_j = 0$.

The form of the quantifier block will be as follows,

$$(\exists \bar{c}.P)(\exists y.N_k)(\exists x_k.M_k)(\exists y.N_{k-1})(\exists x_{k-1}.M_{k-1}) \cdots (\exists y.N_1)(\exists x_1.M_1).$$

The quantifier-free conditions P, N_i and M_i are as follows:

$$P \equiv \bigvee_{i=1}^{k} (\bar{c} = i \;\wedge\; x_i \neq max \;\wedge\; x_{i+1} = x_{i+2} = \cdots = x_k)$$

$$N_i \equiv (i < \bar{c} \land y = x_k) \quad \lor \quad (i = \bar{c} \land \text{SUC}(x_k, y))$$
$$\lor \quad (i > \bar{c} \land y = \max)$$
$$M_i \equiv x_i = y.$$

Thus, we have QB_+, QB_-, and, trivially, QB_0. Observe that the desired QB of Claim 10.19 is a positive boolean combination of these three quantifier blocks. It follows from Lemma 10.15 that QB exists and has $k + 1$ domain variables, as desired. This completes the proof of the Claim and thus of Lemma 10.17. □

Lemma 10.20. $\text{VAR}[k + 1] \subseteq \text{ITER}[\text{arity } k]$.

Proof Here we have a quantifier block of the form

$$QB = [(Q_1 x_{i_1}.M_1)(B_1)\ldots(Q_r x_{i_r}.M_r)(B_r)],$$

where each $i_j \in \{1, \ldots, k+1\}$, the M_i are quantifier-free, and the B_i are blocks of boolean quantifiers over boolean variables $\{b_1, \ldots, b_c\}$. We can convert the iteration of QB into an iterative definition of relations as follows. Let $R_{s,t}$ be a set of k-ary relation symbols, for $1 \leq s \leq r$, and $t \in \{0, 1\}^{\{1,\ldots,c\}}$. Thus, t specifies an assignment to all the boolean variables. Intuitively, the iterative definition for the $R_{s,t}$'s is given as follows:

$$R_{s,t}(x_1, \ldots, \widehat{x_{i_s}}, \ldots, x_{k+1}) \equiv$$
$$(Q_s x_{i_s}.M_s)(B_s) R_{s+1,t'}(x_1, \ldots, \widehat{x_{i_{s+1}}}, \ldots, x_{k+1})$$

Here 1 is added to s modulo r. The notation $R(x_1, \ldots, \widehat{x_i}, \ldots, x_{k+1})$ means that the variable x_i is omited. In the above formula the variable to be quantified next is safely omitted. This is why arity k suffices.

The above formula is a bit misleading in that we must write out boolean quantifier block B_s. For example, the formula $(\forall b_j) R_{s+1,t'}(\bar{x})$ would be expanded to

$$R_{s+1,(t|b_j=0)}(\bar{x}) \land R_{s+1,(t|b_j=1)}(\bar{x}).$$
□

Lemma 10.21. $\text{ITER}[\text{arity } k] \subseteq \text{DSPACE}[n^k]$.

This last inclusion is obvious because $O[n^k]$ bits suffice to record the current meaning of the bounded number of relations of arity k. Each bit of each relation in the next iteration may then be computed by evaluating a fixed first-order formula. This can be done in $\text{DSPACE}[\log n]$ and thus certainly in $\text{DSPACE}[n^k]$. This completes the proof of Theorem 10.16. □

10.4 Using Second-Order Logic to Capture PSPACE

In Theorem 10.13, we saw that PSPACE consists of boolean queries that are describable via first-order quantifier blocks iterated exponentially many times.

Another characterization is that PSPACE is equal to boolean queries that are describable via a second-order quantifier block iterated polynomially many times. First, we define $\mathrm{SO}[t(n)]$ analogously to $\mathrm{FO}[t(n)]$ (Definition 4.24).

Definition 10.22 ($\mathrm{SO}[t(n)]$) A set $S \subseteq \mathrm{STRUC}[\tau]$ is a member of $\mathrm{SO}[t(n)]$ iff there exist quantifier-free formulas M_i, $0 \leq i \leq k$, from $\mathcal{L}(\tau)$, a tuple of constants $\bar{C}$, and a quantifier block

$$\mathrm{QB} = [(Q_1 Z_1.M_1)\ldots(Q_k Z_k.M_k)]$$

such that for all $\mathcal{A} \in \mathrm{STRUC}[\tau]$,

$$\mathcal{A} \in S \quad \Leftrightarrow \quad \mathcal{A} \models \left([\mathrm{QB}]^{t(|\mathcal{A}|)} M_0\right)(\bar{C}) .$$

The only difference between $\mathrm{SO}[t(n)]$ and $\mathrm{FO}[t(n)]$ is that **some** of the variables Z_i may be relation variables: $S_i^{a_i}$, and the corresponding constant C_i must be either an input or numeric relation, or a boolean. Here **false** denotes the empty relation, and **true** denotes the full relations of appropriate arity.

Define $\mathrm{SO}(\mathrm{arity}\ a)[t(n)]$ to be the restriction of $\mathrm{SO}[t(n)]$ allowing quantification only of relations of arity a or less. □

As an example, we prove Proposition 10.25 below. In order to do this, we need,

Exercise 10.23 Let $\mathrm{SO}(\mathrm{arity}\ a)(\mathrm{TC})$ be the language $\mathrm{SO}(\mathrm{arity}\ a)$ extended by the transitive closure operator. Any formula $\Phi \in \mathrm{SO}(\mathrm{arity}\ a)(\mathrm{TC})$ can be written in the following normal form:

$$\Phi \equiv (\mathrm{TC}_{A_1^a\ldots A_k^a, x_1\ldots x_r, A_1'^a\ldots A_k'^a, x_1'\ldots x_r'} \alpha)(\overline{\mathbf{false}}, \overline{\mathbf{true}}) , \qquad (10.24)$$

where α is quantifier-free. [Hint: the proof is similar to that of Theorem 9.20.] □

Proposition 10.25. $\mathrm{SO}(\mathrm{arity}\ a)(\mathrm{TC}) \subseteq \mathrm{SO}(\mathrm{arity}\ a)[n^a]$.

Proof Let $\Phi \in \mathrm{SO}(\mathrm{arity}\ a)(\mathrm{TC})$. We may assume that Φ is in the form of Equation (10.24). We construct a second-order quantifier block similar to the first-order quantifier block in Example 4.22.

Recall there that

$$\mathrm{QB}_{tc} \equiv (\forall z.M_1)(\exists z)(\forall uv.M_2)(\forall xy.M_3) ,$$

where,

$$M_1 \equiv \neg(x = y \vee E(x, y))$$

$$M_2 \equiv (u = x \wedge v = z) \vee (u = z \wedge v = y)$$

$$M_3 \equiv (x = u \wedge y = v)$$

We do almost the same thing here. The difference is that the expression $\overline{A} = \overline{A'}$ is no longer quantifier-free. We must replace it by

$$\forall z.(A_1(z) \leftrightarrow A_1'(z) \wedge \cdots \wedge (A_k(z) \leftrightarrow A_k'(z))) . \qquad (10.26)$$

10.4 Using Second-Order Logic to Capture PSPACE   167

We abbreviate (10.26) by $\forall z.(\overline{A}(z) \leftrightarrow \overline{A'}(z))$. The new quantifier block is as follows:

$$QB_{SO(TC)} \equiv (\forall z.N_1)(\exists \overline{B})(\exists \overline{C}\,\overline{D})(\forall z)(\exists b.N_2)(\exists \overline{A}\,\overline{A'})(\forall z)(\exists b.N_3),$$

where,

$$N_1 \equiv \neg((\overline{A}(z) \leftrightarrow \overline{A'}(z)) \vee \alpha(\overline{A}, \overline{A'}))$$

$$N_2 \equiv (b \wedge \overline{C}(z) \leftrightarrow \overline{A}(z) \wedge \overline{D}(z) \leftrightarrow \overline{B}(z)) \vee$$
$$\quad (\neg b \wedge \overline{C}(z) \leftrightarrow \overline{B}(z) \wedge \overline{D}(z) \leftrightarrow \overline{A'}(z)) \vee$$

$$N_3 \equiv (\overline{A}(z) \leftrightarrow \overline{C}(z) \wedge \overline{A'}(z) \leftrightarrow \overline{D}(z)) \,.$$

Finally, the length of the relevant α-path can be at most 2^{kn^a}. It follows that,

$$\Phi \equiv [QB_{SO(TC)}]^{kn^a}(\mathbf{false})[\mathbf{false}/A, \mathbf{true}/A'] \,,$$

as desired. □

The following theorem summarizes the relationship between deterministic and nondeterministic space and second-order descriptive complexity,

Theorem 10.27. *For $k=1,2,\ldots$*

1. $DSPACE[n^k] = SO(\text{arity }k)(DTC)$,
2. $NSPACE[n^k] = SO(\text{arity }k)(TC)$.

Exercise 10.28 Prove Theorem 10.27. [Hint: note that we have already seen that $REACH_{d\ell}$ is complete for PSPACE and expressible in SO(arity 1)(DTC). In fact, it is complete for DSPACE[n] via first-order, linear-size reductions. You can generalize $REACH_{d\ell}$ to a nondeterministic version $REACH_\ell$ that is complete for NSPACE[n]. Finally, these problems can be generalized to work on relations of arity k.] □

We finish this chapter with a summary of the descriptive characterizations of PSPACE.

Corollary 10.29. *We have the following descriptive characterizations of PSPACE:*

$$PSPACE = FO(PFP) \;\;=\;\; FO[2^{n^{O(1)}}]$$
$$= SO(TC) \;\;=\;\; SO(DTC) \;\;=\;\; SO[n^{O(1)}] \,.$$

From a parallel point of view, we have already seen in Corollary 7.28 that $SO[t(n)]$ is equal to $CRAM\text{-}PROC[t(n), 2^{n^{O(1)}}]$. This suggests that there is a striking tradeoff in parallel time versus hardware,

Corollary 10.30. $PSPACE = CRAM\text{-}PROC[2^{n^{O(1)}}, n^{O(1)}] = CRAM\text{-}PROC[n^{O(1)}, 2^{n^{O(1)}}]$.

Historical Notes and Suggestions for Further Reading

Theorem 10.13 comes from [I89a], although pieces of it were proved earlier. The fact that PSPACE = $FO[2^{n^{O(1)}}]$ was originally shown in [I80]. The equality of PSPACE and FO(PFP) was shown by Vardi in [Var82], although he used "while" instead of PFP. The name PFP comes from Abiteboul and Vianu, [AV91].

Theorem 10.16 is from [I91]. This theorem has been recently generalized by Barrington, Buss, and Immerman to capture all classes DSPACE$[s(n)]$ where $s(n) \geq \log n$ [BBI97]. Theorem 10.27 is from [I83].

Corollary 10.30 is reminiscent of the kind of duality between parallel time and hardware discussed by Hong [Ho86].

11
Uniformity and Precomputation

The first two dimensions of complexity are parallel time and hardware. These can be measured by the descriptive resources quantifier depth *and* number of variables, *respectively. The third dimension is* precomputation: *the work that initially goes into designing the program, formula, or circuit. Precomputation is less well understood than time and hardware.*

The descriptive resource corresponding to precomputation is the numeric predicates. Uniformity is the absence of precomputation: very uniform classes of circuits require minimal precomputation, and their descriptive counterparts require no extra numeric predicates.

We have seen that natural complexity classes admit natural problems that are complete via first-order reductions. Such a complete problem C encapsulates its entire complexity class: a problem S is in the complexity class iff there is a first-order translation of S into C.

In this chapter, we explore just how much computational power reductions need to have for the natural problems to remain complete. The answer is that they may be amazingly weak: We describe first-order and quantifier-free projections and see that they suffice. As a consequence, of complexity classes become remarkably homogeneous. For the same reason, precomputation is usually unnecessary.

We begin this chapter by generalizing the notion of variables so that we may move smoothly between first and second-order logic. In later sections, we explore uniformity and projections. Finally, we see that any boolean query can be understood as a generalized quantifier. For most complexity classes $\mathcal{C}$ and their corresponding complete problems T and generalized quantifiers Q_T, we find that the equation $\mathcal{C} = \text{FO}(Q_T)$ holds.

11.1 An Unbounded Number of Variables

In Theorem 5.22, we proved that for each $t(n)$, uniform circuit class $AC[t(n)]$ is equal to descriptive class $FO[t(n)]$. Thus, the amount of precomputation needed to build circuit n is exactly the writing of a fixed quantifier block $t(n)$ times. The $FO[t(n)]$ formalism requires that there be only a bounded number of variables. As we have seen, this corresponds to a polynomial amount of hardware (Theorem 10.16 and Corollary 5.10).

We now generalize descriptive classes $FO[t(n)]$ to the situation where we have more than a bounded number of variables. The problem is that we still want there to be a bounded number of variables, so that a block of quantifiers can still be iterated. The solution is to introduce variables with a larger range. For example, we can encode $\log n$ first-order variables with one variable that has $\log^2 n$ bits. These longer variables are queried using predicate BIT.

Definition 11.1 For $v(n) > O(1)$, an FO-VAR$[t(n), v(n)]$ formula has two sorts of variables: the domain variables: $x, y, z, \ldots$, ranging over the universe $\{1, 2, \ldots, n\}$, plus extended variables: $X, Y, Z, \ldots$, each of $v(n) \log n$ bits.

The extended variables may be quantified just like domain variables. However, the extended variables do not occur as arguments to any input relations. Their only role is as arguments in predicate BIT. That is, we may assert BIT(Y, x), meaning that bit x of Y is a one. Note that this makes sense only for extended variables with at most n bits. In order to extend the range to a polynomial number of bits, we allow a tuple of first-order variables to encode such a bit position: BIT$(Y, \bar{x})$.

Define FO-VAR$[t(n), v(n)]$, with $v(n) \geq 1$, to be the extension to $FO[t(n)]$ that we get by including a bounded number of $v(n) \log n$-bit extended variables. □

Observe that if $v(n) = n^k / \log n$, then an extended variable Y contains n^k bits and is the same as a second-order variable of arity k. In this special case, the expressions BIT$(Y, x_1, \ldots, x_k)$ and $Y(x_1, \ldots, x_k)$ have the same meaning. Thus, one use of Definition 11.1 is to provide a continuous transformation from first-order logic to second-order logic. The following proposition summarizes this idea:

Proposition 11.2. *For any $t(n)$, SO$[t(n)$, arity $k] =$ FO-VAR$[t(n), n^k/\log n]$. It follows that,* SO$[t(n)] =$ FO-VAR$[t(n), n^{O(1)}]$.

Classes FO-VAR$[t(n), v(n)]$ permit a restatement of Theorem (2.32).

Theorem 11.3. *Let $s(n) \geq \log n$ be space constructible. Then,*

$$\text{NSPACE}[s(n)] \subseteq \text{FO-VAR}[s(n), s(n)/\log n]$$
$$\subseteq \text{ATIME}[s(n)^2] \subseteq \text{DSPACE}[s(n)^2].$$

Proof The last containment was already part of Theorem 2.32. In the first containment, we want to express the existence of a path from start to accept in the computation graph of an NSPACE$[s(n)]$ machine N on some input w, with $n = |w|$. Since each extended variable contains $s(n)$ bits, a bounded number of these variables encode a configuration, i.e, a vertex in the computation graph of $N(w)$.

Since each alternation of quantifiers cuts the distance in half, we need to iterate the quantifier-block $\log(2^{O(s(n))}) = O(s(n))$ times.

In the second inclusion, we simulate the quantification of an extended variable: $\exists Y$ (or $\forall Y$) by the writing down Y's $s(n)$ bits in an existential (or universal) state. Similarly, a standard variable requires $\log n$ bits. Each other part of the FO-VAR$[s(n), s(n)/\log n]$ formula can be simulated in constant time. □

11.1.1 Tradeoffs Between Variables and Quantifier Depth

A fundamental challenge in computational complexity theory is to understand the tradeoff between parallel time and hardware. We know that parallel time is equal to quantifier depth (Theorem 5.2). We also know that there is a tight relationship between the number of variables and deterministic space (Theorem 10.16). There is also a close relationship between number of variables and the amount of hardware in a CRAM (Corollary 5.10).

One instance of the above tradeoff problem seems particularly worthy of study. The following very different characterizations of PSPACE follow from Corollary 10.29 and Proposition 11.2.

$$\text{PSPACE} = \text{FO-VAR}[2^{n^{O[1]}}, O(1)] = \text{FO-VAR}[n^{O[1]}, n^{O(1)}]$$

Thus, the descriptive power of a bounded number of first-order variables (i.e., $O(\log n)$ bits) and exponential quantifier depth is equal to the descriptive power of a bounded number of second-order variables (i.e., $n^{O(1)}$ bits) and polynomial quantifier depth. We would like to know if there is anything in between. For example, what quantifier depth is necessary and sufficient when $\log n$ first-order variables are available?

To make this problem more concrete, recall PSPACE-complete problem REACH$_{d\ell}$ (Equation (10.6)).

Exercise 11.4 Show using Exercise 10.8 that REACH$_{d\ell}$ can be expressed in FO-VAR$[2^n, 2]$ and also in FO-VAR$[n, 3n/\log n]$. □

One interesting instance of the general tradeoff problem is the following,

Open Problem 11.5 What is the tradeoff between number of variables and quantifier-depth for describing REACH$_{d\ell}$?

11.2 First-Order Projections

In this section we define an extremely weak reduction, the first-order projection (fop). For reasons still not completely understood, natural computational problems tend to be complete for standard complexity classes. Furthermore, they remain complete via fops. As we will see, the weakness of these reductions gives a strong characterization of the corresponding complexity classes.

In [Val82], Valiant defined *projection*, an extremely low-level many-one reduction.

Definition 11.6 Let $S, T \subseteq \{0, 1\}^*$. A k-ary *projection* from S to T is a sequence of maps $\{p_n\}$, $n = 1, 2, \ldots$, that satisfy the following properties. First, for all n and for all binary strings s of length n, $p_n(s)$ is a binary string of length n^k and,

$$s \in S \quad \Leftrightarrow \quad p_n(s) \in T.$$

Second, let $s = s_0 s_1 \ldots s_{n-1}$. Then each map p_n is defined by a sequence of n^k literals, $\langle l_0, l_1, \ldots, l_{n^k-1} \rangle$, where

$$l_i \in \{0, 1\} \cup \{s_j, \bar{s}_j \mid 0 \leq j \leq n-1\}.$$

Thus, as s ranges over strings of length n, each bit of $p_n(s)$ depends on at most one bit of s: $p_n(s)[[i]] = l_i(s)$. □

Projections were originally defined as a non-uniform sequence of reductions — one for each value of n. That is, a projection can be viewed as a many-one reduction produced by a family $\{C_n\}$ of circuits of depth one. The circuits consist entirely of wires connecting input bits or negated input bits to outputs. If circuit family $\{C_n\}$ is sufficiently uniform, we arrive at the class of *first-order projections*.

The idea of the definition is that the choice of literals $\langle l_0, l_1, \ldots, l_{n^k-1} \rangle$ in Definition 11.6 is given by a first-order formula in which no input relation occurs. Thus, the formula can talk only about bit positions and not bit values.

Definition 11.7 (First-Order Projections): Let

$$I = \langle \varphi_0, \varphi_1, \ldots, \varphi_r, \psi_1, \ldots, \psi_t \rangle$$

be a k-ary first-order reduction from S to T (Definition 2.11). Suppose further that formula φ_0 defining the universe is entirely numeric, i.e., no input relations occur in φ_0, and for $i \geq 1$, φ_i satisfies the following *projection condition*, as do the ψ_j's.

$$\varphi_i \equiv \alpha_1 \vee (\alpha_2 \wedge \lambda_2) \vee \cdots \vee (\alpha_e \wedge \lambda_e) \tag{11.8}$$

where the α_j's are mutually exclusive formulas in which no input relations occur and each λ_j is a literal, i.e. an atomic formula $P(x_{j_1}, \ldots x_{j_a})$ or its negation.

In this case, predicate $R_i(\langle u_1, \ldots, u_k \rangle, \ldots, \langle \ldots, u_{ka_i} \rangle)$ holds in $I(\mathcal{A})$ if $\alpha_1(\bar{u})$ is true or if $\alpha_j(\bar{u})$ is true for some $1 < j \leq e$ and if the corresponding literal $\lambda_j(\bar{u})$ holds in $\mathcal{A}$. Thus, each bit in the binary representation of $I(\mathcal{A})$ is determined by at most one bit in the binary representation of $\mathcal{A}$. We say that I is a *first-order projection* (fop). Write $S \leq_{\text{fop}} T$ to mean that S is reducible to T via a first-order projection.

If I is a fop and all the φ_i's and ψ_j's are quantifier-free, then we say that I is a *quantifier-free projection* (qfp). Write $S \leq_{\text{qfp}} T$ to mean that S is reducible to T via a quantifier-free projection. Recall that in first-order reductions I, the numeric predicates in $I(\mathcal{A})$ must be first-order definable in $\mathcal{A}$. Similarly, for quantifier-free reductions: the numeric predicates must be quantifier-free definable. In the case of qfps, we allow only the numeric relations $\leq$ and SUC. This is to guarantee transitivity of $\leq_{\text{qfp}}$ (Exercise 11.9). □

The main features of a fop I are that (1) I is a projection (each bit of $I(\mathcal{A})$ is determined by at most one bit of $\mathcal{A}$); and (2) the choice of which bit to look at in $\mathcal{A}$ for the value of each bit of $I(\mathcal{A})$ is determined by a first-order numeric formula. It is important to note that a fop merely copies its input to its output. The first-order numeric formulas α_i in Equation (11.8) tell us which bit of the input to copy and whether or not to negate it.

The difference between a qfp and a fop is that in a qfp, the numeric formulas α_i are actually quantifier-free. We usually concentrate on fops because these seem to enjoy all the nice properties of qfps and fop is a slightly more robust reduction.

Obviously, relations $\leq_{\text{qfp}}$ and $\leq_{\text{fop}}$ are reflexive. Less obviously, that they are transitive:

Exercise 11.9 Show the following concerning fops and qfps:

1. Relation $\leq_{\text{fop}}$ is transitive.
2. Relation $\leq_{\text{qfp}}$ is transitive.
3. When qfp I has $\varphi_0 \equiv \mathbf{true}$, relations $\leq$ and SUC over $I(\mathcal{A})$ are quantifier-free definable.
4. When fop I has $\varphi_0 \equiv \mathbf{true}$, relation BIT over $I(\mathcal{A})$ is first-order definable.

□

As our first example of a problem that is complete via fops, we show,

Proposition 11.10. SAT *is NP-complete via fops and, in fact, via qfps.*

Proof Let B be any boolean query in NP. From the proof of Theorem 7.8, we know that $B = \text{MOD}[\Phi]$, where,

$$\Phi = (\exists S_1^{a_0} \cdots S_g^{a_g})(\forall x_1 \cdots x_s)\gamma ,$$

with γ quantifier-free and in conjunctive normal form: $\gamma \equiv \bigwedge_{j=1}^{r} C_j(\bar{x})$.

We want to write a qfp $I = \langle \mathbf{true}, \varphi_1, \varphi_2 \rangle$ such that for all structures $\mathcal{A} \in \text{STRUC}[\tau]$,

$$\mathcal{A} \models \Phi \quad \Leftrightarrow \quad I(\mathcal{A}) \in \text{SAT} .$$

The formulas φ_1, φ_2 representing the relations P and N of $I(\mathcal{A})$ (Example 2.18) must be quantifier-free and projections (Equation (11.8)).

The boolean variables in the boolean formula $I(\mathcal{A})$ are encoded as tuples $\langle \bar{e}, \bar{c} \rangle$, corresponding to $S_i(\bar{e})$, where $\bar{c}$ is a tuple of first-order, boolean variables encoding the number i and $\bar{e} \in |\mathcal{A}|^a$, and $a = \max\{s, a_j \mid 0 \leq j \leq g\}$. Similarly, the clauses $C_j(\bar{e})$ will be encoded by tuples $\langle \bar{e}, \bar{d} \rangle$, with $\bar{d}$ encoding j. The fact that γ is a fixed, quantifier-free formula makes it trivial to make φ_1 and φ_2 quantifier-free.

In the proof of Theorem 7.8, we can see that no clause contains more than one occurrence of an input relation. The input relations occur exactly in clauses of the form $\neg R(x_4) \vee S_1(0, 0, 0, x_4)$, meaning that if bit x_4 of input relation R is 1, then $S_1(0, 0, 0, x_4)$ holds, meaning that there is a 1 in cell x_4 at time 0. Similarly, there will be a clause $R(x_4) \vee S_0(0, 0, 0, x_4)$ meaning that if bit x_4 of input relation R is 0, then $S_0(0, 0, 0, x_4)$ holds, meaning that there is a "0" in cell x_4 at time 0. This makes it easy to write φ_1 and φ_2 as projection formulas.

As an example, suppose that clauses C_1, C_2, C_3 of γ were the following,

$$C_1 \equiv \neg R(x_4) \vee S_1(0, 0, 0, x_4)$$
$$C_2 \equiv R(x_4) \vee S_0(0, 0, 0, x_4)$$
$$C_3 \equiv S_2(x_2, x_4) \vee x_1 \leq x_2 \vee x_3 = 0 \vee \neg S_2(x_3, x_1).$$

These would contribute the following clauses to the definitions of φ_1, φ_2:

$$\varphi_1(\bar{y}, \bar{d}, \bar{z}, \bar{c}) \equiv (\bar{d} = 1 = \bar{c} \wedge y_1 = y_2 = y_3 = z_1 = z_2 = z_3 = 0 \wedge y_4 = z_4 \wedge R(y_4))$$
$$\vee (\bar{d} = 2 \wedge \bar{c} = 0 \wedge y_1 = y_2 = y_3 = z_1 = z_2 = z_3 = 0 \wedge y_4 = z_4 \wedge \neg R(y_4))$$
$$\vee (\bar{d} = 3 \wedge \neg(y_1 \leq y_2 \vee y_3 = 0) \wedge z_1 = y_2 \wedge z_2 = y_4) \vee \cdots$$
$$\vee \cdots$$

$$\varphi_2(\bar{y}, \bar{d}, \bar{z}, \bar{c}) \equiv (\bar{d} = 3 \wedge \neg(y_1 \leq y_2 \vee y_3 = 0) \wedge z_1 = y_3 \wedge z_2 = y_1) \vee \cdots$$

As usual, a numerical expression involving boolean variables is an abbreviation for a formula saying that the boolean variables are a binary encoding of the number. For example, $\bar{d} = 5$, would be an abbreviation for $d_1 = 1 \wedge d_2 = 0 \wedge d_3 = 1$. In the above, we have the convention that empty clauses are not really clauses and therefore do not have to be satisfied. □

It is easy to understand intuitively why problems such as SAT remain complete via fops: The only place input structure $\mathcal{A}$ is used in output formula $I(\mathcal{A})$ is in placing input bin($\mathcal{A}$) one bit at a time into consecutive tape cells at time 0.

This situation is even more obvious for universal complete problems (Exercise 2.17). Recall that these are essentially universal Turing machines together with complexity bounds. Reductions to universal complete problems consist of writing down a fixed, finite program M and then copying the input string w to the output. In fact, most natural complete problems are disguised universal complete problems. This is why they typically remain complete via fops.

For classes P and above, there is enough space to copy the input onto the work tape. This makes it easy to show that the relevant complete problems remain complete via fops. For L and NL, the situation is more subtle. Consider a typical proof that REACH is complete for NL. Let $S \subseteq \text{STRUC}[\tau]$ be an element of NL and N be an NSPACE[$\log n$] Turing machine accepting S. The reduction maps an input $\mathcal{A} \in \text{STRUC}[\tau]$ to computation graph $G_{N(\text{bin}(\mathcal{A}))}$, with s the initial ID and t the accepting ID. Thus, as desired,

$$\mathcal{A} \in S \quad \Leftrightarrow \quad G_{N(\text{bin}(\mathcal{A}))} \in \text{REACH}.$$

11.2 First-Order Projections

Each possible edge (ID, ID′) in the computation graph depends on the single bit of bin($\mathcal{A}$) that is being read by the read head of ID. Thus the above reduction is a projection, although each bit of the input is copied polynomially many times — for the sake of the polynomially many ID's that read this bit.

An elegant way to prove that REACH is complete for NL via qfps is to use the syntactic proof of Lemma 9.12, which shows that every formula $\varphi \in$ FO(TC) is equivalent to a single application of transitive closure to a quantifier-free formula,

$$\varphi \;\equiv\; (\mathrm{TC}\,\gamma)(\overline{0},\overline{max})\,.$$

An examination of the proof shows that the quantifier-free formula γ resulting from the inductive proof of Lemma 9.12 is in fact a qfp. The same holds for the normal form theorem for FO(DTC) (Lemma 9.15).

Define the alternating transitive closure ATC to be the operator form of problem REACH$_a$. More explicitly, suppose that the formulas $\varphi_1(x_1,\ldots,x_k,x'_1,\ldots,x'_k)$, $\varphi_2(x_1,\ldots,x'_k) \in \mathcal{L}(\tau)$ represent the edge relation and the set of universal nodes, respectively, of an alternating graph. Let $I = \langle \mathbf{true}, \varphi_1, \varphi_2, \overline{0}, \overline{max}\rangle$ be a first-order query from STRUC[τ] to STRUC[τ_{ag}], the vocabulary of alternating graphs. For any structure $\mathcal{A} \in$ STRUC[τ], we have,

$$\mathcal{A} \models (\mathrm{ATC}_{x_1\ldots x_k x'_1 \ldots x'_k}\,\varphi_1, \varphi_2) \quad\Leftrightarrow\quad I(\mathcal{A}) \in \mathrm{REACH}_a\,. \tag{11.11}$$

As we will see in §11.4, this definition of ATC as the operator form of problem REACH$_a$ can be repeated for any other problem. It is not hard to show that a normal form theorem holds for FO(ATC), just as for FO(TC) and FO(DTC).

Exercise 11.12 Show that every problem $A \in$ P is expressible in the form,

$$A = (\mathrm{ATC}_{x_1\ldots x_k x'_1 \ldots x'_k}\,\varphi_1, \varphi_2),$$

where $I = \langle \mathbf{true}, \varphi_1, \varphi_2, \overline{0}, \overline{max}\rangle$ is a qfp. □

Corollary 11.13. *The following problems are complete via quantifier-free projections (qfps):*
REACH$_d$ *for* L, *and* REACH *for* NL, *and* REACH$_a$ *for* P.

In fact, all the problems shown complete so far in this book remain complete via fops. This was not emphasized at the time, but it can be checked by the reader. On the other hand, we can artificially construct a problem that does not remain complete via fops:

Proposition 11.14. *There is a problem S that is NP-complete via first-order reductions, but not via fops.*

Proof Let sq(x) be a string of $|x|^2$ bits, with bit $i|x| + j$ of sq(x) representing the logical AND of bits i and j of x. Let

$$S = \{\mathrm{bin}(\mathcal{A})\mathrm{sq}(\mathrm{bin}(\mathcal{A})) \mid \mathcal{A} \in \mathrm{SAT}\}\,.$$

Obviously S is NP-complete. We now show that S is not NP-complete via projections, even non-uniform projections. Consider the following simple problem in

NP,

$$\text{TWO} = \{x \in \{0,1\}^* \mid x \text{ contains exactly two "1"s}\}$$

Assume for the sake of contradiction that p is a projection from TWO to S, and let $n \in \mathbf{N}$. Let p_n be the part of p that applies to input strings of length n. Obviously, every bit of such an n-bit input string w is relevant to whether $w \in$ TWO. Let w_1 and w_2 be the first two bits of w. These must be copied — or negated and copied — into some bit positions i and j of $p(w)$. Note also that since some such w is an element of TWO the string $p(w)$ must be of length $m + m^2$, for some m depending only on n.

Suppose that $i > m$. For any $w \in \{0,1\}^n$, it must be the case that bit i of $p(w)$ is the conjunction of a designated two bits $i', j' \leq m$. (To be precise, $i = i'm + j'$.) It then follows that at least one of the bits i' and j' depends on w_1. Thus, we may assume that $i, j \leq m$. Now, let $k = im + j$. For all values of w, it must be the case that bit k of $p(w)$ is the conjunction of bits i and j of $p(w)$. This implies that bit k depends on both w_1 and w_2. However, since p is a projection, bit k can depend on at most one of w_1 and w_2. This contradiction proves the proposition. □

First-order and quantifier-free projections have some nice properties. The next corollary follows immediately from Corollary 11.13. We find this result captivating because, for many years, we have often thought that it should not be so hard to prove that there is no quantifier-free projection from SAT to REACH$_d$.

Corollary 11.15. *The conditions* L = NP *and* SAT = REACH$_d$ *are equivalent.*

The following interesting property of first-order projections, which says that there is only one complete problem via first-order projections for each "nice" complexity class.

Fact 11.16. *Let C be one of the complexity classes* NC^1, L, NL, P, NP, PSPACE. *Let A and B be problems complete for C via fops. Then there is a first-order definable isomorphism between A and B.*

11.3 Help Bits

In [KL82], Karp and Lipton gave a uniform definition of non-uniformity.

Definition 11.17 For any complexity class C, define C/poly, the complexity class C with polynomial advice, as follows: For any boolean query S, $S \in C/\text{poly}$ iff there exists a $k \in \mathbf{N}$ and a function $f : \mathbf{N} \to \{0,1\}^*$ such that,

1. For all $n \in \mathbf{N}$, $|f(n)| \leq n^k$, and
2. $S^f \in C$, where, $S^f = \{(\mathcal{A}, f(\|\mathcal{A}\|)) \mid \mathcal{A} \in S\}$.

In other words, $S \in C/\text{poly}$ if polynomially many bits are provided of free advice for the set of inputs of length n, S is recognizable in C. □

In particular, for circuit complexity classes such as NC^1, NC^1/poly is *non-uniform* NC^1. That is, non-uniform NC^1 consists of those problems accepted by a sequence of polynomial-size, $O(\log n)$-depth circuits, with no assumption concerning how hard it is to produce these circuits. The circuits themselves are the advice.

Note that NC^1/poly as well as all other non-uniform problems contain undecidable problems, because advice function f could encode the halting problem. Even though the advice strings $f(n)$ may be arbitrarily complex, non-uniform classes such as NC^1/poly are not all-powerful. This is because a boolean query has exponentially many instances of size n. Polynomially-many advice bits may not suffice to reduce the complexity of all of these. In fact, as we see in Chapter 13, some combinatorial lower bounds on circuit classes apply to the non-uniform case. For example, the proof that PARITY $\notin$ FO (Theorem 13.1) also shows that PARITY $\notin$ FO/poly.

The way to understand Definition 11.17 in the descriptive framework is that advice string $f(n)$ for structures of size n is exactly a new numeric predicate. We have carefully chosen the numeric predicates $\leq$ and BIT, which seem to provide a minimum, robust basis for computation for first-order queries. However, it may turn out that for some applications, more complex numeric predicates help reduce the complexity of computations.

There is a continuum between allowing arbitrarily complex advice as in FO/poly and allowing no advice as in FO. For any complexity classes $\mathcal{C}, \mathcal{C}'$, define $\mathcal{C}$-*uniform* $\mathcal{C}'$ to be the subset of $\mathcal{C}'/\text{poly}$ for which advice function f is an element of $Q(\mathcal{C})$. For example, P-uniform FO is the set of boolean queries that are expressible in first-order logic, using arbitrary, polynomial-time computable numeric predicates.

Exercise 11.18 Show the following concerning uniform complexity classes:
1. If $\mathcal{C}_1 \subseteq \mathcal{C}_2$, then $\mathcal{C}_1$-uniform $\mathcal{C}$ is contained in $\mathcal{C}_2$-uniform $\mathcal{C}$.
2. If $\mathcal{C}_1 \subseteq \mathcal{C}_2$, then $\mathcal{C}_1$-uniform $\mathcal{C}_2$ is equal to $\mathcal{C}_2$. □

Uniformity is in some ways orthogonal to other measures of complexity such as time and space. In particular, the reader should check the following meta-proposition,

Meta-Proposition 11.19. *Every equality or containment of complexity classes proved in this book remains true when, "C-uniform" is placed in front of each complexity class. For example, from Corollary 5.32, we have,*

$$P\text{-}unif\,FO \;=\; P\text{-}unif\,AC^0 \;=\; P\text{-}unif\,CRAM[1] \;=\; P\text{-}unif\,LH \text{ and,}$$

$$FO/\text{poly} \;=\; AC^0/\text{poly} \;=\; CRAM[1]/\text{poly} \;=\; LH/\text{poly}\,.$$

11.4 Generalized Quantifiers

Given any boolean query S, we can define a corresponding generalized quantifier Q_S. Examples are TC, which is the quantifier version of REACH; DTC, which is the quantifier version of $REACH_d$, and ATC, the quantifier version of $REACH_a$ (Equation (11.11)).

Definition 11.20 (**Generalized Quantifier**) Let $S \subseteq \text{STRUC}[\tau]$ be any boolean query. Define the *generalized quantifier* Q_S as follows. Consider any first-order query,

$$I : \text{STRUC}[\sigma] \to \text{STRUC}[\tau], \quad I = \lambda_{x_1^1 \ldots x_a^k} \langle \varphi_0, \ldots, \psi_s \rangle .$$

Then $Q_S(I)$ is a well-formed formula in $\text{FO}(Q_S)$ with the following semantics: For any $\mathcal{A} \in \text{STRUC}[\sigma]$,

$$\mathcal{A} \models Q_S(I) \quad \Leftrightarrow \quad I(\mathcal{A}) \in S .$$

Generalized quantifier Q_S binds variables $x_1^1, \ldots, x_a^k$ of I. □

We give an example using problem REACH. Intuitively, Q_{reach} should be the transitive closure operator. This is nearly true. How should we define I to make the following equation hold?

$$(\text{TC}_{x_1 \ldots x_k x_1' \ldots x_k'} \varphi)(\bar{a}, \bar{b}) \quad = \quad Q_{\text{reach}}(I) .$$

The answer is to define I as follows: $I = \lambda_{x_1 \ldots x_k x_1' \ldots x_k'} \langle \textbf{true}, \varphi, \bar{a}, \bar{b} \rangle$.

The definition of generalized quantifiers allows us to define the notion of first-order, Turing reductions in a natural way.

Definition 11.21 (**First-Order Turing Reductions**) Let A and B be boolean queries. We say that A is *first-order Turing reducible* to B ($A \leq_{\text{fo}}^T B$) iff $A \in \text{FO}(Q_B)$. □

A Turing reduction to a problem B allows arbitrary applications of quantifier Q_B. We have seen that many problems are complete for complexity classes via weaker reductions such as $\leq_{\text{fo}}$, $\leq_{\text{fop}}$, and $\leq_{\text{qfp}}$. Say that a generalized quantifier Q_S has the *first-order normal form property* iff every problem $A \in \text{FO}(Q_S)$ may be expressed in the form,

$$A = Q_S(I) . \tag{11.22}$$

where I is a first-order query and thus has no occurrences of Q_S. Similarly, if I is a first-order projection, then we say that Q_S has the *fop normal form property*; and, if I is a quantifier-free projection, then we say that Q_S has the *qfp normal form property*.

As examples, quantifiers Q_{reach} and TC have the qfp normal form property (Lemma 9.12). The same is true for quantifiers Q_{REACH_d}, DTC (Lemma 9.15), and Q_{REACH_a}, ATC (Exercise 11.12).

The next proposition is obvious. It points out that the completeness of a problem via a weak reduction is equivalent to the problem being flexible enough to absorb repeated applications, as well as quantifications and boolean operations, into itself. (This proposition applies to complexity classes that are closed under first-order Turing reductions — and thus in particular under complementation. For classes such as NP that may not be closed under complementation, a similar proposition holds where all occurrences of the generalized quantifiers are positive.)

Proposition 11.23. *Let S be a boolean query and C a complexity class that is closed under first-order Turing reductions. Then*

1. *S is $\leq_{fo}^t$-complete for C iff $C = FO(Q_S)$.*
2. *S is $\leq_{fo}$-complete for C iff $C = FO(Q_S)$ and Q_S has the first-order normal form property.*
3. *S is $\leq_{fop}$-complete for C iff $C = FO(Q_S)$ and Q_S has the fop normal form property.*
4. *S is $\leq_{qfp}$-complete for C iff $C = FO(Q_S)$ and Q_S has the qfp normal form property.*

Historical Notes and Suggestions for Further Reading

The first measure of descriptive complexity was first-order size (FO-SIZE). This measures how large a first-order formula, i.e., the number of symbols, is needed to describe a given property as a function of the size of the universe of the structure being described. In [I79] it was proved that,

$$\text{NSPACE}[s(n)^2] \subseteq \text{FO-SIZE}[s(n)^2/\log n] \subseteq \text{DSPACE}[s(n)^2] \ . \quad (11.24)$$

Immerman and Landau explored the basic properties of fops and qfps in [IL95]. They described a series of iterated multiplication problems that are complete for a series of complexity classes via fops. A solution to Exercise 11.9 can also be found there. Exercise 11.12 and Corollary 11.13 are from [I83].

The definition of (FO-SIZE) in [I79] used a logspace-uniform sequence of formulas. In this book, we have abandoned FO-SIZE[$s(n)$], replacing it by the syntactically uniform classes FO-VAR[$s(n), v(n)$]. Theorem 11.3 is the translation from Equation 11.24.

Fact 11.16 is due to Allender, Balcázar, and Immerman [ABI97]. It has been extended by Agrawal, Allender, Impagliazzo, Pitassi and Rudich, who showed that complete sets via first-order reductions are (non-uniform) first-order isomorphic [AAI97].

Medina and Immerman showed that it is possible to understand NP-completeness via fops in a syntactic way, i.e., to prove that a problem is NP-complete by looking at the form of its SO∃ description [MI94].

Here is a rare example of a natural problem that is in a slightly non-uniform complexity class and not known to be in the uniform class. Beame, Cook, and Hoover showed that iterated integer multiplication — given n, n-bit natural numbers, compute a specified bit of their product — is in polynomial-time uniform NC^1 [BCH86]. Their proof shows that this problem is in ThC^1-uniform ThC^0 [IL95, Rei87]. However, it is not known to be in uniform ThC^0, or even in L.

The discussion of generalized quantifiers in §11.4 follows [IL95] by Immerman and Landau and [BI97] by Barrington and Immerman. We have adopted the term "generalized quantifier" rather than "operator" in deference to the growing research that uses the former name. In particular, see papers by Makowsky and Pneuli [MP94] and by Kolaitis and Väänänen [KV95]. For a discussion of variants of Proposition 11.23 for NP-complete operators, see Stewart [Ste91].

The *Kolmogorov Complexity* of a binary string w is the length of the minimum binary string p such that $U(p) = w$, where U is a particular, universal Turing machine. This is a robust and fascinating concept. Book [LV93] is an excellent resource on this subject. The Kolmogorov complexity of a string or program measures its essential content in terms of number of bits. An easy counting argument shows that almost all strings are essentially incompressible. These strings are called "Kolmogorov random". When some people hear the term descriptive complexity, they have the mistaken impression that discussing the concept of Kolmogorov complexity is being discussed, i.e., the smallest number of bits needed to describe a fixed, finite concept. However, as the reader by now knows, descriptive complexity is concerned with the richness of a logical language needed to describe a given query not the length of the shortest description in a language of maximal computational complexity.

12
The Role of Ordering

To characterize complexity classes, first-order descriptive complexity requires that each structure has an ordering on its universe. Such an ordering, however, is irrelevant to the properties of graphs or databases that we want to compute. It is easy to use pebble games to prove lower bounds on languages without ordering; on ordered structures these arguments do not work. Furthermore, for databases, the ordering is a low-level implementation issue — how entries are stored in memory or disk — which should be invisible to the person writing queries. For all these reasons, the descriptive characterization of the order-independent queries computable in a given complexity class is a fundamental open problem.

By now we have characterized many complexity classes via extensions of first-order logic. All these characterizations require a total ordering of the universe[1]. A typical example is the fact that P = FO(LFP) (Theorem 4.10).

One way of rephrasing this result is that the set of polynomial-time computable boolean queries concerning finite ordered graphs is exactly characterized by the sentences in FO(LFP) from the language of ordered graphs. But most interesting questions concerning graphs do not depend on the ordering of the vertices.

Recall that a query is order-independent if every pair of isomorphic structures — regardless of their ordering — agree on the query (Definition 1.24). We would very much like to know the answer to the following,

---

[1] Fagin's theorem (Theorem 7.8) does not require an ordering because second-order existential logic is powerful enough to existentially quantify a linear ordering on the universe, which the original proof of Fagin's Theorem does.

Question 12.1. *Is there a recursively enumerable listing of a set of sentences from* FO(LFP) *that describes exactly all the polynomial-time, order-independent boolean queries?*

We know that the language FO(wo$\leq$)(LFP) is not a solution to Question 12.1 — it does not even express the query "Are there an odd number of vertices?" (Proposition 6.14).

In the course of this chapter we suggest some more interesting candidate languages for expressing exactly order-independent P. In Section 12.7, we return to Question 12.1 with new insights; but, we do not answer it.

12.1 Using Logic to Characterize Graphs

Ordering is tied to many issues in computation. We begin with one of the most basic: the problem of characterizing unordered graphs with logic. As we know from Lemma 6.30, the ability to distinguish graphs or other structures is fundamental to the expressive power of the language. To begin with, we are interested in how many variables, and what quantifier-rank are needed to characterize graphs.

Definition 12.2 A useful generalization of graphs that we consider is the set of *colored graphs*. These are structures $G = (E, C_1, \ldots, C_r)$ consisting of a graph together with unary relations, which may be thought of as representing colors on the vertices. Relations $C_1, \ldots, C_r$ encode a range of 2^r possible colors on the vertices. When it is convenient, we assume that each vertex satisfies exactly one color relation, i.e, that there are r colors instead of 2^r. A graph is a colored graph with no color relations. For any graph or colored graph $G = (V, E, C_1, \ldots, C_r)$, a *coloring* of G is an expansion $\widehat{G} = (V, E, C_1, \ldots, C_r, C_{r+1}, \ldots, C_{r'})$ of G that results by adding some new coloring relations. It is convenient to let $\tau_{cg} = \langle E^2, C_1^1, C_2^1, \ldots \rangle$ be the vocabulary with infinitely many coloring relations. We can thus let the set of *colored graphs* be STRUC[τ_{cg}]. For each particular colored graph $G \in$ STRUC[τ_{cg}], only finitely many of the coloring relations may be nonempty. □

We start by saying what it means for a language $\mathcal{L}$ to characterize a colored graph, or a class of colored graphs.

Definition 12.3 Let $\mathcal{L}$ be a language and G be a colored graph. We say that $\mathcal{L}$ *characterizes* G iff for all colorings $\widehat{G}$ of G and for all colored graphs H, if $\widehat{G}$ and H are $\mathcal{L}$-equivalent then they are isomorphic,

$$\widehat{G} \equiv_{\mathcal{L}} H \quad \Rightarrow \quad \widehat{G} \cong H \; .$$

We say that $\mathcal{L}$ *characterizes* a set S of colored graphs iff for all $G \in S$, $\mathcal{L}$ characterizes G. □

Exercise 12.4 Prove that the first-order language of vocabulary τ_{cg} characterizes the set of colored graphs. [Hint: given a graph G on n vertices, we can write a sentence with $n+1$ variables and $n+1$ quantifiers that asserts that there are exactly n vertices and that lists all their properties.] □

Graph isomorphism problem (GRAPH-ISO) is the boolean query that is true of pairs of isomorphic colored graphs.

$$\text{GRAPH-ISO} = \{(G, H) \mid G, H \in \text{STRUC}[\tau_{cg}]\ G \cong H\} \quad (12.5)$$

Usually, when we have an algorithm for GRAPH-ISO for a class of graphs, $\mathcal{C} \subseteq \text{STRUC}[\tau_{cg}]$ we also have a *canonization algorithm*. This is a mapping $f : \mathcal{C} \to \mathcal{C}$ such that,

1. For all $G \in \mathcal{C}$, $f(G) \cong G$, and,
2. For all $G, H \in \mathcal{C}$, $G \cong H \Rightarrow f(G) = f(H)$.

The complexity of graph isomorphism and graph canonization are well studied but remain unresolved. If there is a polynomial-time graph canonization algorithm, then this provides a positive answer to Question 12.1: Let $\varphi \in \text{FO(LFP)}$ be a sentence concerning ordered graphs. We can transform φ into an order-independent query φ' by first describing the canonical image $f(G)$ of input graph G and then applying φ to $f(G)$. The set of all such φ's is thus a listing of exactly all polynomial-time, order-independent graph properties. This provides a positive answer to Question 12.1, using Exercise 3.7.

Graph canonization can be computed in $Q(\Sigma_2^p)$ — the second level of the polynomial-time hierarchy. If P were equal to NP, then the polynomial-time hierarchy would also be equal to P. Thus graph canonization would be in $Q(P)$. It follows that

Corollary 12.6. *If the answer to Question 12.1 is "No", then* $P \neq NP$.

As the following exercise shows, the first-order characterization of all colored graphs leads to inefficient graph isomorphism and canonization algorithms. Below, we develop more efficient isomorphism and canonization algorithms by restricting the number of variables used to characterize certain classes of graphs.

Exercise 12.7 Use the first-order characterization of all colored graphs (Exercise 12.4) to give exponential-time isomorphism and canonization algorithms for the class of all colored graphs. □

12.2 Characterizing Graphs Using $\mathcal{L}^k$

Recall that $\mathcal{L}^k$ is the restriction of first-order logic to formulas in which only variables $x_1, x_2, \ldots, x_k$ occur (Definition 6.9). We saw for example in Exercise 6.26 that three variables are necessary and sufficient to express the connectivity of graphs. In this section, we see some simple classes of colored graphs that can

184   12. The Role of Ordering

be characterized in $\mathcal{L}^k$, for some value of k. Consider the following restriction on colored graphs.

Definition 12.8 (Color-Class Size) We say that a colored graph G has *color-class size* k iff no set of more than k vertices from G agree on all the color predicates. Write CC_k to denote the set of graphs of color-class size k. □

For example, the graphs in Figure 6.6 have color-class size 2. Babai showed that for every fixed k, there is a polynomial-time algorithm for graph isomorphism for graphs of color-class size k. The following proposition shows that graphs of color-class size 3 are quite simple: they are characterized by $\mathcal{L}^3$. Contrast this with the graphs of Figure 6.6 which are $\mathcal{L}^2$-equivalent but not isomorphic. Thus, $\mathcal{L}^2$ does not characterize the graphs of color-class size 2.

Proposition 12.9. $\mathcal{L}^3$ *characterizes graphs of color-class size at most three.*

Proof Let G and H be $\mathcal{L}^3$-equivalent colored graphs. We will build an isomorphism $f : G \to H$. We first refine the colorings of the vertices of G and H to correspond to $\mathcal{L}^3$ types. For $A, B \in \{G, H\}$, vertices $a \in A$ and $b \in B$ will have the same $\mathcal{L}^k$-refined color iff they satisfy the same $\mathcal{L}^k$ formulas, i.e.,

$$\{\varphi \in \mathcal{L}^k \mid (A, a/x_1) \models \varphi\} = \{\varphi \in \mathcal{L}^k \mid (B, b/x_1) \models \varphi\}.$$

The following lemma says that we may assume that the color types of G and H are already refined.

Lemma 12.10. *Let the finite, colored graphs G and H be $\mathcal{L}^k$-equivalent and let G' and H' be the $\mathcal{L}^k$ color refinements of G and H. Then G' and H' are $\mathcal{L}^k$-equivalent.*

Proof Since G and H are finite, each refined color class C'_i is determined by the conjunction $\psi_i \in \mathcal{L}^k$ of a finite set of formulas. That is, for all i, G' and H' satisfy

$$\forall x_1(C'_i(x_1) \leftrightarrow \psi_i(x_1))$$

Thus any occurrence of $C'_i(x_1)$ may be replaced by the equivalent $\psi_i(x_1)$. Now, for any formula $\alpha \in \mathcal{L}^3(C'_1, C'_2, \ldots)$ we may replace each occurrence of $C'_i(x_j)$ by $\psi_i(x_j)$ to obtain an equivalent formula $\alpha' \in \mathcal{L}^3(C_1, \ldots C_r)$. □

By the above lemma we may assume that the color classes of G and H correspond exactly to the $\mathcal{L}^3$-types of the vertices. Let red and blue be two colors and consider the edges between red and blue vertices in G or H. Note that this is a regular bipartite graph because we can express in $\mathcal{L}^3$ that a red vertex has 0, 1, 2, or all blue vertices as neighbors. Note also that for color classes of size at most 3, the only regular bipartite graphs representing nontrivial relationships between vertices are the 1:1 correspondence graphs and their complements. Let us then change such bipartite graphs as follows: replace the complete bipartite graph by its complement and replace the graphs of degree two whose complements are 1:1 correspondence graphs by these complements. When we perform these changes on G and H, the new graphs are still $\mathcal{L}^3$-equivalent, and they are isomorphic now iff they were before.

Let the *color valence* of a graph be the maximum number of edges from any vertex to vertices of a fixed color. We have reduced the problem to constructing an isomorphism between $\mathcal{L}^3$-equivalent graphs G and H when these graphs have color valence one.

We construct isomorphism f as follows: Begin by choosing $g \in G$, $h \in H$, a pair of vertices of the same color, and letting $f(g) = h$. Next, while there is a vertex g_1 in the domain of f with a (unique) neighbor g_2 of color C_i not yet in the domain of f, do the following. Let h_2 be the neighbor of $f(g_1)$ of color C_i and let $f(g_2) = h_2$.

We have defined f on the connected component that contains g. If some vertex $g' \in G$ is not yet in the domain of f, then there must be an $h' \in H$ of the same color that is not yet in the range of f. Let $f(g') = h'$ and continue in the same way to define f on the connected component containing g'.

We claim that the function f constructed above is an isomorphism from G to H. If not, then there is a loop of a certain color sequence in one of the graphs but not the other. For example, suppose we chose $g_1, g_2, \ldots, g_j$ and $h_1, h_2, \ldots, h_j$ so that g_1 and h_1 are color C_1, and for $i < j$, g_{i+1} and h_{i+1} are the unique neighbors of g_i and h_i, respectively, of color C_{i+1}. However, suppose now that the neighbor of h_j of color C_1 is h_1, but that g_1 is not a neighbor of g_j. In this case there is a certain easily describable loop in H but not in G. This means that G and H disagree on the following $\mathcal{L}^3$ formula:

$$C_1(x_1) \wedge \exists x_2 \big(C_2(x_2) \wedge E(x_1, x_2) \wedge \exists x_3 (C_3(x_3) \wedge E(x_2, x_3) \wedge$$
$$\wedge \exists x_2 (C_4(x_2) \wedge E(x_3, x_2) \wedge \ldots \wedge \exists x_i (C_j(x_i) \wedge E(x_i, x_1)) \ldots)$$

Since $G \equiv_{\mathcal{L}^3} H$, they must agree on the above formula. Therefore f is an isomorphism as claimed. □

Languages $\mathcal{L}^k$ have the serious weakness that they cannot express counting (Proposition 6.14). In the next section we add counting to first-order logic, forming the languages $\mathcal{C}^k$. We see that $\mathcal{C}^2$ suffices to characterize almost all graphs. We will see in Theorem 13.26 that Proposition 12.9 cannot be easily extended to graphs of color-class size 4 or greater, even when counting is added.

12.3 Adding Counting to First-Order Logic

When we remove ordering from first-order logic, the languages become too weak to capture general computation. These languages become too weak to even count mod 2 (Proposition 6.14). In this section, we show how to add counting to languages without ordering. The reason to do this is so that the languages retain their order-independent nature — the expressible properties are properties of the graph or database in question, not of its representation — and they can express basic counting properties.

As we have seen, an ordering on the n-element universe lets us treat these elements as the numbers $\{0, 1, \ldots, n-1\}$ with the usual ordering and arithmetic operations. In the following definition, we separate these two concepts into two domains and then add counting quantifiers to rejoin them.

Definition 12.11 (Number and Counting) Let *first-order logic with numbers* FO(NUMBER) be first-order logic interpreted over two-sorted structures. Let $\mathcal{A} = \langle |\mathcal{A}|, R_1^{\mathcal{A}} \ldots R_r^{\mathcal{A}}, c_1^{\mathcal{A}} \ldots c_s^{\mathcal{A}} \rangle$ be any structure of vocabulary $\tau = \langle R_1^{a_1}, \ldots, R_r^{a_r}, c_1, \ldots, c_s \rangle$. Let $n = \|\mathcal{A}\|$. Define the following two-sorted structure.

$$\mathcal{A}(\text{NUMBER}) = (\{0, 1, \ldots, n-1\}, \leq, \text{BIT}, 0, 1, max; |\mathcal{A}|, R_1^{\mathcal{A}} \ldots R_r^{\mathcal{A}}, c_1^{\mathcal{A}} \ldots c_s^{\mathcal{A}})$$

The idea is that the numeric predicates and constants refer to the number domain and the input predicates and constants refer to the universe $|\mathcal{A}|$. For simplicity, in language FO(NUMBER) we let variables $i, j, k, \ldots$ range over the number domain, while variables $x, y, z, u, v, w, \ldots$ range over $|\mathcal{A}|$.

Once numbers are available it is nice to be able to count. To do this, we can add counting quantifiers. Let the meaning of the formula:

$$(\exists i \, x)\varphi(x)$$

be that there exist at least i distinct elements x such that $\varphi(x)$. Note that this quantifier binds x and leaves i free. We will also use the quantifiers $(\exists ! i \, x)$, meaning that there exists exactly i x's:

$$(\exists ! i \, x)\varphi(x) \equiv (\exists i \, x)\varphi(x) \wedge \neg(\exists i+1 \, x)\varphi(x).$$

Let FO(COUNT) denote first-order logic over structures with numbers and counting quantifiers. □

As an example — in contrast to Proposition 6.14 — we show,

Proposition 12.12. *The boolean query on graphs that is true iff there are an odd number of vertices is expressible in* FO(NUMBER). *The query that there are an odd number of directed edges is not expressible in* FO(NUMBER), *but it is expressible in* FO(COUNT).

Proof The two queries are expressible as follows. The first sentence says that $n - 1$ is even, so n is odd. Recall that addition is first-order expressible using BIT (Theorem 1.17). The last sentence says that there are an odd number of edges by saying that there are an odd number of vertices of odd degree.

$$\text{ODD-VERTICES} \equiv (\exists i)(i + i = max)$$
$$\text{ODD-DEGREE}(x) \equiv (\exists i)(\forall j)((j + j \neq i) \wedge (\exists ! i \, y)E(x, y))$$
$$\text{ODD-EDGES} \equiv (\exists i)(\forall j)((j + j \neq i) \wedge (\exists ! i \, x)(\text{ODD-DEGREE}(x)))$$

The reason that ODD-EDGES is not expressible in FO(NUMBER) is that without counting quantifiers, there is no relation between the number domain and the vertex domain except that they have the same cardinality. Let G_n and H_n be graphs with $4n$ vertices each such that $2n$ of G_n's vertices have loops, $2n - 1$ of H_n's vertices have loops and there are no other edges. Then $G_n(\text{NUMBER}) \sim^{2n-1} H_n(\text{NUMBER})$. In the pebble game, the number domains of $G_n(\text{NUMBER})$ and $H_n(\text{NUMBER})$ are identical, so it is never to Samson's advantage to play there. □

Corollary 12.13. *Suppose Delilah has a winning strategy on the k-pebble, m-move game on $\mathcal{A}, \mathcal{B}$, i.e., $\mathcal{A} \sim^k_m \mathcal{B}$ and $\|\mathcal{A}\| = \|\mathcal{B}\|$. Then Delilah has a winning strategy on the same structures with numbers, i.e., $\mathcal{A}(\text{NUMBER}) \sim^k_m \mathcal{B}(\text{NUMBER})$.*

Exercise 12.14 Show that the boolean query that an undirected graph has an odd number of edges is expressible in FO(COUNT). [Hint: the difficulty is that the trick used in Proposition 12.12 will not work because each edge is counted twice. Furthermore, you cannot count only the edges to larger numbered vertices because the vertices are not ordered. Try counting the number of directed edges mod 4.] □

In language FO(COUNT) each reasonably sized number has a very short unique description. It is simpler when proving upper and lower bounds to ignore the length of these descriptions by adding constant symbols for each possible number. Define $\mathcal{C}^k$ to be the extension of $\text{FO}^k(\text{COUNT})$ in which we have a constant symbol for each number. Thus, $\mathcal{C}^k$ is the extension of $\mathcal{L}^k$ in which we have numeric constants and counting quantifiers. In the next section, we introduce the appropriate pebble game for $\mathcal{C}^k$ and show that $\mathcal{C}^2$ is powerful enough to characterize all trees and almost all graphs.

Exercise 12.15 Show that in FO(NUMBER), each number i has a unique description using three variables and quantifier-rank $O(\log \log i)$. □

In this chapter, we are interested in what can be expressed in language FO(COUNT) without ordering. Ordered structures already can think of their elements as numbers. On the other hand, counting quantifiers are equivalent to the majority quantifier, which we have seen previously (Theorem 5.27). Thus,

Proposition 12.16. *Over ordered structures the following equalities hold,*

$$\text{FO(NUMBER)} = \text{FO}$$
$$\text{FO(COUNT)} = \text{FO}(M) = \text{Th}\mathcal{C}^0 \;.$$

Exercise 12.17 Prove Proposition 12.16. [Hint: see Exercise 5.28.] □

12.4 Pebble Games for $\mathcal{C}^k$

Definition 12.18 We now define the *counting game* $\mathcal{G}^k_{C,m}$, which is the generalization of the Ehrenfeucht-Fraïssé game $\mathcal{G}^k_m$ (Definition 6.2) to language $\mathcal{C}^k$. The new game is the same as the old except that each move consists of two parts. Suppose the current game position is $(\mathcal{A}, \alpha_r, \mathcal{B}, \beta_r)$. Move $r+1$ consists of the following two steps:

188   12. The Role of Ordering

1. Samson picks up a pair of pebbles, say pair i. He then chooses a set E of vertices from one of the structures. Delilah must now answer with a set F of vertices from the other structure. F must have the same cardinality as E.
2. Samson places his pebble i on some vertex $f \in F$. Delilah answers by placing the other pebble i on some $e \in E$. Thus, the new configuration $\alpha_{r+1}, \beta_{r+1}$ is the same as α_r, β_r except that $\{\alpha_{r+1}(x_i), \beta_{r+1}(x_i)\} = \{e, f\}$.

The definition for winning is as before. We use the notation $\mathcal{A} \sim^k_{\mathcal{C},m} \mathcal{B}$ to mean that Delilah has a winning strategy for game $\mathcal{G}^k_{\mathcal{C},m}$ on $\mathcal{A}, \mathcal{B}$. □

What is going on in the two-step move is that Samson is asserting that there exist $|E|$ elements in one of the structures satisfying some property, and Delilah responds with the same number of elements in the other structure. Samson then challenges with an element f in the set chosen by Delilah. Delilah must answer with an element e from Samson's chosen set. For Delilah to win, e and f must be indistinguishable in the remaining moves of the game

Example 12.19 As an example, consider game $\mathcal{G}^2_{\mathcal{C}}$ on graphs G, H of Figure 6.6. We claim that $G \sim^2_{\mathcal{C}} H$. The proof is inductive. Suppose that after move r, the configuration is G, α_r, H, β_r and Samson has not yet won. This means that the vertices pebbled by x_i in each graph are the same color, for $i = 1, 2$. Suppose now that Samson starts a counting move by picking up the pair of x_2 pebbles, and let red be the color of the vertices on which the x_1 pebbles are sitting. Define a 1:1 correspondence ρ from the vertices of G to the vertices of H as follows: ρ maps the red pebbled vertex in G to the red pebbled vertex in H and the red non-pebbled vertex in G to the red non-pebbled vertex in H. For each other color, there is one vertex of that color adjacent to $\alpha_r(x_1)$ and one vertex not adjacent. Map these four vertices accordingly. Now, whatever set E chooses, Delilah answers with $\rho(E)$ if $E \subseteq |G|$ and with $\rho^{-1}(E)$ otherwise. When Samson chooses $f \in F$, Delilah answers with the corresponding point $\rho(f)$ or $\rho^{-1}(f)$. Much as in Proposition 6.7, this is a winning strategy for Delilah. □

The next theorem shows the unsurprising fact that game $\mathcal{G}^k_{\mathcal{C}}$ characterizes language $\mathcal{C}^k$. In the following, $\mathcal{A} \equiv^k_{\mathcal{C},m} \mathcal{B}$ means that $\mathcal{A}$ and $\mathcal{B}$ agree on all sentences from $\mathcal{C}^k$ of quantifier-rank m.

Theorem 12.20. *Let $\mathcal{A}$ and $\mathcal{B}$ be finite structures of the same vocabulary and let α_0, β_0 be a k-configuration of $\mathcal{A}, \mathcal{B}$. Then the following are equivalent:*

1. $(\mathcal{A}, \alpha_0) \sim^k_{\mathcal{C},m} (\mathcal{B}, \beta_0)$,
2. $(\mathcal{A}, \alpha_0) \equiv^k_{\mathcal{C},m} (\mathcal{B}, \beta_0)$.

Exercise 12.21 Prove Theorem 12.20. [Hint: you must prove a version of Exercise 6.11 for $\mathcal{C}^k_m$. In this case, you need that the two structures are finite, so that only a finite number of number constants are available.] □

Delilah's strategy in Example 12.19 involved constructing at each step a bijection from $|\mathcal{A}|$ to $|\mathcal{B}|$. Surprisingly, this is necessary as well as sufficient for Delilah to win:

Definition 12.22 Define the *bijection game* $\mathcal{G}_{\mathcal{B},m}^k$ to be similar to game $\mathcal{G}_{\mathcal{C},m}^k$ (Definition 12.18). Each move of $\mathcal{G}_{\mathcal{B},m}^k$ consists of the following three steps.

1. Samson picks up a pair of pebbles, say pebble pair i.
2. Delilah chooses a bijection $\rho : |\mathcal{A}| \to |\mathcal{B}|$.
3. Samson places the x_i pebbles on some elements $a \in |\mathcal{A}|$ and $\rho(a) \in |\mathcal{B}|$.

□

Theorem 12.23. *The bijection and counting games are equivalent. That is, for all k, m, and k-configurations $(\mathcal{A}, \alpha_0, \mathcal{B}, \beta_0)$, the following are equivalent:*

1. $(\mathcal{A}, \alpha_0) \sim_{\mathcal{C},m}^k (\mathcal{B}, \beta_0)$
2. $(\mathcal{A}, \alpha_0) \sim_{\mathcal{B},m}^k (\mathcal{B}, \beta_0)$

Exercise 12.24 Prove Theorem 12.23. [Hint: for the more difficult direction, suppose that (1) holds and that Samson has started a bijective move by picking up pebble pair i. Partition the universes of $\mathcal{A}$ and $\mathcal{B}$ according to what formulas $\varphi(x_i)$ from $\mathcal{C}_{m-1}^k$ they satisfy.] □

Exercise 12.25 Show that Hanf's Theorem (Theorem 6.21) remains true when the language includes counting quantifiers. [Hint: the proof follows the same outline. The fact that there are exactly the same number of each 2^r-type gives Delilah a way to define the bijections that she needs to win the r-round bijection game.] □

12.5 Vertex Refinement Corresponds to $\mathcal{C}^2$

We now show that the expressive power of $\mathcal{C}^2$ is characterized by the well-known method of vertex refinement (see [Ba81, HT72]). Let $G = (V, E, C_1, \ldots, C_r)$ be a colored graph in which every vertex satisfies exactly one color relation. Let $f : V \to \{1 \ldots n\}$ be given by $f(v) = i$ iff $v \in C_i$. We then define the refinement f' of f as follows: The new color of each vertex v is defined to be the tuple $\langle f(v), n_1, \ldots, n_r \rangle$, where n_i is the number of vertices of color i that v is adjacent to. We sort these new colors lexicographically and assign $f'(v)$ to be the number of the new color class that v inhabits. Thus two vertices are in the same new color class just if they were in the same old color class, and they were adjacent to the same number of vertices of each color. We keep refining the coloring until at some level $f^{(k)} = f^{(k+1)}$. We let $\bar{f} = f^{(k)}$ and call $\bar{f}$ the *stable coloring* of f. See Figure 12.26 for an example of a stable coloring.

The equivalence of stable colorings and $\mathcal{C}^2$ equivalence is summed up by the following theorem.

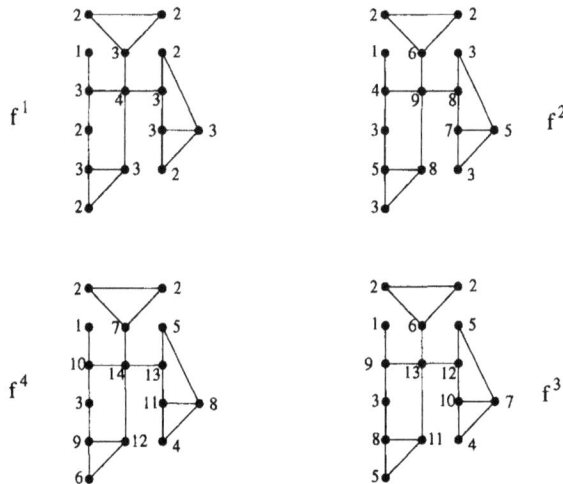

Figure 12.26: Stable Coloring of a Graph: $\bar{f} = f^4$

Theorem 12.27. *Given a colored graph $G = (V, E, C_1, \ldots, C_r)$, vertices $g_1, g_2 \in V$, and a number m, the following are equivalent:*

1. $f^m(g_1) = f^m(g_2)$,
2. $(G, g_1/x_1) \equiv^2_{C,m} (G, g_2/x_1)$,
3. $(G, g_1/x_1) \sim^2_{C,m} (G, g_2/x_1)$.

Proof We already know from Theorem 12.20 that (2) and (3) are equivalent. We prove their equivalence with (1) by induction on m. In the base case, $f^0(g_1) = f^0(g_2)$ iff g_1 and g_2 have the same initial color. This holds iff $(G, g_1/x_1) \sim^2_{C,0} (G, g_2/x_1)$.

Inductively, assume the equivalence for all quantifier ranks less than m.

($\neg 1 \Rightarrow \neg 3$): Suppose that $f^m(g_1) \neq f^m(g_2)$ but they agree on f^{m-1}. It follows that g_1 and g_2 have a different number of neighbors of some $f^{(m-1)}$ color class i. Let N be the maximum of these two numbers and suppose they are neighbors of g_1. In the first move of $\mathcal{G}^2_{C,m}$, Samson picks up the second pebble pair and chooses the set of N neighbors of g_1 of color i. Delilah must respond with N vertices on the other side. If one of these is not a neighbor of g_2, then Samson can choose it and win immediately. Otherwise, at least one of the vertices chosen by Delilah is not color i. Samson chooses this vertex and by induction wins the remaining $m-1$ move game.

($1 \Rightarrow 3$): Suppose $f^m(g_1) = f^m(g_2)$. We show that $(G, g_1/x_1) \sim^2_{B,m} (G, g_1/x_1)$. On the first move, Samson picks up the second pebble pair — otherwise he loses immediately, since the two graphs are equal. Since $f^m(g_1) = f^m(g_2)$, g_1 and g_2 have the same number of neighbors of each f^{m-1} color. Delilah builds the bijective map ρ that maps neighbors of g_1 of some color to neighbors of g_2 of the same color. Since we are playing on two copies of the same graph, it follows that there are also the same number of non-neighbors of each color on both sides.

Thus, ρ maps the non-neighbors of g_1 of each color to non-neighbors of g_2 of the same color. Whatever choice $(h_1, \rho(h_1))$ Samson makes, we have that $f^{m-1}(h_1) = f^{m-1}(\rho(h_1))$. It follows by induction that Samson wins the remaining $m-1$ moves of the bijective game. □

All Trees and Almost All Graphs

Theorem 12.27, combined with some facts about stable colorings provides us with several corollaries concerning graphs characterized by C^2. First, we show that trees are characterized by C^2.

Proposition 12.28. *The set of undirected, finite trees is characterized by C^2.*

Proof The standard linear-time tree isomorphism algorithm is a stable coloring in which first the leaves are colored, then vertices of height one — those that have at most one neighbor of height greater than zero, then vertices of height two — those that have at most one neighbor of height greater than one, etc. The *children* of a vertex v are the neighbors of v of height less than v's height. The color of a vertex is the tuple consisting of its height and the number of children it has of each color. This is part of the stable coloring of a vertex and is thus expressible in C^2. See Figure 12.29 which shows the coloring of a tree. Notice that two vertices have the same color iff their subtrees are isomorphic.

The subtree rooted at v is the tree consisting of v together with its children, grandchildren, etc. We show by induction on the height of v that if v and w are the same color, then their subtrees are isomorphic. In the base case, all leaves of the same color are isomorphic. Inductively, suppose that this holds for all vertices of height less than h, and let v and w be vertices of height h of the same color. Thus, v and w have the same number of children of each color. By induction, the subtrees rooted at same colored children are isomorphic. It follows that mapping v to w and the subtrees of children of v to subtrees of children of the same color of w is an isomorphism. □

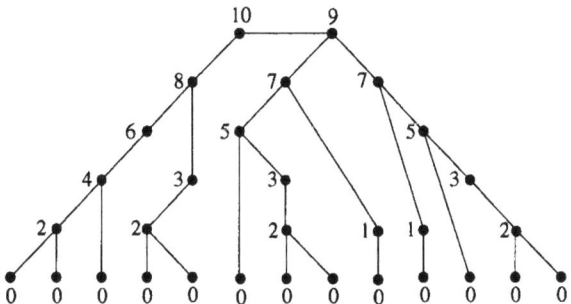

Figure 12.29: Linear-Time Coloring of a Tree

Exercise 12.30 Extend Proposition 12.28 to show that it holds for undirected, colored forests. Show that it also holds for colored forests of directed trees in which edges are directed from root to leaves. □

The following notation makes it easier to discuss the values of k for which $\mathcal{L}^k$ or $\mathcal{C}^k$ characterizes a set of graphs.

Definition 12.31 Let Σ be a set of finite graphs. Define var(Σ, n) (resp. vc(Σ, n)) to be the minimum k such that $\mathcal{L}^k$ (resp. $\mathcal{C}^k$) characterizes the graphs in Σ with at most n vertices. Let var(n) = var($GRAPHS$, n) and vc(n) = vc($GRAPHS$, n). When var(Σ, n) or vc(Σ, n) is bounded, write var(Σ) = $\max_n$ var(Σ, n), and vc(Σ) = $\max_n$ vc(Σ, n). □

By combining various results obtained so far, we know that var(GRAPHS, n) = $n + 1$, $var(CC_1) = 2$, and $var(CC_2) = var(CC_3) = 3$. Proposition 12.28 says that vc(TREES) = 2. The following fact shows that counting was crucial in Proposition 12.28.

Fact 12.32. *Let T_k be the set of finite trees such that each node has at most k children, and let S_k be the subset of T_k in which each non-leaf has exactly k children. The number of variables v so that $\mathcal{L}^v$ characterizes these classes of trees is*

$$var(T_k) = \begin{cases} 2 & \text{if } k = 1 \\ 3 & \text{if } 2 \leq k \leq 3 \\ k & \text{if } k > 3 \end{cases}$$

$$var(S_k) = \begin{cases} 2 & \text{if } 1 \leq k \leq 2 \\ 3 & \text{if } 3 \leq k \leq 6 \\ \lceil k/2 \rceil & \text{if } k > 6 \end{cases}.$$

Babai and Kučera have proved the following result about stable colorings of random graphs:

Fact 12.33. *There exists a constant $\alpha < 1$ such that if G is chosen randomly from the set of all labeled graphs on n vertices then*

$$Prob\{G \text{ has two vertices of the same stable color}\} < \alpha^n.$$

Fact 12.33 implies that for almost all graphs, each vertex has a unique description in $\mathcal{C}^2$. It follows that,

Corollary 12.34. *Almost all finite graphs are characterized by $\mathcal{C}^2$.*

It is easy to see that Fact 12.33 fails for regular graphs: all regular graphs of degree d on n vertices are $\mathcal{C}^2$-equivalent. Kučera has given an expected linear-time algorithm for canonization of regular graphs of a given, fixed degree [Ku87]. It follows from his results that:

Corollary 12.35. *For all d and for sufficiently large n, C^3 characterizes more than $1 - O[1/n]$ of the regular graphs of degree d on n vertices.*

12.6 Abiteboul-Vianu and Otto Theorems

In this section, we show how to describe a total ordering of the $\mathcal{L}^k$ and $\mathcal{C}^k$-types of vertices in the unordered languages FO(LFP) and FO(LFP, COUNT), respectively. It follows that if two sets of graphs are distinguishable in FO(PFP), then they are already distinguishable in FO(LFP). Similarly if two sets of graphs are distinguishable in FO(PFP, COUNT), then they are already distinguishable in FO(LFP, COUNT). These imply the theorems of Abiteboul-Vianu (Theorem 12.50) and Otto (Theorem 12.57), respectively. We begin with C^2, because this corresponds to the familiar notion of stable coloring (Theorem 12.27).

Theorem 12.36. *There is a formula $\Phi_2(x, y) \in$ FO(LFP, COUNT) such that for all graphs G and vertices g, h,*

$$G \models \Phi_2(g, h) \quad \Leftrightarrow \quad \bar{f}(g) < \bar{f}(h)$$

where $\bar{f}(v)$ refers to the stable coloring of vertex v.

Proof We write an inductive definition $\varphi_2(R, x, y)$ so that $\Phi_2 = (\text{IFP}\varphi_2)$ is the desired ordering. We use inflationary fixed point to avoid the requirement that R occur positively (Corollary 9.8).

The base coloring of all vertices is the same, so R is initially false. (For colored graphs, we would just order vertices according to their input colors.)

Inductively, the new color of each vertex is its old color together with how many of its neighbors have each old color. Thus, relation $R(x, y)$ holds now if it held before, or if x and y were in the same color class last time, but now for some vertex z, z's color class is the first on which x and y differ, and y has more neighbors than x of z's color. The following formula expresses this condition. We use the subformula $v(x, i, w)$ meaning that x has at least i neighbors of the same old color as w. To say that v is the same old color as w we write $\neg R(v, w) \land \neg R(w, v)$.

$$v(x, i, w) \equiv (\exists i \, v)(\neg R(v, w) \land \neg R(v, w) \land E(x, v))$$

$$\varphi_2(R, x, y) = R(x, y) \lor$$
$$\left(\neg R(y, x) \land (\exists z)(\forall w. R(w, z))((\forall i)(v(x, i, w) \leftrightarrow v(y, i, w)) \right.$$
$$\left. \land (\exists j)(v(y, j, z) \land \neg v(x, j, z))) \right)$$

□

Exercise 12.37 Show how to modify the ordering of C^2-types defined in Theorem 12.36 when the inputs are colored graphs. □

In a similar way, the ordering of $\mathcal{L}^2$ types is expressible in FO(LFP),

Corollary 12.38. *There is a formula $\Psi_2(x, y) \in$ FO(LFP) that describes a total ordering on the $\mathcal{L}^2$ types of vertices of any graph.*

Exercise 12.39 Prove Corollary 12.38. You should also do this in the case that the inputs are colored graphs. [Hint: the only change from Theorem 12.36 is that instead of counting how many neighbors x and y have of a certain color, we need to say whether or not they have any neighbors of this color and whether all vertices of this color are their neighbors.] □

Stable Coloring of k-tuples

Theorem 12.27 shows that $\mathcal{C}^2$-equivalence is determined by stable coloring of vertices. Similarly, $\mathcal{C}^k$-equivalence is determined by stable coloring of $k-1$-tuples. However, it turns out simpler to talk about the $\mathcal{C}^k$-types of k-tuples rather than $k-1$-tuples and this is what we now do. At the same time, we become more general by talking about arbitrary vocabularies rather than just colored graphs.

Definition 12.40 ($\mathcal{C}^k$-**Type of k-tuples**) For any structure $\mathcal{A} \subseteq$ STRUC$[\tau]$, let $n = \|\mathcal{A}\|$. For $t \in \mathbf{N}$, define the mappings $\mathcal{C}_t^k : |\mathcal{A}|^k \to \mathbf{N}$ as follows. The initial type $\mathcal{C}_0^k(v_1, \ldots, v_k)$ indicates the isomorphism type of the k-tuple. Another way of putting this is that for any two tuples $g_1, \ldots, g_k$ and $h_1, \ldots, h_k$, $\mathcal{C}_0^k(\bar{g}) = \mathcal{C}_0^k(\bar{h})$ iff the map $\alpha : g_i \mapsto h_i, i = 1 \ldots k$, is an isomorphism. This is equivalent to $\bar{g}$ and $\bar{h}$ agreeing on all quantifier-free formulas.

Inductively, type $t+1$ of a tuple $\mathcal{C}_{t+1}^k(g_1, \ldots, g_k)$ is determined by $\mathcal{C}_t^k(\bar{g})$ together with the number of elements of the universe that result in a certain $\mathcal{C}_t^k$-type when substituted for a certain entry of $\bar{g}$. More formally, for $1 \leq i \leq k$ and $c \in \mathcal{C}_t^k((\mathcal{A})^k)$, let

$$n_{t,c}^i(\bar{g}) = \left|\{v \in |\mathcal{A}| \mid c = \mathcal{C}_t^k(g_1, \ldots, g_{i-1}, v, g_{i+1}, \ldots, g_k)\}\right|$$

Then $\mathcal{C}_{t+1}^k(\bar{g}) = \mathcal{C}_{t+1}^k(\bar{h})$ iff $\mathcal{C}_t^k(\bar{g}) = \mathcal{C}_t^k(\bar{h})$ and for all i and c, $n_{t,c}^i(\bar{g}) = n_{t,c}^i(\bar{h})$. We assign numerical values to $\mathcal{C}_{t+1}^k(\bar{g})$ by sorting the n^k tuples,

$$\langle \mathcal{C}_t^k(\bar{g}), n_{t,0}^1(\bar{g}), \ldots, n_{t,c_m}^1, \ldots, n_{t,c_m}^k \rangle, \quad \bar{g} \in |\mathcal{A}|^k$$

where c_m is the maximum element of $\mathcal{C}_t^k(|\mathcal{A}|^k)$. Finally, define the $\mathcal{C}^k$-type of a k-tuple, $\mathcal{C}^k(\bar{g})$ to be $\mathcal{C}_T^k(\bar{g})$ where T is minimum such that $\mathcal{C}_T^k = \mathcal{C}_{T+1}^k$. Observe that $T < n^k$ because each iteration that makes progress increases the number of $\mathcal{C}^k$-types and there cannot be more $\mathcal{C}^k$-types than k-tuples. □

The following theorem, which follows immediately from Definition 12.40, says that it is correct to call $\mathcal{C}^k(\bar{g})$ the $\mathcal{C}^k$-type of $\bar{g}$.

Theorem 12.41. *For any $\mathcal{A} \in$ STRUC$[\tau]$, any $g_1, \ldots, g_k, h_1, \ldots, h_k \in |\mathcal{A}|$ and any t, the following are equivalent.*

1. $\mathcal{C}_t^k(\bar{g}) = \mathcal{C}_t^k(\bar{h})$,
2. $(\mathcal{A}, g_1/x_1, \ldots, g_k/x_k) \equiv_{C,t}^k (\mathcal{A}, h_1/x_1, \ldots, h_k/x_k)$.

In particular,

$$C^k(\bar{g}) = C^k(\bar{h}) \quad \Leftrightarrow \quad (\mathcal{A}, g_1/x_1, \ldots, g_k/x_k) \equiv_C^k (\mathcal{A}, h_1/x_1, \ldots, h_k/x_k).$$

Exercise 12.42 Let the language $FO^k(COUNT)$ be $FO(COUNT)$ restricted to have k domain variables $x_1, \ldots, x_k$. Number variables are left unrestricted, and quantification of number variables does not count toward the quantifier rank of formulas. In Theorem 12.41 the equivalence is stated in terms of language C^k. Show that this theorem remains true when the C_m^k-equivalence is replaced by $FO_m^k(COUNT)$ equivalence. □

In analogy to Definition 12.40, we define the $\mathcal{L}^k$-types of k-tuples. $\mathcal{L}_t^k(\bar{g})$ is defined exactly as $C_t^k(\bar{g})$ except that instead of the numbers $n_{t,c}^i(\bar{g})$ are replaced by booleans defined as follows,

$$b_{t,c}^i(\bar{g}) = \text{true} \quad \Leftrightarrow \quad (\exists v \in |\mathcal{A}|)(c = \mathcal{L}_t^k(g_1, \ldots, g_{i-1}, v, g_{i+1}, \ldots, g_k)).$$

The following theorem, which is similar to Theorem 12.41, follows directly from the definition.

Theorem 12.43. *For any $\mathcal{A} \in STRUC[\tau]$, any $g_1, \ldots, g_k, h_1, \ldots, h_k \in |\mathcal{A}|$ and any t, the following are equivalent.*

1. $\mathcal{L}_t^k(\bar{g}) = \mathcal{L}_t^k(\bar{h})$,
2. $(\mathcal{A}, g_1/x_1, \ldots, g_k/x_k) \equiv_t^k (\mathcal{A}, h_1/x_1, \ldots, h_k/x_k)$.

In particular,

$$\mathcal{L}^k(\bar{g}) = \mathcal{L}^k(\bar{h}) \quad \Leftrightarrow \quad (\mathcal{A}, g_1/x_1, \ldots, g_k/x_k) \equiv^k (\mathcal{A}, h_1/x_1, \ldots, h_k/x_k)$$

Just as in the case of Theorem 12.36, total orderings of C^k-types and $\mathcal{L}^k$-types are expressible in $FO(LFP, COUNT)$ and $FO(LFP)$, respectively.

Theorem 12.44. *For all k, there is a formula $\Phi_k(x_1, \ldots, x_k, y_1, \ldots, y_k)$ in $FO(LFP, COUNT)$ such that for all structures $\mathcal{A}$ and k-tuples of elements $\bar{g}, \bar{h} \in |\mathcal{A}|^k$,*

$$\mathcal{A} \models \Phi_k(\bar{g}, \bar{h}) \quad \Leftrightarrow \quad C^k(\bar{g}) < C^k(\bar{h})$$

Proof This proof is similar to that of Theorem 12.36. We write an inductive definition $\varphi_k(R, x_1, \ldots x_k)$ so that $\Phi_k = (\text{IFP}\varphi_k)$ is the desired ordering.

In the base case, $C_0^k(\bar{g})$ is determined by the quantifier-free properties of $\bar{g}$ and there are only a bounded number of possibilities.

Inductively, $C_{t+1}^k(\bar{g}) < C_{t+1}^k(\bar{h})$ iff $C_t^k(\bar{g}) < C_t^k(\bar{h})$ or $C_t^k(\bar{g}) = C_t^k(\bar{h})$ and for some i, c, $n_{t,c}^i(\bar{g}) < n_{t,c}^i(\bar{h})$, and for all (i', c') lexicographically less than (i, c), $n_{t,c'}^{i'}(\bar{g}) = n_{t,c'}^{i'}(\bar{h})$.

To write this inductive definition in $FO(COUNT)$, note that we express $C_t^k(\bar{g}) = C_t^k(\bar{h})$ as $\neg R(\bar{g}, \bar{h}) \wedge \neg R(\bar{h}, \bar{g})$. We express a color c by quantifying a k-tuple $c_1, \ldots, c_k$ of that color. Thus, to express the property that $m = n_{t,c}^i(\bar{g})$ we would write the following,

$$(\exists! m \, x_i)(\mathcal{C}_t^k(g_1, \ldots, g_{i-1}, x_i, g_{i+1}, \ldots, g_k) = \mathcal{C}_t^k(c_1, \ldots, c_k))$$

□

In a similar way, we can express the ordering of the $\mathcal{L}^k$-types.

Theorem 12.45. *For all k, there is a formula $\Phi_k(x_1, \ldots, x_k, y_1, \ldots, y_k) \in$* FO(LFP) *such that for all structures $\mathcal{A}$ and k-tuples of elements $\bar{g}, \bar{h} \in |\mathcal{A}|^k$,*

$$\mathcal{A} \models \Phi_k(\bar{g}, \bar{h}) \quad \Leftrightarrow \quad \mathcal{L}^k(\bar{g}) < \mathcal{L}^k(\bar{h}).$$

We are now ready to prove the Abiteboul-Vianu theorem. The main insight is that in FO(LFP) we can express the transformation from any structure $\mathcal{A}$ to $E^k(\mathcal{A})$, the ordered structure of $\mathcal{A}$'s $\mathcal{L}^k$-types.

Definition 12.46 Let $\mathcal{A} \in \text{STRUC}[\tau]$ where $\tau = \langle R_1^{a_1}, \ldots, R_r^{a_r}, c_1, \ldots, c_s \rangle$. Assume that $a_i \leq k$, for $1 \leq i \leq r$. Define the structure $E^k(\mathcal{A})$, the $\mathcal{L}^k$-*invariant* of $\mathcal{A}$, as follows,

$$(|\mathcal{A}|^k / \mathcal{L}^k(\bar{x}) = \mathcal{L}^k(\bar{y}), \leq, =', R_1', \ldots, R_r', C_1, \ldots, C_r, X_1, \ldots, X_k, P_{s^1}, \ldots, P_{s^{k^k}})$$

where the universe is the set of $\mathcal{L}^k$-types from $\mathcal{A}$, the ordering is the ordering of these $\mathcal{L}^k$-types as in Theorem 12.45. The s^j's are k-tuples of indices, $s^j \in \{1, \ldots, k\}^k$. The other relations are defined as follows:

$$\begin{aligned}
E^k(\mathcal{A}) &\models =' ([b_1, \ldots, b_k]) & &\Leftrightarrow & b_1 &= b_2 \\
E^k(\mathcal{A}) &\models R_i'([b_1, \ldots, b_k]) & &\Leftrightarrow & \mathcal{A} &\models R_i(b_1, \ldots, b_{a_i}) \\
E^k(\mathcal{A}) &\models C_i([b_1, \ldots, b_k]) & &\Leftrightarrow & c_i^{\mathcal{A}} &= b_1 \\
E^k(\mathcal{A}) &\models X_i([b_1, \ldots, b_k], [d_1, \ldots, d_k]) & &\Leftrightarrow & b_j &= d_j, j \in \{1, \ldots, k\} - \{i\} \\
E^k(\mathcal{A}) &\models P_s([b_1, \ldots, b_k], [d_1, \ldots, d_k]) & &\Leftrightarrow & d_i &= b_{s_i}, 1 \leq i \leq k.
\end{aligned}$$

Note that the above are well defined as long as the meaning for X_i and P_s is that this relation holds on the right hand side for *some* choice of representatives. □

In the following theorem, we see that everything we can say about $\mathcal{A}$ we can translate to a statement about $E^k(\mathcal{A})$.

Theorem 12.47. *Let $k \in \mathbb{N}$, let τ be any vocabulary, and let φ be any formula of this vocabulary in $\text{FO}^k(\text{LFP})$ or $\text{FO}^k(\text{PFP})$. There exists an "equivalent" formula φ' also in FO(LFP), FO(PFP), respectively. Conversely, for any φ' there exists a corresponding formula φ, where in either case,*

$$\text{for all } \mathcal{A} \in \text{STRUC}[\tau], \quad \mathcal{A} \models \varphi \quad \Leftrightarrow \quad E^k(\mathcal{A}) \models \varphi' \tag{12.48}$$

Proof Let $\varphi \in \mathcal{L}^k$. We inductively define formula φ' such that for all structures $\mathcal{A}$ and all elements $b_1, \ldots, b_k \in |\mathcal{A}|$, the following equivalence is maintained,

$$(\mathcal{A}, b_1/x_1, \ldots, b_k/x_k) \models \varphi \quad \Leftrightarrow \quad (E^k(\mathcal{A}), [b_1, \ldots, b_k]/x) \models \varphi'. \tag{12.49}$$

For atomic formulas,

$$\varphi \equiv R_i(x_{s_1}, \ldots, x_{s_a}) \Rightarrow \varphi' \equiv (\exists y)(P_s(x, y) \wedge R'_i(y)),$$
$$\varphi \equiv (x_{s_1} = x_{s_2}) \Rightarrow \varphi' \equiv (\exists y)(P_s(x, y) \wedge =' (y)).$$

Inductively,

$$\varphi \equiv \alpha \wedge \beta \Rightarrow \varphi' \equiv \alpha' \wedge \beta',$$
$$\varphi \equiv \neg\alpha \Rightarrow \varphi' \equiv \neg\alpha',$$
$$\varphi \equiv (\exists x_i)\alpha \Rightarrow \varphi' \equiv (\exists y)(X_i(x, y) \wedge \alpha'(y/x)),$$
$$\varphi \equiv (\forall x_i)\alpha \Rightarrow \varphi' \equiv (\forall y)(X_i(x, y) \to \alpha'(y/x)).$$

It is easy to see by induction that Equation (12.49) holds. Furthermore, least fixed points and partial fixed points remain unchanged except that the arity of the new relation symbol in φ' is reduced to one:

$$\varphi \equiv R(x_{s_1}, \ldots, x_{s_a}) \Rightarrow \varphi' \equiv (\exists y)(P_s(x, y) \wedge R'(y)),$$
$$\varphi \equiv (\text{LFP}_{R, x_1 \ldots x_k} \alpha) \Rightarrow \varphi' \equiv (\text{LFP}_{R', x} \alpha'),$$
$$\varphi \equiv (\text{PFP}_{R, x_1 \ldots x_k} \alpha) \Rightarrow \varphi' \equiv (\text{PFP}_{R', x} \alpha').$$

In the converse case, the transformation from φ' to φ is essentially defined by the rules of Definition 12.46. A quantification $(\exists y)$ in φ' is mapped to $(\exists y_1 \ldots y_k)$ in φ. □

In Chapter 13, we see some separations of languages without ordering that do not seem to imply any separation of their corresponding complexity classes. A corollary of Theorem 12.47 is that this situation changes when both complexity classes are at least P.

Theorem 12.50. (**Abiteboul-Vianu Theorem**) *The following two conditions are equivalent.*

1. FO(wo≤)(LFP) = FO(wo≤)(PFP),
2. P = PSPACE.

Proof (1) implies (2) because if FO(wo≤)(LFP) = FO(wo≤)(PFP) then FO(LFP) = FO(PFP), since input languages may include ordering relations. Thus, P = PSPACE.

Conversely, suppose P = PSPACE. Let $\varphi \in$ FO(wo≤)(PFP) and let k be the number of distinct variables occurring in φ. By Theorem 12.47, φ' describes an equivalent PSPACE computation on the ordered structures $E^k(\mathcal{A})$. Thus, φ' is also a PTIME computation. By Theorem 4.10, this PTIME computation is described by a formula $\gamma \in$ FO(LFP). But ordered structure $E^k(\mathcal{A})$ is definable from $\mathcal{A}$ in FO(wo≤)(LFP), so the condition $E^k(\mathcal{A}) \models \gamma$ is definable in FO(wo≤)(LFP). Thus φ is definable in FO(wo≤)(LFP). □

Exercise 12.51 In Definition 12.46 we assumed that no relation symbol had arity greater than k. Show how to modify the definition of $E^k(\mathcal{A})$ to the case where this does not hold. Of course, your generalization must maintain the truth of Theorem 12.47. [Hint: we are interested only in formulas from $\mathcal{L}^k$.] □

Exercise 12.52 Recall that in the proof of the zero-one laws for $\mathcal{L}^k$ (Theorem 6.40) we used the complete extension axiom γ_k (Equation 6.37). Show using the proof of Lemma 6.39 that if $G \models \gamma_k$ and $\bar{a}, \bar{b}$ are k-tuples of vertices from G, then,

$$(G, \bar{a}) \equiv_0 (G, \bar{b}) \quad \Rightarrow \quad (G, \bar{a}) \equiv^k (G, \bar{b}) \,.$$

That is, if $\bar{a}$ and $\bar{b}$ satisfy the same quantifier-free formulas, then they have the same $\mathcal{L}^k$-type. Let b_k be the number of isomorphism types of undirected graphs on at most k vertices. Conclude that for any graph G that satisfies γ_k, the $\mathcal{L}^k$-invariant of G has bounded size: $G \models \gamma_k \Rightarrow \|E^k(G)\| \leq b_k$. □

Exercise 12.52 tells us that for a randomly chosen structure $\mathcal{A}$, almost surely the $\mathcal{L}^k$-invariant of $\mathcal{A}$ has bounded size. This is a special case of the following theorem of Tyszkiewicz. Tyszkiewicz's theorem implies that when a zero-one law holds, for almost all structures, the language involved can express only a bounded number of inequivalent sentences. In the following, say that a *mixed zero-one law* holds for a class of structures K iff there is a finite partition $K = K_1 \cup \cdots \cup K_t$ where the zero-one law holds for each K_i.

Fact 12.53. *Let K be any recursive class of finite structures. The following conditions are equivalent:*

1. *K has a mixed zero-one law for $\mathcal{L}^k$.*
2. *There is a fixed bound b such that for a structure $\mathcal{A}$ chosen at random from K, almost surely $\|E^k(G)\| \leq b$.*

Facts 12.53 and 12.33 expose a large difference between languages FO(LFP) and FO(LFP, COUNT) over random, unordered structures. Almost surely, the first language expresses only a bounded number of sentences, and the parity of the size of the universe is not one of them. On the other hand, there is almost surely an ordering expressible in FO(LFP, COUNT), so all polynomial-time computable queries are expressible.

In fact, Hella, Kolaitis and Luosto have observed that a weaker construction than counting already exposes this difference [HKL97]. Let *parity quantifier* $(\oplus x)\varphi$ mean that there are an odd number of x's such that φ. Karp has shown that Fact 12.33 remains true when, instead of counting how many neighbors each vertex has, we count the parity of the number of neighbors [Ka79]. It follows that almost surely an ordering, and thus all polynomial-time queries, are expressible in FO(LFP, $\oplus$).

We close this section by proving Otto's theorem, which is a version of the Abiteboul-Vianu theorem for FO(LFP, COUNT). The first step is to revise structure $E^k(\mathcal{A})$, which codes the $\mathcal{L}^k$-types of k-tuples from $\mathcal{A}$ (Definition 12.46) to $E_C^k(\mathcal{A})$, which codes the $\mathcal{C}^k$-types.

Definition 12.54 Let $\mathcal{A} \in \text{STRUC}[\tau]$ be as in Definition 12.46 with $\tau = \langle R_1^{a_1}, \ldots, R_r^{a_r}, c_1, \ldots, c_s \rangle$, and let $n = \|\mathcal{A}\|$. Assume that $a_i \leq k$, for $1 \leq i \leq r$. Define structure $E_C^k(\mathcal{A})$ — the $\mathcal{C}^k$-invariant of $\mathcal{A}$ — as follows,

$$E_C^k(\mathcal{A}) = ((|\mathcal{A}|^k / \mathcal{C}^k(\bar{x}) = \mathcal{C}^k(\bar{y})) \times n,$$
$$\leq, =' R'_1, \ldots, R'_r, C_1, \ldots, C_r, Y_1, \ldots, Y_k, P_{s^1}, \ldots, P^{s_{k^k}})$$

where the universe is the product of the set of numbers $n = \{0, 1, \ldots, n-1\}$ with (essentially) the universe of $E^k(\mathcal{A})$. All the relations are as before — with the number component ignored — except for the Y_i's. The number components are used to indicate how many neighbors of a certain color each color class contains. Let $Y_i([b_1, \ldots, b_k], m_1, [d_1, \ldots, d_k], m_2)$ hold iff $d_j = b_j$, for $j \neq i$, and there exists at least m_2 x_i's such that $(C^k(b_1, \ldots, b_{i-1}, x_i, b_{i+1}, \ldots, b_k) = C^k(d_1, \ldots, d_k))$. □

As in Theorem 12.47 we have

Theorem 12.55. *Let $k \in \mathbf{N}$, let τ be any vocabulary and let φ be any formula of this vocabulary in* $\mathrm{FO}^k(\mathrm{LFP}, \mathrm{COUNT})$ *or* $\mathrm{FO}^k(\mathrm{PFP}, \mathrm{COUNT})$. *Then there exists a formula φ' also in* $\mathrm{FO}(\mathrm{LFP}, \mathrm{COUNT})$, $\mathrm{FO}(\mathrm{PFP}, \mathrm{COUNT})$, *respectively. Conversely, given any φ' there is a corresponding φ. In each case,*

$$\text{for all } \mathcal{A} \in \mathrm{STRUC}[\tau], \quad \mathcal{A} \models \varphi \quad \Leftrightarrow \quad E_C^k(\mathcal{A}) \models \varphi' . \tag{12.56}$$

Proof The construction of φ' is done inductively, as in the proof of Theorem 12.47. The difference is in the case of quantifiers. Let

$$\varphi \equiv (\exists m\, x_i)\alpha \quad \Rightarrow \quad \varphi' \equiv (\exists m\, y)(Y_i(x, y) \wedge \alpha'(y/x)) .$$

The definition of the Y_i's guarantees that (12.56) holds as desired.

Conversely, consider a formula φ'. Since $E_C^k(\mathcal{A})$ is an ordered structure, every formula $\varphi' \in \mathrm{FO}(\mathrm{LFP}, \mathrm{COUNT})$ may be translated to a corresponding formula $\varphi'' \in \mathrm{FO}(\mathrm{LFP})$. We can then translate φ'' to φ essentially as in Theorem 12.47. □

Using, Theorem 12.55, we can now prove the following theorem in the same way as Theorem 12.50.

Theorem 12.57. (**Otto's Theorem**) *The following two conditions are equivalent.*

1. $\mathrm{FO}(\mathrm{wo}\leq)(\mathrm{LFP}, \mathrm{COUNT}) = \mathrm{FO}(\mathrm{wo}\leq)(\mathrm{PFP}, \mathrm{COUNT})$
2. $\mathrm{P} = \mathrm{PSPACE}$

An interesting open question related to C^k-invariants is the following:

Open Problem 12.58 Is there a polynomial-time mapping from $E_C^k(\mathcal{A})$ to a canonical structure $\mathcal{A}'$ with $E_C^k(\mathcal{A}') = E_C^k(\mathcal{A})$?

12.7 Toward a Language for Order-Independent P

We still do not have an answer to Question 12.1. We see in the next chapter that the language FO(LFP, COUNT) does not expressible all of order-independent polynomial time (Theorem 13.26). Dawar has shown that Question 12.1 is robustly formulated, [Daw93]:

Theorem 12.59. *The following conditions are equivalent:*

1. *There is a recursively enumerable listing of a set of sentences from FO(LFP) that describe exactly all the polynomial-time, order-independent boolean queries. That is, the answer to Question 12.1 is, "Yes."*
2. *There is a first-order query whose range is a listing of a set of sentences from FO(LFP) that describe exactly all the polynomial-time, order-independent boolean queries.*
3. *There is a boolean query that is complete for order-independent P via first-order reductions that do not use ordering or any other numeric predicate.*

Proof The implication, 3. $\Rightarrow$ 2., is clear: Let $\gamma \in$ FO(LFP) be the complete problem. The desired listing consists of all formulas $\widehat{Q}(\gamma)$, where Q is an FO(wo$\leq$) query and $\widehat{Q}$ is its dual (Definition 3.3).

The implication 2. $\Rightarrow$ 1. is immediate.

To prove 1. $\Rightarrow$ 3., let M be a total Turing machine whose output $M(0) = \varphi_0$, $M(1) = \varphi_1, \ldots$, consists of sentences in FO(LFP) describing all polynomial-time, order-independent boolean queries. Define the problem $C \subset \{0, 1, \#\}^*$ as follows,

$$C = \left\{ \varphi_i \#^{t_i} \text{code}(\mathcal{A}) \#^r \;\middle|\; \begin{array}{l} M \text{ outputs } \varphi_i \text{ within } t_i \text{ steps, and} \\ \mathcal{A} \models \varphi_i \text{ and this can be checked in } r \text{ steps.} \end{array} \right\}$$

C is a universal complete problem for order-independent P, cf. Exercise 2.17. C is acceptable in linear time by construction: We can simulate M to check that in t_i steps it outputs φ_i. Note that fixed constant t_i depends only on i. Next, we are given time r to check that $\mathcal{A} \models \varphi_i$. We know by Theorem 4.10 that it is sufficient to have $r = \|\mathcal{A}\|^{c_i}$ where c_i is a constant depending only on i.

By 1., every order-independent polynomial-time boolean query is of the form

$$G_i = \{\mathcal{A} \in \text{STRUC}[\tau_i] \mid \mathcal{A} \models \varphi_i\} \; .$$

G_i is reducible to C via the reduction

$$r_i : \mathcal{A} \mapsto \varphi_i \#^{t_i} \text{code}(\mathcal{A}) \#^r \; .$$

Reduction r_i is a first-order query, but it is not clear how to write r_i without using ordering. For example, the typical way we encode structures is as bin($\mathcal{A}$), which uses ordering and addition (Exercise 2.3).

A simple solution is to modify the complete problem C. Note that from number t_i φ_i can be computed, and we may assume that $t_i > c_i$. Let the reduction $r'_i : \text{STRUC}[\tau_i] \to \text{STRUC}[\tau_s]$ be a t_i-ary reduction. All that r'_i needs to do is to encode $\mathcal{A}$; its arity will encode t_i.

For example, if $\tau_i = \langle R_1^1, R_2^2, c_1, c_2 \rangle$, then let $r'_i = \lambda_{x^1 \ldots x^{t_i}} \langle \text{true}, \varphi_1 \rangle$ where the formula φ_1 — encoding the single unary relation S for the string $r'_i(\mathcal{A})$ — is the following:

$$\varphi_1 \equiv x^1 = x^2 = \cdots = x^{t_i - 1} \; \vee$$
$$x^1 = x^2 \wedge x^3 = x^4 \wedge x^5 \neq x^6 \wedge R_1(x^{t_i}) \; \vee$$
$$x^1 = x^2 \wedge x^3 \neq x^4 \wedge x^5 = x^6 \wedge R_2(x^{t_i - 1}, x^{t_i}) \; \vee$$

$$x^1 = x^2 \wedge x^3 \neq x^4 \wedge x^5 \neq x^6 \wedge x^{t_i} = c_1 \vee$$
$$x^1 \neq x^2 \wedge x^3 = x^4 \wedge x^5 = x^6 \wedge x^{t_i} = c_2 \ .$$

The reason for the first line in the definition of φ_1 is to encode the number $\|\mathcal{A}\|$. The definition of the relevant complete problem can now be reconstructed:

$$C' = \left\{ r_i'(\mathcal{A}) \mid i \in \mathbf{N}, \mathcal{A} \in \text{STRUC}[\tau_i], \mathcal{A} \models \varphi_i \right\} \ . \qquad \square$$

Most people who have studied the issue believe that every correct proof in every mathematics paper or textbook can — in principle — be formalized in Zermelo-Fraenkel set theory, plus the axiom of choice (ZFC). This indicates that the following is an ultimate candidate for a set of boolean queries satisfying Question 12.1,

$$Z = \left\{ \varphi \in \text{FO(LFP)} \mid \text{ZFC} \vdash \text{``}\varphi\text{ is order-independent''} \right\}$$

If any solution to Question 12.1 can be proved correct in ZFC, then Z is such a solution. However, there is no obvious Ehrenfeucht-Fraïssé game for class Z, so we get little insight into the inherent role of ordering.

Question 12.1 can be generalized to any other complexity class smaller than P. It may be the case that some of these questions are more tractable.

Historical Notes and Suggestions for Further Reading

The discussion of $\mathcal{C}^k$ including Definition 12.3, Proposition 12.9, Definition 12.18, Theorems 12.20 and 12.27 are from [IL90].

The bijective game (Definition 12.22) and Theorem 12.23 are due to Hella [He96]. The characterization of trees in $\mathcal{C}^2$ (Proposition 12.28) follows from Theorem 12.27 and the standard linear time isomorphism algorithm for trees [AHU74]. Fact 12.32 is from Immerman and Kozen [IK87]. Fact 12.33 is from Babai and Kučera [BK80]. Babai showed that for every fixed k, there is a polynomial-time algorithm for graph isomorphism for graphs of color-class size k. Recently, Grohe has proved that the class of finite planar graphs can be characterized in FO(LFP) and thus in $\mathcal{L}^k$ for a fixed k, [Gro97b].

It is not hard to see that for any fixed k, testing whether two graphs are L_k equivalent or C_k equivalent can be done in polynomial time, although it may be exponential in k. Grohe proved that for $k \geq 2$, testing L_k equivalence or C_k equivalence of two input structures is complete for polynomial time [Gro96].

The fact that Hanf's Theorem remains true in the presence of counting quantifiers (Exercise 12.25) is due to Hella, Libkin, and Nurmonen [HLN97].

Theorem 12.50 is due to Abiteboul and Vianu, [AV91] and Theorem 12.57 is due to Grädel and Otto [GO93, Ott96]. Our presentation of this material uses notation and ideas from Dawar, Lindell, and Weinstein [DLW95].

If an ordering is definable on a structure $\mathcal{A}$, then $\mathcal{A}$ is *rigid*, i.e., it has no nontrivial automorphisms. For a language $\mathcal{L}$, we say that $\mathcal{A}$ is $\mathcal{L}$-rigid iff for all $a_1, a_2 \in |\mathcal{A}|$,

$$(\mathcal{A}, a_1) \equiv_{\mathcal{L}} (\mathcal{A}, a_2) \quad \Rightarrow \quad a_1 = a_2 \ .$$

Intuitively, $\mathcal{A}$ is $\mathcal{L}$-rigid if $\mathcal{L}$ has a unique name for each element in the universe of $\mathcal{A}$. Let $\mathcal{A}$ be an ordered structure and let $\mathcal{L}$ be at least as powerful as FO(DTC) or FO-VAR[$\log n$, 3]. Then $\mathcal{A}$ is $\mathcal{L}$-rigid.

Dawar conjectured that for every finitely axiomatizable class of rigid structures, K, an ordering of all the structures in K is definable by a formula in FO(LFP) [Daw93]. Gurevich and Shelah refuted this conjecture by constructing a finitely axiomatizable class of rigid structures that are not C^k-rigid for any k [GS96]. It follows that no formula in FO(LFP, COUNT) can order all the structures in K. The Gurevich and Shelah construction is closely related to the proof of Theorem 13.26, which we give in the next chapter.

For more information about the set theory system ZFC see Cohen's book [Coh66]. After presenting the axioms of ZFC, he writes, "The reader should be reasonably convinced even on a first reading that the axioms easily encompass all of traditional mathematics."

Fact 12.53 is due to Tyszkiewicz [Tys97].

Problem 12.58 was considered by Otto in [Ott97]. He proved there that there is such a canonical form when $k = 2$. Otto also showed that if there is a polynomial-time algorithm to produce such canonical forms from the C^3-invariants, then there is a logic for $C^k \cap \mathrm{P}$, for all k. The corresponding problem for $\mathcal{L}$-invariants was settled in the negative by Grohe who proved that there is no recursive bound on the size of the smallest structure $\mathcal{A}$ with invariant $E^k(\mathcal{A})$ [Gro97].

Recently Grohe showed that is very unlikely that there is a polynomial-time algorithm that given a structure $\mathcal{A}$ will find a canonical structure of the same L^k type [Gro97a].

Question 12.1 remains quite open at this writing. One possible direction for solving this may be to add to FO(LFP, COUNT) some algorithms to manipulate permutation groups via their generators, cf. [BL83]. An interesting step in this direction was taken by Gire and Hoang in [GH97].

13
Lower Bounds

The very simple problem PARITY is too hard for first-order logic, no matter what numeric predicates we add. When we add counting, but remove ordering, PARITY is expressible. However, a different sort of parity problem becomes inexpressible. A related lower bound suggests that complete problems for P are inherently sequential.

13.1 Håstad's Switching Lemma

Recall boolean query PARITY, which is true of boolean strings that have an odd number of ones. Using pebble games, we have shown that PARITY is not first-order in the absence of the numeric predicate BIT (Proposition 6.14, Proposition 6.45). This theorem is much more subtle with the inclusion of BIT.

Theorem 13.1. *Query* PARITY *is not first-order expressible:* PARITY $\notin$ FO.

The known proofs of Theorem 13.1 all prove the stronger result that PARITY is not in the non-uniform class AC^0/poly or, equivalently, PARITY is not first-order, no matter what numeric predicates are available (Proposition 11.19). The proof we present here is via the Håstad Switching Lemma, following the treatment in [Bea96].

Let f be a boolean function, with boolean variables $V_n = \{x_1, \ldots, x_n\}$. A restriction on V_n is a map $\rho : V_n \to \{0, 1, \star\}$. The idea is that some of the variables are set to "0" or "1" and the others — those assigned "$\star$" — remain variables.

Restriction ρ applied to function f results in function $f|_\rho$ in which value $\rho(x_i)$ is substituted for x_i in f, for each x_i such that $\rho(x_i) \neq \star$. Thus, $f|_\rho$ is a function of the variables that have been assigned "$\star$". Let $\mathcal{R}_n^r$ be the set of all restrictions on V_n that map exactly r variables to "$\star$".

We state and prove the switching lemma using decision trees. Given a formula F in disjunctive normal form (DNF)[1] define the *canonical decision tree $T(F)$ for F* as follows: Let $C_1 = \ell_1 \wedge \cdots \ell_i$ be the first term of F, so $F = C_1 \vee F'$. The top of $T(F)$ is a complete binary decision tree on the variables in C_1. Each leaf of the tree determines a restriction ρ that assigns the appropriate value to the variables in C_1 and assign "$\star$" to all the other variables. There is a unique leaf that makes C_1 true and this should remain a leaf and be labeled "1". To each other leaf, determining restriction ρ, we attach the canonical decision tree $T(F'|_\rho)$.

Let $h(T)$ be the height of tree T. We now show that for any formula F in DNF, if F has only small terms, then when randomly choosing a restriction ρ from $\mathcal{R}_n^r$, with high probability the height of the canonical decision tree of the resulting formula, $h(T(F|_\rho))$, is small.

It then follows that the negation of $F|_\rho$ can also be written in DNF — as the disjunction of the conjunction of each branch in the tree that leads to "0". Thus, with high probability, a random restriction switches a DNF formula that has only small terms to a conjunctive normal form (CNF) formula.

Lemma 13.2. (**Håstad Switching Lemma**) *Let F be a DNF formula on n variables, such that each of its terms has length at most k. Let $p \leq 1/7$, $r = pn$, and $s \geq 0$. Then,*

$$\frac{|\{\rho \in \mathcal{R}_n^r \mid h(T(F|_\rho)) \geq s\}|}{|\mathcal{R}_n^r|} < (7pk)^s \ .$$

Proof The proof of Lemma 13.2 is a somewhat intricate counting argument. Let Stars(k, s) be the set of all sequences $w = (S_1, S_2, \ldots, S_t)$ where each S_i is a nonempty subset of $\{1, 2, \ldots, k\}$ and the sum of the cardinalities of the S_i's equals s. We use the following upper bound on the size of Stars(k, s).

Lemma 13.3. *For $k, s > 0$, $|\text{Stars}(k, s)| \leq (k/\ln 2)^s$.*

Proof We show by induction on s that $|\text{Stars}(k, s)| \leq \gamma^s$, where γ is such that $(1 + 1/\gamma)^k = 2$. Since $(1 + 1/\gamma) < e^{1/\gamma}$, we have $\gamma < k/\ln 2$ and thus the lemma will follow.

Suppose that the lemma holds for any $s' < s$. Let $\beta \in \text{Stars}(k, s)$. Then $\beta = (S_1, \beta')$, where $\beta' \in \text{Stars}(k, s - i)$ and $i = |S_1|$. Thus,

$$|\text{Stars}(k, s)| = \sum_{i=1}^{\min(k,s)} \binom{k}{i} |\text{Stars}(k, s - i)|$$

Thus, by the induction hypothesis,

$$|\text{Stars}(k, s)| \leq \sum_{i=1}^{k} \binom{k}{i} \gamma^{s-i}$$

---

[1] A DNF formula is an "or" of "and"s. This is the dual of CNF.

13.1 Håstad's Switching Lemma

$$= \gamma^s \sum_{i=1}^{k} \binom{k}{i} (1/\gamma)^i$$

$$= \gamma^s[(1 + 1/\gamma)^k - 1] = \gamma^s.$$

□

Let $R \subseteq \mathcal{R}_n^r$ be the set of restrictions ρ such that $h(T(F|_\rho)) \geq s$. We will define a 1:1 map,

$$\alpha : R \rightarrow \mathcal{R}_n^{r-s} \times \text{Stars}(k, x) \times 2^s. \tag{13.4}$$

Once we show that α is one to one, it will follow that

$$\frac{|R|}{|\mathcal{R}_n^r|} \leq \frac{|\mathcal{R}_n^{r-s}|}{|\mathcal{R}_n^r|} \cdot |\text{Stars}(k, s)| \cdot 2^s. \tag{13.5}$$

Observe that $|\mathcal{R}_n^r| = \binom{n}{r} 2^{n-r}$, so,

$$\frac{|\mathcal{R}_n^{r-s}|}{|\mathcal{R}_n^r|} = \frac{(r)(r-1)\cdots(r-s+1)}{(n-r+s)(n-r+s-1)\cdots(n-r+1)} \cdot 2^s \leq \left(\frac{2r}{n-r}\right)^s.$$

Substituting this into Equation (13.5) and using Lemma 13.3, we have,

$$\frac{|R|}{|\mathcal{R}_n^r|} \leq \left(\frac{2r}{n-r}\right)^s \cdot (k/\ln 2)^s \cdot 2^s$$

$$= \left(\frac{4rk}{(n-r)\ln 2}\right)^s$$

$$= \left(\frac{4pk}{(1-p)\ln 2}\right)^s$$

when $r = pn$. This is less than $(7pk)^s$ when $p < 1/7$.

It thus suffices to construct 1:1 map α (Equation (13.4)). Let $F = C_1 \vee C_2 \vee \cdots$. Let $\rho \in R$, and let C_{i_1} be the first term of F that is not set to "0" in $F|_\rho$.

Let b be the first s steps of the lexicographically first branch in $T(F|_\rho)$ that has length at least s. Let V_1 be the set of variables in $C_{i_1}|_\rho$. Let a_1 be the assignment to V_1 that makes $C_{i_1}|_\rho$ true. Let b_1 be the initial segment of b that assigns values to V_1. If b ends before all the values of V_1 are defined, then let $b_1 = b$, and shorten a_1 so that it assigns values only to the variables that b_1 does. See Figure 13.6.

Define the set $S_1 \subseteq \{1, 2, \ldots, k\}$ to include those j such that the jth variable in C_{i_1} is set by a_1. S_1 is nonempty. Note that from C_{i_1} and S_1 we can reconstruct a_1.

If $b \neq b_1$, then $(b - b_1)$ is a path in $T(F|_{\rho b_1})$. Let C_{i_2} be the first term of F not set to "0" by ρb_1. As above, we generate b_2, a_2, and S_2. Repeat this until the whole branch b is used up. We have $b = b_1 b_2 \cdots b_t$, and let $a = a_1 a_2 \cdots a_t$. Define the map $\delta : \{1, \ldots, s\} \rightarrow \{0, 1\}$ such that $\delta(j) = 1$ if a and b assign the same value at their step j, and $\delta(j) = 0$ if a and b assign different values to variable j. We finally define the map α as,

$$\alpha(\rho) = \langle \rho a, (S_1, S_2, \ldots, S_t), \delta \rangle.$$

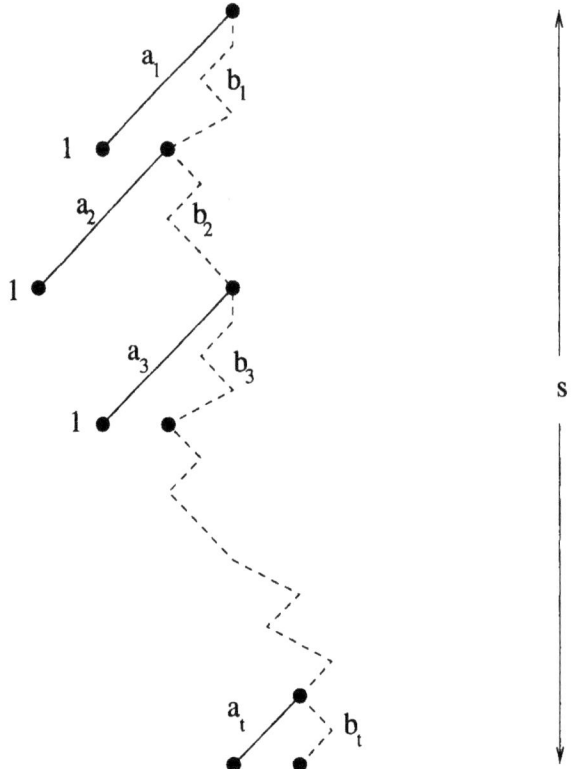

Figure 13.6: Decision tree $T(F|_\rho)$ with path of length s, $b = b_1 b_2 \cdots b_t$.

From $\alpha(\rho)$ we can reconstruct ρ as follows: C_{i_1} is the first clause that evaluates to "1" using ρa. From C_{i_1} and S_1 we reconstruct a_1. Then, using δ, we can compute the restriction $\rho' = \rho b_1 a_2 \cdots a_t$. Next, C_{i_2} is the first clause evaluating to "1" using ρ'. From this and S_2, we can compute a_2, and so on. Thus α is 1:1. This completes the proof of Håstad's Switching Lemma. □

A striking consequence of the switching lemma is that AC^0 circuits have restrictions on which they are constant even though many variables are assigned to "$\star$":

Theorem 13.7. *Let C be an unbounded fan-in circuit with n inputs, having size s and depth d. Let $r \leq n/(14^d (\log s)^{d-1}) - (\log(s) - 1)$. Then there is a restriction $\rho \in \mathcal{R}_n^r$ for which $C|_\rho$ is constant.*

Proof We show inductively from the leaves up, that there is a restriction that turns all the gates into DNF or CNF formulas all of whose terms have length at most $\log s$.

Assume that level one of the circuit — the nodes sitting above the inputs and their negations — consists of "or" gates. Thus, each of these gates g is a DNF formula whose maximum term size is one. By Lemma 13.2, with $p = 1/14$, $n_1 = n/14$, $k = 1$, we have,

$$|\{\rho \in \mathcal{R}_n^{n_1} \mid h(T(g|_\rho)) \geq \log s\}| < (2)^{-\log s} \cdot |\mathcal{R}_n^{n_1}|.$$

Since there are at most s gates at level one, the number of restrictions ρ such that $h(T(g|_\rho)) \geq \log s$ for some g is less than,

$$s \cdot (2)^{-\log s} \cdot |\mathcal{R}_n^{n_1}| = |\mathcal{R}_n^{n_1}|.$$

Thus, there is at least one restriction $\rho_1 \in \mathcal{R}_n^{n_1}$ under which all the gates at level one are CNF formulas with terms of size less than $\log s$. It follows that the "and" gates at level two are CNF formulas with terms of size less than $\log s$.

Let $g_2 = g|_{\rho_1}$ be any such gate. Using Lemma 13.2, with $k = \log s$, $p = 1/(14 \log s)$, $n_2 = n_1/(14 \log s)$, we have,

$$|\{\rho \in \mathcal{R}_{n_1}^{n_2} \mid h(T(g_2|_\rho)) \geq \log s\}| < (2)^{-\log s} \cdot |\mathcal{R}_{n_1}^{n_2}|.$$

Thus, there is a restriction $\rho_2 \in \mathcal{R}_{n_1}^{n_2}$ under which every gate at level two is a DNF formula all of whose terms have length less than $\log s$.

Repeating this argument through all d levels, we have a restriction $\rho = \rho_1 \rho_2 \cdots \rho_d \in \mathcal{R}_{n_d}^n$ such that the height $T(C|_\rho)$ of the decision tree of the root of the circuit is less than $\log s$. Observe that $n_d = n/(14^d (\log s)^{d-1})$. Let b be the restriction corresponding to any branch of the decision tree. It follows that $C|_{\rho b}$ is constant and has at least $r = n_d - (\log(s) - 1)$ inputs. $\square$

Suppose that circuit C in Theorem 13.7 computes the parity of its n inputs. Then any restriction of C also computes the parity of its remaining inputs. Thus, if $1 \leq r$ in Theorem 13.7, then C must not compute PARITY. It follows that if C is a size s, depth d circuit computing parity on n inputs, then the following inequalities hold,

$$1 > n/(14^d (\log s)^{d-1}) - (\log(s) - 1)$$
$$\log s > n/(14^d (\log s)^{d-1})$$
$$(\log s)^d > n/(14^d)$$
$$s > 2^{\frac{1}{14} n^{\frac{1}{d}}}.$$

We thus have the following lower bound on the number of iterations of a first-order quantifier block needed to compute PARITY. This corollary is optimal by Exercise 4.19.

We use the "big omega" notation for lower bounds. The "equation" $f(n) = \Omega(g(n))$ is equivalent to $g(n) = O(f(n))$. It means that for almost all values of n, $f(n)$ is at least some constant multiple of $g(n)$.

Corollary 13.8. *If* PARITY $\in$ FO[$s(n)$], *then* $s(n) = \Omega(\log n / \log \log n)$, *and this holds even in the presence of arbitrary numeric predicates.*

Exercise 13.9 Show that PARITY is first-order reducible to REACH. Conclude that the same lower bound as in Corollary 13.8 holds for REACH. $\square$

13.2 A Lower Bound for REACH$_a$

In this section, we prove a lower bound (Theorem 13.11) on the quantifier-rank needed to express the P-complete problem REACH$_a$ (Definition 3.24), when ordering and the other numeric predicates are not available. If the same result were proved for the language with ordering, it would imply that NC is strictly contained in P, and in fact that $\bigcup_k$ DSPACE[$(\log n)^k$] does not contain P.

Exercise 13.10 Show that REACH$_a$ is expressible in FO-VAR(wo$\leq$)[$n, 2$]. [Hint: just write down the natural inductive definition of the alternating path relation.] □

In the remainder of this section, we prove the following lower bound:

Theorem 13.11. *Boolean query* REACH$_a$ *is not expressible in quantifier rank* $2\sqrt{\log n - 1}$ *in the language without ordering.*

To prove Theorem 13.11, we construct graphs G_m and H_m with the following properties:

$$\begin{aligned}&1. \quad \|G_m\| = \|H_m\| < m^{1+\log m} \\ &2. \quad G_m \sim_m H_m \\ &3. \quad G_m \in \text{REACH}_a; \ H_m \notin \text{REACH}_a .\end{aligned} \quad (13.12)$$

Note that condition (1) implies that for $n = \|G_m\|$, $\log n < (1 + \log m)(\log m)$, so $\sqrt{\log n} < 1 + \log m$, so $2\sqrt{\log n - 1} < m$. Thus, Equation (13.12) implies Theorem 13.11.

The first step in producing G_m and H_m is to introduce the building block out of which they will be constructed.

Lemma 13.14. *Let X_d be the alternating graph pictured in Figure 13.13. Then X_d has automorphisms that switch any two of the pairs (a_1, b_1), (a_2, b_2), and (a_3, b_3), leaving the other pair fixed.*

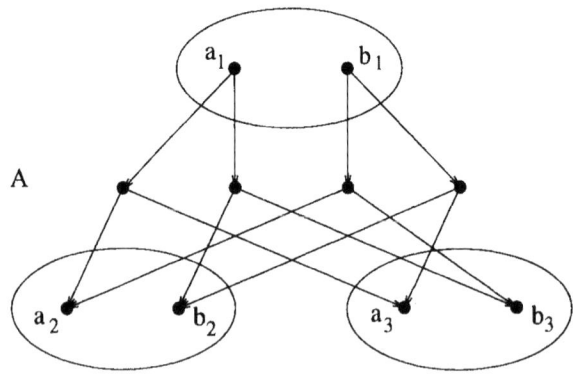

Figure 13.13: The Directed Switch X_d

Proof The idea is that when X_d is placed in a graph, each of the pairs will consist of one point that can reach d and one point that cannot. Note that the four points at the middle of X_d are "and"-nodes and the other points are "or"-nodes. The boolean formulas corresponding to alternating graph X_d are the following:

$$a_1 \equiv (a_2 \wedge a_3) \vee (b_2 \wedge b_3),$$
$$b_1 \equiv (a_2 \wedge b_3) \vee (b_2 \wedge 2_3).$$

The proof of the lemma is an easy computation. □

Before we construct the graphs G_m and H_m satisfying Equation (13.12), we build exponential size graphs as a warm-up.

Let T_m be a complete binary tree of height m, with root r and edges directed from root to leaves. Define $X_d(T_m)$ to be the graph obtained by replacing each vertex v from T_m by a copy $X_d(v)$ of the switch X_d. Let y and z be the left and right children of v, respectively. Then $X_d(T_m)$ contains the edges $(a_2(v), a_1(y))$, $(b_2(v), b_1(y))$ and $(a_3(v), a_1(z))$, $(b_3(v), b_1(z))$. Furthermore, add an additional vertex t and draw the edges $(a_2(\ell), t)$ and $(a_3(\ell), t)$ for each leaf ℓ of T_m. Finally, interpret constant symbol s as $a_1(r)$. Define $\tilde{X}_d(T_m)$ to be the same as $X_d(T_m)$ except that s is interpreted as $b_1(r)$. Thus $X_d(T_m) \in \text{REACH}_a$, but $\tilde{X}_d(T_m) \notin \text{REACH}_a$. See Figure 13.15 for a diagram of $\tilde{X}_d(T_m)$.

The following observation about $X_d(T_m)$ and $\tilde{X}_d(T_m)$ leads to $\mathcal{D}$'s winning strategy in $\mathcal{G}_m(X_d(T_m), \tilde{X}_d(T_m))$:

Observation 13.16. *Suppose that in $\tilde{X}_d(T_m)$, we take any pair of edges, $(a_i(v), a_i(w))$, $(b_i(v), b_i(w))$, and switch them, i.e., replace them by $(a_i(v), b_i(w))$, $(b_i(v), a_i(w))$. Then the resulting graph is isomorphic to $X_d(T_m)$. In fact, switching*

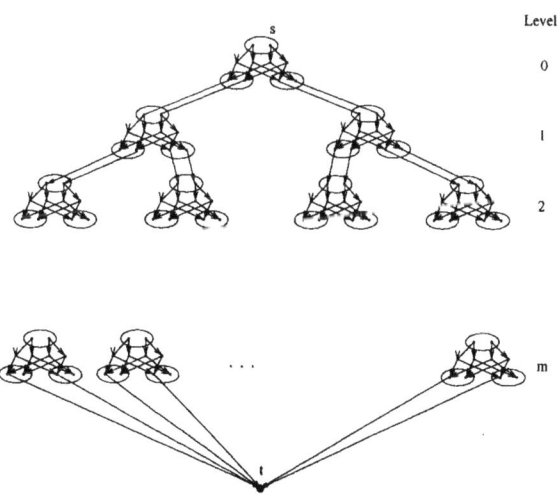

Figure 13.15: The graph $\tilde{D}_d(T_m)$.

210   13. Lower Bounds

any odd number of edge pairs in $X_d(T_m)$ ($\tilde{X}_d(T_m)$) yields a graph that is isomorphic to $\tilde{X}_d(T_m)$ ($X_d(T_m)$).

Proof The proof follows from Lemma 13.14. First, take any single edge-pair-switch in $\tilde{X}_d(T_m)$. By Lemma 13.14 there is an automorphism of the graph that flips the pairs $a_i(v)$, $b_i(v)$ and $a_1(v)$, $b_1(v)$. The result thus pushes the edge switch up one level in the tree. When the top is reached, $a_1(r)$ and $b_1(r)$ have been switched i.e., $\tilde{X}_d(T_m)$ has been changed to $X_d(T_m)$. If there is more than one pair of switched edges, then in this way they can be pushed to the root one by one. Each time $\tilde{X}_d(T_m)$ is changed to $X_d(T_m)$ or vice-versa. □

Lemma 13.17. *For $m = 1, 2, \ldots,$ $X_d(T_m) \sim_m \tilde{X}_d(T_m)$.*

Proof By induction on m. This is clear for $m = 0$. Assume that the lemma for m and consider the game $\mathcal{G}_{m+1}(X_d(T_{m+1}), \tilde{X}_d(T_{m+1}))$. Suppose that Samson's first move is to place a pebble on a vertex a in $X_d(v)$ for some $v \in T_m$. It does not matter whether a is in $X_d(T_{m+1})$ or $\tilde{X}_d(T_{m+1})$. Either $v = r$ is the root of T_{m+1} or it is in the left or right subtree of r. If v is in the left subtree, then Delilah should answer according to the isomorphism σ provided by Observation 13.16 between $X_d(T_{m+1})$ and $\tilde{X}_d(T_{m+1})$ with the edge pair $(a_3(r), a_1(w))$, $(b_3(r), b_1(w))$ switched, where w is the right child of r. Notice that w is now the root of a copy of $\tilde{X}_d(T_m)$. Any further moves in the right subtree should be answered according to Delilah's inductive winning strategy in the game $\mathcal{G}_m(X_d(T_m), \tilde{X}_d(T_m))$. Any further moves in the other part of the tree should be answered by the isomorphism σ. Thus, this strategy is always a win for Delilah. If the first move was in the right subtree, then Delilah's answer is similar. If v is the root of T_{m+1}, then Delilah may arbitrarily place the imaginary edge switch in the right subtree and answer according to the isomorphism σ. □

Exercise 13.18 Show that Samson wins the game $\mathcal{G}^3_{m+2}(X_d(T_m), \tilde{X}_d(T_m))$.

[Hint: Suppose that Samson place his first two pebbles on $a_1(v_1)$ and $a_2(v_2)$, where v_1 and v_2 are the children of the root of T_m. Delilah cannot answer with $a_1(v)$ and $a_2(v)$ on the other side or she will lose in two more moves. Thus, in two moves and three pebbles, Samson can push the difference between the two graphs one level down the tree.] □

We continue with the proof of Theorem 13.11. The last step of the proof is the introduction of graphs D_k. $D_{\log m}$ will replace the binary tree T_m in the construction. $D_{\log m}$ has about $m^{\log m}$ vertices but it has essentially $\log m$ degrees of freedom — enough to let Delilah win the m-move game.

The graph $D_k = (V_k, E_k)$ is defined as follows. See Figure 13.19 for the graph D_2. The vertices V_k consist of $k + 1$-tuples, $\langle x_1, \ldots, x_k, r \rangle$ where $x_1, \ldots, x_k$ can be thought of as coordinates and r is the row number. As we move from each row to the one below, one of the coordinates is expanded by one. Call a *block* of D_k, k consecutive rows. So from the top of one block to the top of the next, each coordinate is expanded by one.

13.2 A Lower Bound for REACH$_a$

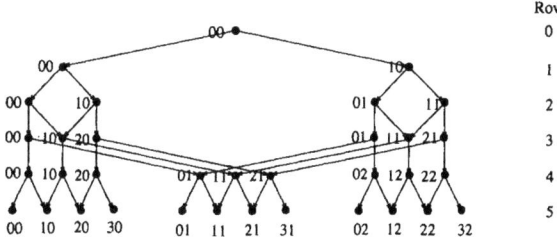

Figure 13.19: The graph D_2.

$$V_k = \{\langle x_1,\ldots,x_k,r\rangle \mid r = ak + j, a < 2^k, 0 \le x_i \le a+1, x_i \le a \text{ for } i > j\}$$

$$E_k = \{(\langle \bar{x}, r-1\rangle, \langle \bar{x}, r\rangle) \mid \text{for all } r < k2^k\} \cup$$
$$\{(\langle x_1,\ldots,x_k,r-1\rangle, \langle x_1,\ldots x_i, x_i+1, x_{i+1}\ldots x_k, r\rangle) \mid i \equiv r \pmod{k}\}$$

Let $G_m = X_d(D_{\log m})$ and $H_m = \tilde{X}_d(D_{\log m})$. See Figure 13.20 for a drawing of part of $X_d(D_2)$. Notice that in the directed tree, all internal nodes have in-degree one, but in D_k, some vertices have in-degree two. For such vertices v, there are several incoming edges to $a_1(v)$ and $b_1(v)$.

We show that the three conditions of Equation 13.12 hold. Conditions 1. and 3. are immediate. We must show that Delilah wins the m-move game on $X_d(D_{\log m})$ and $\tilde{X}_d(D_{\log m})$.

Think of a round of the game as labeling a vertex v in $D_{\log m}$: it is labeled "0" if Delilah answers with the same point as Samson, e.g., they both choose $a_1(v)$. It is labeled "1" if Delilah answers with the opposite point. The *labeling rule* is that if the two children of v are labeled, then v must be labeled with the "exclusive or" of its children's labels. Delilah wins the game as long as she never breaks the labeling rule and never labels a bottom vertex "1" or the root "0".

The crucial property of D_k is stated in the following claim. It says that it does Samson no good in the 2^k-move game, to choose a vertex more than k levels below where he has forced a vertex to be labeled "1". Thus, in m moves, Samson cannot force a vertex at the bottom level to be labeled "1".

Claim 13.22. *Suppose row r of D_k is entirely labeled and let any $2^k - 1$ vertices on or below row $r + k$ be chosen. If the chosen vertices are all labeled "0", then there is still a labeling of the rest of D_k that is consistent with row r.*

Proof By induction on k. For $k = 1$, let v be the chosen vertex on or below row $r + 1$. Suppose v is on row $r + 1$ of D_1. Let ℓ be a labeling of row $r + 1$ that generates the required labeling of row r. Observe that $\bar{\ell}$ — the complement of ℓ — generates the same labeling of row r. Clearly one of ℓ and $\bar{\ell}$ labels v "0" as desired. If v is below row $r + 1$, then take an arbitrary labeling of row $r + 1$ and proceed down to the row above v and then use the same argument.

Inductively, assume the claim is true for all $k' < k$. Let row r be fixed and suppose that $2^k - 1$ vertices of D_k have been chosen on or below row $r + k$. Let i be the coordinate that is expanded as we pass from row r to row $r + 1$. That is,

212   13. Lower Bounds

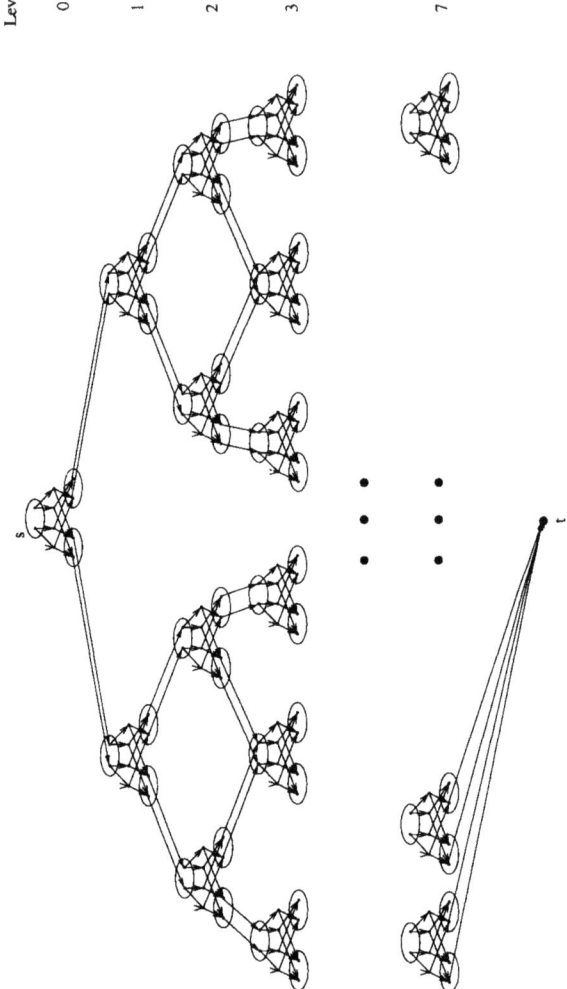

Figure 13.20: The graph $X_d(D_2)$.

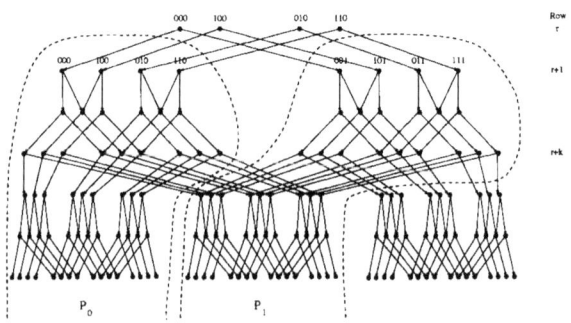

Figure 13.21: Proof of Claim 13.22, $k = 3, r = 2, i = 3$.

$i \equiv r + 1 \pmod{k}$. Let t be the maximum coordinate i occurring in row $r + k$. Let $P_0, P_1, \ldots, P_t$ be the subsets of D_{k+1} below row r, projected onto values $0, 1, \ldots, t$ of coordinate i,

$$P_j = \{\langle x_1, \ldots, x_{k+1}, s\rangle \mid s > r, x_i = j\}.$$

(See Figure 13.21.)

Observe that each of the P_j's is a copy of D_{k-1} except that each row $k - 1$ is repeated. There can be at most one of the P_j's — call it P_{j_0} — that contain at least 2^{k-1} chosen vertices. Assume that all the vertices in P_{j_0} have been labeled "0". By induction, we can label row $r + 1$ of the rest of the P_j's as we please. Thus, we can label row r of D_k as we please. □

This completes the proof of Theorem 13.11. Observe that Delilah's winning strategy in the game $\mathcal{G}_m(X_d(D_{\log m}), \tilde{X}_d(D_{\log m}))$ is in fact a winning strategy in the bijection game, $\mathcal{G}^k_{B,m}(X_d(D_{\log m}), \tilde{X}_d(D_{\log m}))$ (Definition 12.22). At each move, for each vertex $v \in D_k$, Delilah decides whether she would label this vertex "0" or "1". In the former case, she maps every vertex in $X(v) \subset X_d(X_{\log m}))$ to the same vertex in $\tilde{X}_d(D_{\log m})$. In the latter case, she maps the vertices according to one of the automorphisms that switch $a_1(v)$ and $a_2(v)$ as given in Lemma 13.14.

Thus, we have proved,

Corollary 13.23. *Boolean query* REACH$_a$ *is not expressible in quantifier rank* $2^{\sqrt{\log n} - 1}$ *even in language* FO(COUNT).

For large n, the function $2^{\sqrt{\log n}}$ dominates $(\log n)^k$, for any value of k. Recall that class NC is equal to FO[$(\log n)^{O(1)}$] (Corollary 5.26). Thus, if Theorem 13.11 went through with ordering, we would have proved that NC is strictly contained in P. This would mean that polynomial-time complete problems are *inherently sequential* in that they cannot be computed in parallel time $(\log n)^{O(1)}$ using polynomially many processors.

Of course, problem REACH$_a$ is expressible in FO(wo$\leq$)(LFP). In the next section, we present a different use of switch X (Figure 13.13). We prove that the language FO(wo$\leq$)(LFP, COUNT) is strictly contained in order-independent P.

13.3 Lower Bound for Fixed Point and Counting

That the language with fixed point and counting, FO(wo$\leq$)(LFP, COUNT), we now prove falls far short of capturing order-independent polynomial-time.

The argument is similar to the lower bound on REACH$_a$ (Theorem 13.11). Let switch X be the graph shown in Figure 13.24. This is the undirected version of switch X_d of Figure 13.13. Notice that the a vertices are circled in Figure 13.24 to distinguish them in the drawing from their companion b vertices. Note that the four central vertices have the property that each of them is connected to an even number of circled vertices. Just as in Lemma 13.14, we have that,

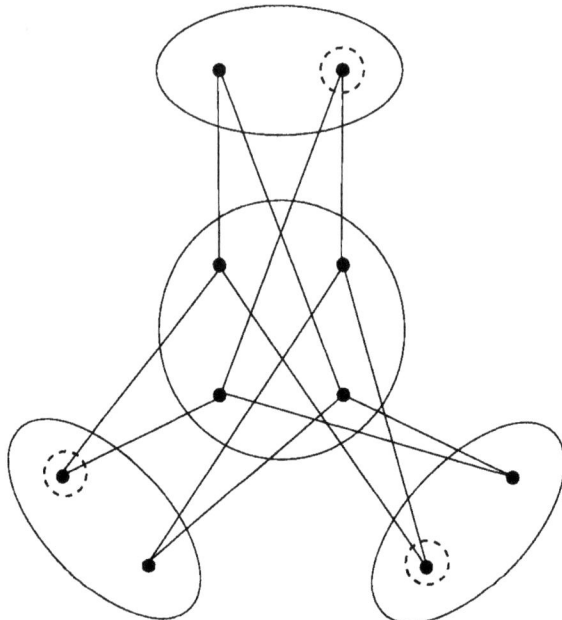

Figure 13.24: Switch X drawn with the a-vertices circled.

Lemma 13.25. *Let X be the graph pictured in Figure 13.24. Then X has automorphisms that switch any two of the pairs of a and b vertices, leaving the other pair fixed.*

Using switch X, we construct a sequence of pairs of graphs that are computationally simple to distinguish but require a linear number of variables to distinguish, even in the presence of counting:

Theorem 13.26. *There exists a sequence of pairs of graphs $\{A_n, \tilde{A}_n\}$, $n \in \mathbf{N}$, admitting a linear time canonical labeling algorithm and having the following additional properties:*

1. A_n and $\tilde{A}_n$ have $O(n)$ vertices.
2. A_n and $\tilde{A}_n$ have degree three and color class size four.
3. $A_n \equiv_{C^n} \tilde{A}_n$.
4. A_n is not isomorphic to $\tilde{A}_n$.

Before we prove Theorem 13.26 note a few of its consequences:

Corollary 13.27. *FO(wo$\leq$)(LFP, COUNT) is strictly included in order-independent polynomial-time.*

Proof By Theorem 13.26, the problem of distinguishing A_n from $\tilde{A}_n$ is in order-independent P. Suppose that there were a sentence $\varphi \in$ FO(wo$\leq$)(LFP, COUNT) that distinguished A_n from $\tilde{A}_n$. Let k be the number of variables in φ.

Thus there is a sentence $\varphi' \in C^k$ such that $\langle A_n, \tilde{A}_n \rangle \models \varphi$ and $\langle A_n, A_n \rangle \models \neg\varphi$. However, we know that $A_n \sim_{C^n} \tilde{A}_n$, sot $\langle A_n, A_n \rangle \sim_{C^n} \langle A_n, \tilde{A}_n \rangle$ because Delilah's winning strategy in the first game carries over to the second by using her strategy in the second components and the identity map in the first components. This is a contradiction when $n \geq k$. □

The following corollary of Theorem 13.26 is in sharp contrast to Proposition 12.9, which says that three variables suffice to identify graphs of color class size three.

Corollary 13.28. *A linear number of variables is required to identify graphs of color class size 4, even in the presence of counting. In symbols,* $\text{var}(CC_4) = \Omega(n)$ *and* $\text{vc}(CC_4) = \Omega(n)$.

Proof of Theorem 13.26: Let G be an undirected graph that is regular of degree three. Define $X(G)$ to be the graph in which each vertex of G is replaced by $X(v)$, a copy of the switch X. For each edge (u, v) of G, a pair of vertices denoted by $a(u, v), b(u, v)$ is selected from $X(u)$, and similarly, the pair of vertices $a(v, u), b(v, u)$ is selected from $X(v)$. Edges $(a(u, v), a(v, u))$ and $(b(u, v), b(v, u))$ are drawn. See Figures 13.29 and 13.30 for a sample degree-three graph H and the corresponding $X(H)$.

If G has an ordering on its vertices, then $X(G)$ inherits a partial ordering. Call the four central vertices of $X(v)$, $c_i(v)$, $1 \leq i \leq 4$. Then vertices $a(u, v), b(u, v)$, and $c_i(v)$ are partially ordered according to the lexicographic ordering of $\langle u, v \rangle$, and $\langle v, v \rangle$. Observe that if G is ordered, then $X(G)$ has color class size 4 in the sense that the partial ordering distinguishes all vertices except the pairs $a(u, v), b(u, v)$ and the quadruples $c_1(v), \ldots, c_4(v)$.

From now on in this section, G will be a regular, degree-three graph with an ordering on its vertices. Define $\tilde{X}(G)$ to be the graph $X(G)$ except that the edges are flipped between $X(v_1)$ and $X(v_2)$, for (v_1, v_2) the lexicographically first

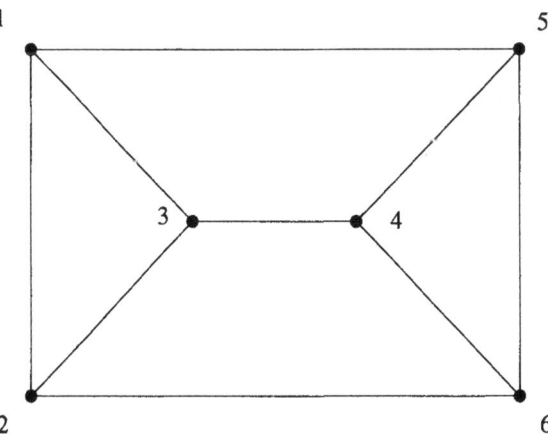

Figure 13.29: H is a regular, degree-three graph.

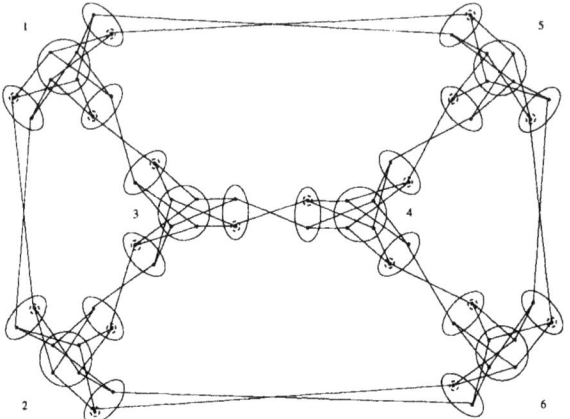

Figure 13.30: The graph $X(H)$.

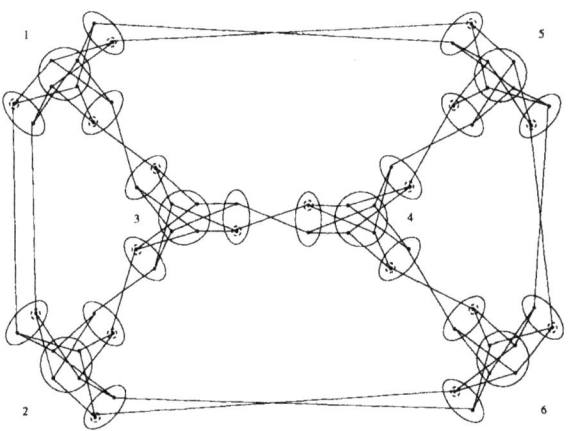

Figure 13.31: The graph $\tilde{X}(H)$.

edge in G. By "flipped" we mean that instead of the edges $(a(v_1, v_2), a(v_2, v_1))$, $(b(v_1, v_2), b(v_2, v_1))$, $\tilde{X}(G)$ has the edges, $(a(v_1, v_2), b(v_2, v_1)), (b(v_1, v_2), a(v_2, v_1))$.

Compare the drawing of $\tilde{X}(H)$ in Figure 13.31 with the drawing of $X(H)$ in Figure 13.30.

The following observation is similar to Observation 13.16,

Observation 13.32. *Let G be any regular, degree-three, connected graph. Let $\hat{X}(G)$ be like $X(G)$ except that exactly t pairs of edges are flipped. Then $\hat{X}(G)$ is isomorphic to $X(G)$ iff t is even and $\hat{X}(G)$ is isomorphic to $\tilde{X}(G)$ iff t is odd.*

The following are amusing and useful exercises.

Exercise 13.33 Prove Observation 13.32. The main subtlety is in proving that $X(G)$ is not isomorphic to $\tilde{X}(G)$. □

Let STRAIGHT be the set of graphs Z such that Z is equal to $X(G)$ for some ordered, regular, degree-three graph G. Similarly, let FLIP be the set of such graphs equal to $\tilde{X}(G)$ for such a G. In the next exercise you are asked to show that mutually exclusive boolean queries STRAIGHT and FLIP are each computable in linear time. Let $\oplus$ be the counting-mod-2 quantifier:

$$(\oplus x)\varphi(x) \quad \equiv \quad \text{"there are an odd number of } x\text{'s satisfying } \varphi\text{."}$$

In fact, STRAIGHT and FLIP are expressible — over ordered graphs — in FO($\oplus$). The class FO($\oplus$) is equivalent to circuit class AC^0 extended by counting-mod-two gates in addition to the usual "and", "or" and "not" gates. This very small complexity class is strictly contained in ThC^0 = FO(M) (Fact 13.37).

Exercise 13.34 Prove that STRAIGHT and FLIP are computable in linear time on a RAM. Prove also that they are expressible in the language FO($\oplus$).
[Hint: if Z is equal to $X(G)$ or $\tilde{X}(G)$ for an ordered graph G, then its ordering relation groups together each pair $a(u, v), b(u, v)$. Using the ordering of Z, we can label the first element in this pair a and the second b. This is the key point: the ordering gives us a global labeling of all these pairs. Now, each pair of edges between a, b-pairs is *straight* if the edge is from a to a and b to b and *flipped* otherwise. Similarly, each $X(v)$ is straight if the c_i's are each connected to an even number of a's and flipped if they are each connected to an odd number of a's. (If neither, then Z is not in FLIP or STRAIGHT.) The algorithm has only to count the number of flips mod two.] □

Let $G = (V, E)$ be a connected graph. Define a *separator* of G to be a subset $S \subset V$ such that the induced subgraph on $V - S$ has no connected component with more than $|V|/2$ vertices. A probabilistic construction shows that there exist regular, degree-three graphs whose separators all have size $\Omega(n)$ [Ajt87].

Let $T_1, T_2, \ldots$ be a sequence of regular, degree-three graphs such that T_n has $O(n)$ vertices and its smallest separator has size at least $n + 1$. The fact that T_n has only large separators means that it is "very connected". For this reason, the flip in $\tilde{X}(T_n)$ can be almost anywhere. Even after we have pinned down a set of n vertices, S, the largest connected component of $T_n - S$ still includes over half the vertices of T_n. The flip can hide inside this largest connected component and never be pinned down by Samson if he has only n pebbles. The following is the key idea in the proof of Theorem 13.26.

Lemma 13.35. *As above, let T_k be a regular, degree-three graph whose smallest separator has size at least $k + 1$. Then,*

$$X(T_k) \sim^k_C \tilde{X}(T_k).$$

Proof We show that Delilah wins bijection game $\mathcal{G}^k_B(X(T_k)\tilde{X}(T_k))$ (Definition 12.22).

We know by Observation 13.32 that if we flip any edge pair in $X(T_k)$, then the resulting graph is isomorphic to $\tilde{X}(T_k)$. After move r, let Q_r be the largest connected component in $T_k - P_r$, where P_r is the set of all vertices $v \in T_k$ such that just after move r, some pebble is placed on a vertex in $X(v)$. Since T_k has no separator of size less than $k + 1$, Q_r contains over half of the vertices of T_k.

Delilah's winning strategy is to maintain the following invariant,

> For each vertex $v \in Q_r$, let $X^v(T_k)$ be $X(T_k)$
> with an edge pair adjacent to $X(v)$ flipped.
> There is an isomorphism $\eta_{r,v}$ from $X^v(T_k)$ to $\tilde{X}(T_k)$, such that
> for all $x_i \in \text{dom}(\alpha_r)$, $\eta_{r,v}(\alpha_r(x_1)) = \beta_r(x_i)$. (13.36)

Clearly Invariant 13.36 holds before the first move. Suppose that it holds just after move r, and in move $r + 1$, let Samson pick up pebble pair i. Delilah responds with the bijection ρ_{r+1} defined as follows:

For each $v \in T_k$, let S_v consist of v together with all vertices w from T_k such that a pebble — not including x_i — is currently on a vertex in $X(w)$. Let C_v be the largest connected component of $T_k - S_v$. Since S_v consists of at most k vertices, C_v contains more than half of the vertices in T_k as does Q_r. Let z be a vertex in $Q_r \cap C_v$. Define ρ_{r+1} to be equal to $\eta_{r,z}$ on $X(v)$.

At the end of move $r + 1$, Samson places the x_i pebbles on a vertex g in some $X(v)$, and $\rho_{r+1}(g) = \eta_{r,z}(g)$. Delilah has not lost because $\eta_{r,z}$ is an isomorphism that maps the currently pebbled points to pebbled points.

After move $r + 1$, the new Q_{r+1} is equal to S_v. Define $\eta_{r+1,z} = \eta_{r,z}$. For all other vertices $w \in Q_{r+1}$, consider a path from z to w that stays within Q_{r+1}. Define $\eta_{r+1,w}$ to be the modification of $\eta_{r,z}$ that arises by pushing the flip from z to w. This affects no points outside Q_{r+1}.

Thus, Invariant 13.36 holds after move $r + 1$. □

Finally, choosing $A_n = X(T_n)$, and $\tilde{A}_n = \tilde{X}(T_n)$ we have proved Theorem 13.26. □

Historical Notes and Suggestions for Further Reading

Theorem 13.1 was originally proved by Ajtai [Ajt83] and independently by Furst, Saxe, and Sipser [FSS84]. The paper [Ajt83] is very rich. It also proves, among other things, that there is a strict arity hierarchy in SO∃. Sipser also proved that there is a strict depth hierarchy for AC^0, [Sip83]. That is, there are first-order boolean queries of alternation depth $k + 1$ that are not in non-uniform, alternation depth k first-order.

Yao improved the bounds of Theorem 13.1, showing that a bounded depth AC^0 type circuit must have exponential size to express parity [Yao85]. Håsted simplified Yao's proof and strengthened the bounds, thus proving Lemma 13.2 and Corollary 13.8.

Finally, Razborov used an algebraic argument to show that PARITY $\notin FO(\oplus_3)$, where $\oplus_3$ is the counting mod 3 quantifier [Raz87]. This was extended by Smolensky to show the following [Smo87]:

13.3 Lower Bound for Fixed Point and Counting

Fact 13.37. *For distinct primes p and q, $FO(\oplus_p)$ is not contained in $FO(\oplus_q)$, and these are thus both strictly contained in $FO(M) = ThC^0$.*

Theorem 13.26 was proved by Cai, Fürer and Immerman [CFI92]. It provides an $\Omega(n)$ lower bound on the number of variables needed to identify graphs on n vertices (Corollary 13.28). Before this theorem was proved, it was believed that a constant number of variables might suffice. This would have led to a polynomial-time algorithm for general graph isomorphism. There had been considerable work in this direction by Weisfeiler, Lehman, and others [Wei76].

Hella introduced the Bijection Game (Definition 12.22) and proved that it is equivalent to the counting game (Theorem 12.23). He used this result to show that the lower bound of Theorem 13.26 proves something much stronger: adding unary generalized quantifiers to FO(LFP, COUNT) never captures order-independent P [He96].

There are many other descriptive lower bounds without ordering that are worth noting. Grädel and McColm proved lower bounds on logics with transitive closure operators [Grä92, GM96, GM95]. Etessami and Immerman proved upper and lower bounds on transitive closure logics concerning tree isomorphism and canonization [EI95a]. This was part of Etessami's thesis on the power of local orderings [Ete95a, Ete95].

Grohe proved the following very strong theorem that there is a strict arity hierarchy for transitive closure, least fixed point, and partial fixed point logics [Gro96a]:

Fact 13.38. (**Arity Hierarchy Theorem**) *For every $k > 1$, there is a formula $\varphi_k(x_1, \ldots, x_k, x'_1, \ldots, x'_k)$ that is expressible in $FO(TC(arity\ k))$ but is not expressible in $FO(PFP(arity\ k-1))$, and thus not in $FO(LFP(arity\ k-1))$ or $FO(TC(arity\ k-1))$.*

14
Applications

The largest application of descriptive complexity is to the theory of databases. A relational database is a finite logical structure. This chapter begins with a discussion of the expressibility of relational queries. Other applications covered are Dynamic Complexity and Computer Aided Verification.

14.1 Databases

A *relational database* is a finite logical structure. Recall the genealogical database from Example 1.2. There, a genealogical database was any structure $\mathcal{B}$ of vocabulary $\langle F^1, P^2, S^2 \rangle$. A relational database is typically thought of as a set of tables. In this case we would have tables called FEMALE, PARENT, and SPOUSE. Examples of such a database are shown in Figure 14.1. Notice that each column is given a heading, which are often used in database query languages.

The definition of first-order query (Definition 1.26) that we have been using throughout this book is consistent with the standard notion of a first-order query to a relational database. Thus, for example, we can write the query "Daughter-in-law" as follows:

$$\text{D-in-Law}(x, y) \equiv (\exists z)(\text{PARENT}(x, z) \land \text{SPOUSE}(z, y))$$

The answer to this query is a new binary relation consisting of all pairs (x, y) such that y is the daughter-in-law of x. In database query languages, we may name elements of the domain in our queries: we have constants for every current element of the domain. For example, we could particularize the daughter-in-law

FEMALE	Woman
	Leah
	Rachel
	Rebekah
	Sarah
	...

PARENT	DadOrMom	Child
	Abraham	Isaac
	Isaac	Jacob
	Rebekah	Jacob
	Sarah	Isaac
	...	...

SPOUSE	Husband	Wife
	Abraham	Sarah
	Isaac	Rebekah
	Jacob	Leah
	Jacob	Rachel
	...	...

Figure 14.1: A genealogical relational database.

query by asking for the daughters-in-law of a particular person, e.g., the query D-in-Law(Isaac, y) would produce the set {Leah, Rachel}. Similarly, the boolean query D-in-Law(Isaac,Rachel) would produce the answer **true**.

14.1.1 SQL

Typical users of database systems are presented with a convenient query language so that they are not forced to write out formulas in first-order logic. Here we sketch one such popular language called SQL. An SQL query equivalent to Daughter-in-Law is the following:

SQL Query 14.2 (Daughter-in-Law)
    select   DadOrMom, Wife
    from   PARENT, SPOUSE
    where   Child = Husband

Observe that the select-from-where query captures existential quantification. If we want to define a relation, or set of relations, that we are going to reuse, we can define it in SQL as a *view*. This is done as follows:

SQL Query 14.3 (D-in-Law View)
    create view   D-in-Law **as**
        select   DadOrMom, Wife
        from   PARENT, SPOUSE
        where   Child = Husband

SQL contains boolean connectives and it also allows nested queries. For example, suppose that we wanted to express the following query:

$$\varphi(z) \equiv (\exists x y_1 y_2)(y_1 \neq y_2 \land \text{PARENT}(z, x) \land \text{D-in-Law}(x, y_1) \land \text{D-in-Law}(x, y_2))$$

returning those people who have children (x) having at least two daughters-in-law. This would be done as follows:

SQL Query 14.4 (Child-has-two-Daughters-in-law)
```
select   DadOrMom
from     PARENT
where    Child    in
                  (select  D1.DadOrMom
                   from    D-in-Law D1, D-in-Law D2
                   where   D1.DadOrMom = D2.DadOrMom
                           and not D1.Child = D2.Child)
```

We have seen that SQL contains existential quantification and boolean operations, and that these may be nested. Thus the part of SQL that we have described so far can express almost all first-order queries. The "almost" is because the structure of an SQL query requires that it computes some subset of a projection of the cross product of given relations. For example, SQL Query 14.2 can be computed by performing the following steps:

1. Take the cross product of relations PARENT and SPOUSE.
2. Select just those tuples for which the second component is equal to the third component.
3. Project the resulting relation onto its first and fourth components.

This query can be written algebraically as,

$$\pi_{1,,4}\sigma_{2=3}(\text{PARENT} \times \text{SPOUSE}) \,.$$

Similarly, Query 14.4 is a projection of a subset of relation PARENT.

SQL also has a "union" operation to build queries that have a component that is not a subset of the elements occurring in a particular column of one other relation. To take the union of two relations, they must be of the same type, i.e., the same arity and having the same labels. We may rename the labels for this purpose. For example, consider the following SQL query, which computes all people currently occurring in relation PARENT. Note that the labels "DadOrMom" and "Child" are both changed to "Name" so that the union may be taken.

SQL Query 14.5 (Human)
```
(select  DadOrMom Name
 from    PARENT
 where   true)
union
(select  Child Name
 from    PARENT
 where   true)
```

A query is called *safe* iff every element of every tuple occurs in at least one of the database relations. Database query languages restrict themselves to safe queries. One reason for this is that the values occurring in tuples of real database relations are usually from some basic type like **integer**, **real**, or **string**. We certainly should not write the query "**not** Human", which might be expected to return the set of all character strings that are not names in our current relation Human. In descriptive complexity, the concept of "safe query" is not an important restriction. In all plausible cases, the universe of our structure is equal to set of elements occurring in some tuple in some relation. Of course, this is immediate for an ordered structure.

The proof of the following theorem should now be clear from the above discussion.

Theorem 14.6. *All safe, first-order queries are expressible in SQL.*

Since real databases include numbers, SQL includes a limited amount of arithmetic, $+, -, \times, /$ and some aggregate operators, **sum, average, count, min** and **max**. For example, we can ask how many parents are in a database as follows. (The use of command **distinct** causes the selection to consider each DadOrMom only once. If omitted, this query would return the number of tuples in the Parent relation.)

SQL Query 14.7 (How Many Parents)
    select   count(**distinct** DadOrMom)
    from   Parent
    where   true

Similarly, the following **group by** instruction would tell us how many children each parent has.

SQL Query 14.8 (How Many Kids per Parent)
    select   DadOrMom, **count**(Child)
    from   Parent
    group by DadOrMom

The aggregate operators of SQL (and other commercial, relational database languages) take us beyond the first-order queries. As we have seen, these operators are all expressible in language FO(COUNT) (Theorem 5.27, Exercise 5.29, Proposition 12.16). As a consequence, we have the following,

Theorem 14.9. *The queries computable in SQL are exactly the safe queries in $Q(\text{FO}(\text{COUNT}))$.*

Exercise 14.10 Write SQL queries to the genealogical database expressing the following:

1. The "Cousin" relation.
2. The "Niece" relation.
3. The maximum number of children of any one DadOrMom.
4. The average number of children per married couple.
5. The average number of children per family that has at least one child.

□

14.1.2 Datalog

Some database computations need to go beyond the descriptive power of SQL. For this purpose, we may turn to Datalog, a stripped down version of the logic programming language Prolog.

In Datalog, we begin as usual with a vocabulary τ. A Datalog program consists of a set of propositional Horn clauses $C_1, \ldots, C_k$ from the language of τ together with some new relation variables, A_i^j. Each clause $C_i = \{\ell_1, \ldots, \ell_r\}$ is a set of literals. At most one of these literals is positive, and the positive literal must be an occurrence of one of the relation variables.

As an example, let τ_g be the vocabulary of graphs. Consider a Datalog program D_{tc} consisting of the following two clauses:

$$C_1 = \{A(x, y), \neg A(x, z), \neg A(z, y)\}; \qquad C_2 = \{A(x, y), \neg E(x, y)\} .$$

D_{tc} can be drawn more evocatively as follows,

Datalog Query 14.11 (D_{tc})

$$A(x, y) \leftarrow A(x, z), A(z, y)$$
$$A(x, y) \leftarrow E(x, y)$$

Datalog treats the program as a set of simultaneous inductive definitions of the relation variables occurring on the left. Any variable occurring on the right but not the left is existentially quantified. If a relation variable occurs as the left side of two or more rules, then its definition is the disjunction of these rules. [1]

Continuing our example, Datalog Query D_{tc} is equivalent to the following positive inductive definition:

$$A(x, y) \equiv (\exists z)(A(x, z) \wedge A(z, y)) \vee E(x, y) .$$

Thus, it describes the transitive closure of relation E.

Note that the semantics of Datalog is different from that of Prolog. If one submitted Datalog Program 14.11 to a Prolog interpreter, it might go into an infinite loop, trying to satisfy successive subgoals $A(x, z), A(x, z_1). A(x, z_2), \ldots$.

It is clear from the definition that Datalog corresponds to existential first-order logic plus the least-fixed-point operator. What is missing is the ability to express the negation of the input and numeric relations. Let Datalog* be the enhancement of Datalog to allow negative input and numeric literals on the right side of rules (but still not the negation of any of the new relation symbols). Thus the following is not hard to show:

Proposition 14.12.

$$\text{Datalog}^* = \text{FO}\exists(\text{LFP}) .$$

In the presence of an ordering relation, Datalog can simulate a universal quantifier. Then, by Theorem 4.10 and Corollary 4.11, we have the following:

Theorem 14.13. *Over ordered structures, the following equalities hold,*

$$\text{Datalog}^* = \text{FO}(\text{LFP}) = P .$$

---

[1] If a clause has no left side, then this is an auxiliary rule. If the whole right side of such a clause ever holds, then Datalog would return **false** as the answer to this query.

Exercise 14.14 Prove Theorem 14.13. [Hint: Show how to simulate a universal quantifier in FO∃(LFP), using the ordering. The ordering allows you to examine item 0, then item 1, and so on.] □

Exercise 14.15 Express the following as Datalog queries.

1. The set of descendants of Abraham.
2. The Ancestor-Chain-Length relation: this expresses the length of a path from ancestor to progeny. For example, A-C-L includes the following tuples: (Abraham, Abraham, 0), (Abraham, Isaac, 1), (Abraham, Jacob, 2), (Sarah, Joseph, 4), etc.

□

14.2 Dynamic Complexity

Traditional complexity classes are not completely appropriate for database systems. Two important differences between database complexity and traditional complexity are:

1. Databases are *dynamic*. The work to be done consists of a long sequence of small updates and queries to a large database. Each update and query should be performed very quickly in comparison to the size of the database.
2. Computations on databases are for the most part *disk access bound*. The cost of computing a request is usually tied closely to the number of disk pages that must be read or written to fulfill the request.

Many modern uses of computers have the above two features. In this section, we discuss dynamic complexity. Traditional complexity classes are *static*. This means that they have the following paradigm:

Read the entire input, $\mathcal{A}$, of size $n = \|\mathcal{A}\|$. Then test whether $\mathcal{A}$ satisfies a particular boolean query, Q. The static complexity of Q is the complexity needed to do this checking. All the complexity classes studied so far in this book: FO, NC, L, NL, P, NP, PSPACE, etc., are static.

The fundamental issue in static complexity is, "What is the fastest (or least complex) way, upon reading the entire input, to compute the query?" Static complexity is important, but it is not always the best point of view. In fact, the static point of view comes very much from the days in which we submitted our jobs to the mainframe as a deck of cards. When our turn came, the computer read our deck of cards, which included our program and data. Then it computed its response and charged us for the amount of computation time taken.

A more common approach nowadays is to be working on a large problem over a long period of time, with the current draft stored and being manipulated on line. Suppose for example that we are working on a large program. We would like to make changes in a few lines of the program and recompile it. We do not want the compiler to have to read and recompile the entire program each time we make a small change. The area of incremental compilation is well developed.

In fact, most applications are dynamic. Here is a list of examples:

- Databases
- Spread-sheets
- Texing a file
- Compiling a program
- Performing a calculation
- Processing a visual scene
- Understanding a natural language
- Verifying a circuit

The vision example is evocative: Suppose a speaker enters a room. In a second or so, she can see that she is in a lecture hall, with many chairs, many people sitting in the chairs, decorations on the walls, etc. If during her lecture a member of the audience raises his hand, she can recognize that event very rapidly. She does not have to process the entire scene from scratch in order to understand the change.

In the dynamic paradigm the object in question may undergo a long series of inserts, deletes, changes, and, queries. On each query, we must *very quickly* compute the query of the current database.

Below, we give a general definition of dynamic complexity class Dyn-$\mathcal{C}$, which corresponds to any static complexity class $\mathcal{C}$. The fundamental issue in dynamic complexity is no longer how to answer the query upon reading the entire database. Instead, we would like to know: What auxiliary information should be maintained in order to to compute the query of the current object very quickly?

A classic example of this sort of auxiliary structure is the index. If we know that we are going to be frequently asked to find customer records given their name, we will maintain a dynamic index of these names. This will enable us to find a record in time $\log n$ or less, rather than the linear time that would be needed if we had no such index.

14.2.1 Dynamic Complexity Classes

Let $\tau = \langle R_1^{a_1}, \ldots R_s^{a_s}, c_1, \ldots, c_t \rangle$ be a vocabulary, and let $\mathcal{Q} \subseteq \text{STRUC}[\tau]$ be a boolean query over τ. We can then think of Q as a dynamic problem in the following sense.

The *initial structure of size n*,

$$\text{init}(\tau, n) = \langle \{0, 1, \ldots, n-1\}, \emptyset, \ldots, \emptyset, 0, \ldots, 0 \rangle$$

contains a potential universe of size n, with all relations empty and all constants set to 0. Let the *basic requests* $\mathcal{R}_{n,\tau}$ for vocabulary τ and size n consist of the following:

- **ins**($\bar{a}, R_i$): insert tuple $\bar{a}$ into relation R_i, each $a_j < n$.
- **del**($\bar{a}, R_i$): delete tuple $\bar{a}$ from relation R_i, each $a_j < n$.

- **set**(c_j, b):   set constant symbol $c_j := b, b < n$.

For example, recall problem PARITY (Example 2.12), the boolean query over $\tau_s = \langle S^1 \rangle$ that is true of the boolean strings with an odd number of one's.

The basic requests for problem PARITY consist of single changes of bits from 0 to 1 (insert into S) or from 1 to 0 (delete from S). Clearly this problem is easy to compute dynamically: In addition to the string itself, we maintain an additional bit b, initially set to 0, which we use to remember the current parity of the string. Any request to insert a bit that is already on or delete a bit that is already off is ignored. A request that changes a bit causes us to change that bit and toggle b. The answer to query PARITY is just the current value of b, see Figure 14.17. Even before we make the formal definition of Dyn-$\mathcal{C}$, the following proposition should be intuitively obvious.

Proposition 14.16. *PARITY is an element of dynamic complexity classes* Dyn-FO *and* Dyn-RAMTIME[1].

We now define the dynamic complexity classes. Refer to Figure 14.19. Let $S \subseteq \text{STRUC}[\sigma]$ be any boolean query. The view of S as a dynamic problem is as follows: We begin with $\mathcal{A}_0 = \mathbf{init}(\tau, n)$, where n is an estimate of the potential size universe that we will need to handle.

Next, we are given a series of basic requests $r_1, r_2, \ldots, r_t \in \mathcal{R}_{n,\tau}$. Inductively, define $\mathcal{A}_{i+1}$ to be the result of applying request r_{i+1} to structure $\mathcal{A}_i$. Interspersed with the requests, are queries as to whether the current $\mathcal{A}_i$ satisfies query S.

Note that even though we are only asking one boolean query we may use constants to parameterize a whole set of queries. For example, boolean query REACH is parameterized by the constants s and t. If we want to know if there is a path from vertex 17 to vertex 102, then we can set s to 17, set t to 102, and ask if REACH holds.

In order to tell whether $\mathcal{A}_i$ satisfies S, we maintain a data structure $\mathcal{B}_i$. The key properties of our data structure sequence $\mathcal{B}_0, \mathcal{B}_1, \ldots, \mathcal{B}_t$ are the following: (1) We can compute $\mathcal{B}_0$ efficiently from $\mathcal{A}_0$; (2) the $\mathcal{B}_i$'s are at most polynomially larger

Structure	Request	Data Structure
0000000		0000000 0
	ins(3, S)	
0010000		0010000 1
	ins(7, S)	
0010001		0010001 0
	del(3, S)	
0000001		0000001 1

Figure 14.17: The dynamic PARITY problem.

than the $\mathcal{A}_i$'s; (3) We can compute $\mathcal{B}_{i+1}$ efficiently from $\mathcal{B}_i$ and r_{i+1}; and (4) Given $\mathcal{B}_i$, we can test efficiently whether $\mathcal{A}_i$ satisfies S. Here is the formal definition.

Definition 14.18 (Dynamic Complexity Classes) Let $\mathcal{C}$ be any (static) complexity class. Dynamic complexity class Dyn-$\mathcal{C}$ is the set of boolean queries $S \subseteq \text{STRUC}[\sigma]$ that satisfy the following conditions:

There exists another boolean query, $T \subseteq \text{STRUC}[\tau]$ in (static) $\mathcal{C}$ that serves as the data structure for the dynamic algorithm. There exist mappings $f, g \in Q(\mathcal{C})$ and a constant $k \in \mathbf{N}$ such we have the following. Let $n \in \mathbf{N}$ and let $r_1, r_2, \ldots, r_t$ be any sequence of requests from $\mathcal{R}_{n,\tau}$. Let $\mathcal{A}_0, \ldots, \mathcal{A}_t$ be defined as above. Then the maps f and g determine a sequence $\mathcal{B}_0, \ldots, \mathcal{B}_t \in \text{STRUC}[\tau]$ such that for all $i \leq t$,

1. $\mathcal{B}_0 = f(n)$
2. $\|\mathcal{B}_i\| \leq n^k$
3. $\mathcal{B}_i = g(\mathcal{B}_{i-1}, r_i)$, and
4. $\mathcal{A}_i \in S \quad \Leftrightarrow \quad \mathcal{B}_i \in T$

Thus, we can efficiently, i.e., in static $\mathcal{C}$ for each request and query, maintain the data structure $\mathcal{B}_i$, and test whether $\mathcal{A}_i \in S$ by testing whether $\mathcal{B}_i \in T$. □

As our first example, we prove Proposition 14.16:

Proof To show that PARITY is an element of Dyn-FO and Dyn-RAMTIME[1] we construct boolean query T as follows. T consists of those elements of $\text{STRUC}[\langle S^1, b \rangle]$ such that the boolean constant b is equal to "1". Clearly this can be checked in FO and in TIME[1] on a random access machine.

Function f on input n produces the structure,

$$f(n) = \langle \{0, \ldots, n-1\}, \emptyset, 0 \rangle$$

consisting of the all-zero binary string of length n, and the constant "0".

Similarly, the following formulas define the query g, which on input the structure $\mathcal{B}_{i-1} = \langle \{0, \ldots, n-1\}, S, b \rangle$, and a request r_i produces the new structure, $\mathcal{B}_i =$

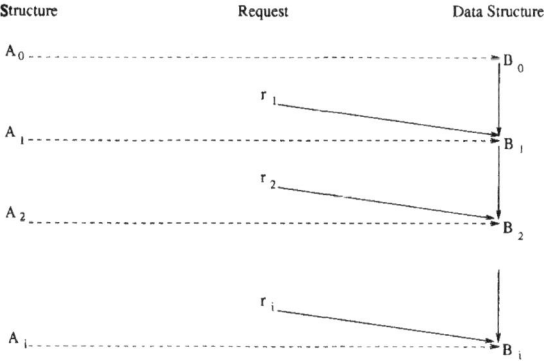

Figure 14.19: Definition of Dyn-$\mathcal{C}$.

$\langle\{0,\ldots,n-1\}, S', b'\rangle$. There are two cases, depending on whether r_i is an insert or a delete:

$$\mathbf{ins}(a, S): \quad S' := S(x) \vee x = a$$
$$b' := (b \wedge S(a)) \vee (\neg b \wedge \neg S(a))$$

$$\mathbf{del}(a, S): \quad S' := S(x) \wedge x \neq a$$
$$b' := (b \wedge \neg S(a)) \vee (\neg b \wedge S(a))$$

We maintain string S in our data structure so that we can ignore silly requests to set a bit that is already one or unset a bit that is already zero. If we can trust the user not to make such a request, then our data structure can consist of the single bit b.

Since bit b in $\mathcal{B}_i$ maintains the parity of the boolean string $\mathcal{A}_i$, we have,

$$\mathcal{A}_i \in \text{PARITY} \quad \Leftrightarrow \quad \mathcal{B}_i \in T .$$

Furthermore, functions f and g are first-order and constant time. Thus, PARITY $\in$ Dyn-FO and PARITY $\in$ Dyn-RAMTIME[1] as desired. □

Class Dyn-FO is the set of boolean queries that we can maintain in a first-order way if we plan ahead. For example, while PARITY is not in (static) FO, it is in Dyn-FO, because we can allocate an extra bit b to keep track of the parity of a certain string. Surprisingly, as we now see, some reachability problems are also in Dyn-FO.

Let REACH(acyclic) be the dynamic problem of testing whether the current directed, acyclic graph has a path from s to t. We assume that the graph remains acyclic during its entire history.

Theorem 14.20. *REACH(acyclic)* $\in$ *DynFO.*

Proof As our auxiliary data structure, we maintain the path relation P, the transitive closure of edge relation E. Boolean query T just checks whether $P(s, t)$ holds,

$$T = \left\{ \mathcal{B} \in \text{STRUC}[\langle E^2, P^2, s, t\rangle] \mid \mathcal{B} \models P(s, t) \right\} .$$

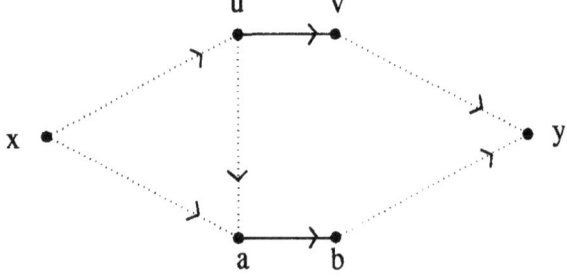

Figure 14.21: Maintaining P when deleting edge (a, b).

The first-order query to update P when an edge is inserted into E is the following:
ins(a, b, E) : $P'(x, y) := P(x, y) \vee (P(x, a) \wedge P(b, y))$.

The query to update P when an edge is deleted is more subtle and depends on the fact that the graph is acyclic. The interesting case is how to determine if there is still a path from x to y after deleting edge (a, b), assuming that there was a path before that used the edge, i.e., both $P(x, a)$ and $P(b, y)$ hold. Consider any path from x to y that does not use edge (a, b). Let u be the last vertex along this path from which a is reachable. Note that u is not equal to y because the graph was acyclic before the deletion of (a, b). Thus, edge (u, v) described in the following formula must exist. Furthermore, acyclicity insures that the path from x to u to v to y does not involve edge (a, b), see Figure 14.21.

$$\mathbf{del}(a, b, E): P'(x, y) \equiv P(x, y) \wedge \Big[\neg(P(x, a) \wedge P(b, y)) \vee$$
$$(\exists uv)\Big(P(x, u) \wedge E(u, v) \wedge P(v, y)$$
$$\wedge P(u, a) \wedge \neg P(v, a) \wedge (a \neq u \vee b \neq v)\Big)\Big] \quad \square$$

Next, we show that undirected reachability is also in Dyn-FO.

Theorem 14.22.   REACH$_u \in$ Dyn-FO .

Proof The idea is to dynamically maintain a spanning forest for the current undirected graph $G = (V, E)$. The data structure maintains the following three relations:

$$E(a, b) \equiv \text{``}\langle a, b \rangle \text{ is an edge''}$$
$$F(a, b) \equiv \text{``edge } \langle a, b \rangle \text{ is in the forest''}$$
$$V(a, b, m) \equiv \text{``there is a forest path from } a \text{ to } b \text{ via } m\text{''}$$

$V(a, b, m)$ means that $a \neq b$ and the unique path in the forest from a to b passes through m, either as an intermediate point or as an endpoint. The first-order property T in this case is,

$$T = \{\mathcal{A} \in \text{STRUC}[\langle E^2, F^2, V^2, s, t \rangle] \mid \mathcal{A} \vDash V(s, t, t)\} .$$

In the definitions of the queries corresponding to insertion and deletion of an edge, we use the following first-order abbreviations:

$$Eq(x, y, a, b) \equiv (x = a \wedge y = b) \vee (x = b \wedge y = a)$$
$$P(x, y) \equiv x = y \vee V(x, y, y) .$$

We handle the request to insert the edge $\langle a, b \rangle$ into E as follows: First, make $E'(a, b)$ and $E'(b, a)$ true. Next, if there had not been a path from a to b, then add the undirected edge (a, b) to forest, F, and update the path-via relation V, accordingly:
ins(a, b, E):

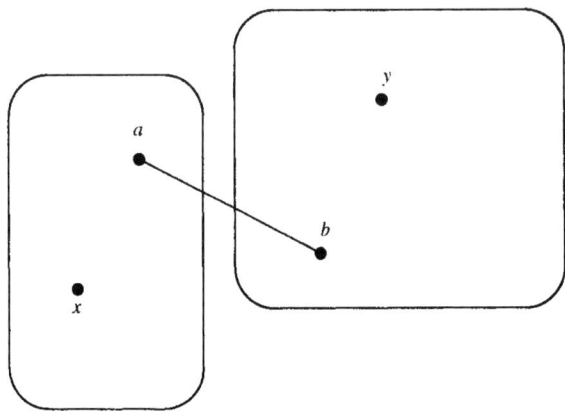

Figure 14.23: Dynamic REACH$_u$: insertion of edge (a, b).

if $\neg E(a, b)$
then $E'(x, y) := E(x, y) \lor Eq(x, y, a, b)$
  if $\neg P(a, b)$
  then $F'(x, y) := F(x, y) \lor Eq(x, y, a, b)$
    $V'(x, y, z) := V(x, y, z) \lor \big[\neg P(x, y)$
    $\land (\exists uv)(Eq(u, v, a, b) \land P(x, u) \land P(v, y)$
    $\land (z = a \lor z = b \lor V(x, u, z) \lor V(v, y, z)))\big]$

We handle the request to delete edge (a, b) from the graphs as follows, see Figure 14.24:

If edge (a, b) is not in the forest ($\neg F(a, b)$), then the updated relations are unchanged, except that $E'(a, b)$ and $E'(b, a)$ are set to false. Otherwise, first identify the vertices of the two trees in the forest created by the deletion, and then pick an edge, say e, out of all the edges (if any) that run between the two trees and *insert* e into the forest, updating relations V and F appropriately.

Define a temporary relation T to denote relation V after (a, b) is deleted, before the new edge, e, is inserted.

$$T(x, y, z) := V(x, y, z) \land \neg(V(x, y, a) \land V(x, y, b))$$

Using T, pick the new edge that must be added to the spanning forest. New(x, y) is true iff edge (x, y) is the minimum edge that connects the two disconnected components. (Here "minimum" could mean in terms of the ordering on the universe, or we could remember the order in which edges have been inserted.)

$$\text{New}(x, y) := E(x, y) \land T(a, x, a) \land T(b, y, b) \land$$
$$(\forall uv)\big[(E(u, v) \land T(a, u, a) \land T(b, v, b))$$
$$\to (x < u \lor (x = u \land y \leq v))\big]$$

E', F' and V' are then defined as follows:

$$E'(x, y) := E(x, y) \land \neg Eq(x, y, a, b)$$

Remove (a, b) from the forest and add the new edge.

$$F'(x, y) := (F(x, y) \land \neg Eq(x, y, a, b)) \lor \text{New}(x, y) \lor \text{New}(y, x).$$

The paths in the forest, from x to y via z, that did not pass through a and b, are valid. Also, new paths have to be added as a result of the insertion of a new edge in the forest.

$$V'(x, y, z) := T(x, y, z) \lor [(\exists u, v)(\text{New}(u, v) \lor \text{New}(v, u)) \land T(x, u, x) \\ \land T(y, v, y) \land (T(x, u, z) \lor T(y, v, z))] \qquad \square$$

In summary, the response to the request to delete edge (a, b) is the following:

if $F(a, b)$

then Remove (a, b) from F;

Form $T(a)$ and $T(b)$;

if $(\exists (c, d) \in E)(c \in T(a) \land d \in T(b))$

then insert least such edge into F;

compute V'

Exercise 14.25 The data structure in the proof of Theorem 14.22 contains relation V of arity three. Show that instead of V it suffices to maintain the binary relations F as a directed spanning forest, F^*, the transitive closure of F, and the unary relation $R(x)$ meaning that x is currently the root of its tree in F. [Hint: show that

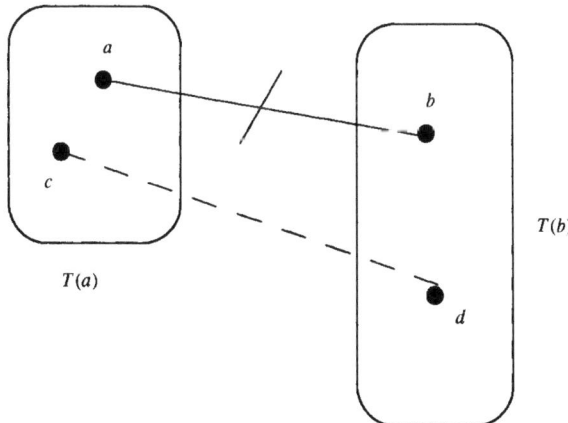

Figure 14.24: Dynamic REACH$_u$: deletion of edge (a, b).

there is a first-order query: Changeroot(r, r') that updates R,F, F* so that the tree of F rooted at r is now rooted at r' for some vertex r' for which $F^*(r, r')$ held. Then, to handle the insertion or deletion of edge (a, b), first move the root to b.] □

Exercise 14.26 Let R be any regular language. Show that $R \in$ Dyn-FO. [Hint: think of the input as a sequence of n slots that initially all contain the empty string, ϵ. Think of a complete binary tree of height $\log n$ sitting above the string. At each node r of the tree, dynamically maintain the function $\delta^*(\cdot, w_r)$ computing the state resulting from starting at any given state and then reading w_r. Here, w_r is the string sitting below the subtree rooted at r.] □

14.3 Model Checking

An exciting application of Descriptive Complexity is to a part of computer-aided verification called Model Checking. For a long time, the area of formal methods had a reputation of claiming more than it could deliver. The promise of proving large programs and systems correct was held out for decades as a technology that was right around the corner.

Now, model checking has come along as an extremely modest approach to formal methods. In fact, it is so modest that it is practical and useful. Many hardware design companies, including Intel, have adopted model checking as part of their basic design method, and the area is taking off at this writing.

The idea is simple. When we design a program, distributed protocol, or circuit, we do so in a formal language. This might be a programming language such as Java or a circuit design language such as VHDL.

Such a formal design determines a logical structure, $\mathcal{K} = \langle S, p_1, \ldots, p_k, R\rangle$, called a Kripke structure or transition system. Here S consists of the set of possible global states of the design, the set of all possible gate values for the circuit, or all possible assignments of values to the variables of a program, or all possible assignments of states to each processor in the distributed protocol. Binary relation R is the next move relation: $R(s, s')$ means that the move from s to s' is a possible atomic transition of the circuit or program. We assume that Kripke structures have the property that every state has some successor state. This can be achieved by putting a loop $R(s, s)$ on any state that has no other successors.

Finally, unary relations $p_1, \ldots, p_k$ express properties of the global states, for example being an initial state, being an accepting state, or that a particular variable has a special value.

The reader will notice that the size of S is often exponential in the size of the design. This phenomenon is typically referred to as the *state explosion problem*. For this reason, structure $\mathcal{K}$ is often represented symbolically. This practice is sometimes called *symbolic model checking*.

Once we have our formal design, a representation of the transition system can be automatically generated. We may now wish to write some simple correctness conditions in a formal language. For example, we might want to express the following:

1. If the Restart button is pressed, we eventually restart.
2. Doors are not opened between stations.
3. Division is performed correctly.
4. The motor is not turned on during maintenance.

These conditions express the fact that (1) if we press the Restart button on a personal computer it will eventually restart — without our having to unplug it and plug it in again; (2) the subway train controller meets some simple safety conditions; (3) Intel's new processor does its arithmetic correctly; and (4) the mixer in a nuclear waste storage container (which is designed to keep its contents mixed and thus less volatile) cannot turn on (and thus cause injury) during maintenance.

Now that we have formally expressed our design and our statement concerning its behavior, we can simply press a button and be told whether or not $\mathcal{K} \models \varphi$, that is, does the design satisfy the desired property?

This is the model checking problem: given $\mathcal{K}$ and φ, test whether $\mathcal{K}$ satisfies φ. Model checkers are currently being hailed as great debugging aides. This they are. We can write a simple property that our system should satisfy and press the button. If it does not satisfy the property then we will be told so. Typical model checking programs will present a counter-example run of the system, i.e., they produce the explicit bug. If the model does satisfy the property, then we will be told so. Note that this is better than just not finding a bug. Our model checker has automatically proved that our design has a certain desirable property.

14.3.1 Temporal Logic

Not surprisingly, there is a tradeoff between the descriptive power of the language that describes the properties to be checked, and the computational complexity of checking these properties. In a nutshell, this is how Descriptive Complexity relates to Model Checking.

A popular and quite expressive language for Model Checking is computation tree logic CTL*. CTL* is a version of temporal logic that combines linear and branching time. CTL* has two kinds of formulas: *state formulas*, which are true or false at each state, and *path formulas*, which are true or false with respect to an infinite path through $\mathcal{K}$. The following is an inductive definition of the state and path formulas of CTL*.

Definition 14.27 (Syntax of CTL*) State formulas $\mathcal{S}$ and path formula $\mathcal{P}$ of CTL* are the smallest sets of formulas satisfying the following:

State Formulas, $\mathcal{S}$:

1. The boolean constants **true** and **false** are elements of $\mathcal{S}$.
2. If p_i is a unary predicate, then $p_i \in \mathcal{S}$.
3. If $\varphi \in \mathcal{P}$, then $\mathrm{E}\varphi \in \mathcal{S}$

Intuitively, $E\varphi$ means that there exists an infinite path starting at the current state and satisfying φ.

Path Formulas, $\mathcal{P}$:

1. If $\alpha \in \mathcal{S}$ then $\alpha \in \mathcal{P}$.
2. If $\varphi, \psi \in \mathcal{P}$, then $\neg\varphi, \varphi \wedge \psi, \mathbf{X}\varphi$, and $\varphi \mathbf{U} \psi$ are in $\mathcal{P}$.

Intuitively, $\mathbf{X}\varphi$ means that φ holds at the next time and $\varphi \mathbf{U} \psi$ means that at some time now or in the future, ψ holds, and from now until then, φ holds. □

Several other operators can be defined in terms of the ones included in Definition 14.27. For each of these operators, we include its intuitive meaning.

$$\begin{array}{rcll} \varphi \vee \psi & \equiv & \neg(\neg\varphi \wedge \neg\psi) & \text{``}\varphi \text{ or } \psi \text{ holds.''} \\ \mathbf{A}\varphi & \equiv & \neg \mathbf{E} \neg \varphi & \text{``}\varphi \text{ holds on all paths.''} \\ \mathbf{F}\varphi & \equiv & \mathbf{true}\mathbf{U}\varphi & \text{``}\varphi \text{ holds at some time in the future''} \\ \mathbf{G}\varphi & \equiv & \neg \mathbf{F} \neg \varphi & \text{``}\varphi \text{ holds at all times in the future.''} \\ \varphi \mathbf{B} \psi & \equiv & \neg(\neg\varphi \mathbf{U} \neg\psi) & \text{``}\varphi \text{ holds before } \psi \text{ fails.''} \end{array}$$

Next, formally define the semantics of the above operators. A path ρ is a mapping of the natural numbers to states in $\mathcal{K}$ such that for all i, $\mathcal{K} \models R(\rho(i), \rho(i+1))$. We use the notation $\rho|i$ for the tail of ρ, with states $0, 1, \ldots, i-1$ removed.

Definition 14.28 (Semantics of CTL*) The following are inductive definitions of the meaning of CTL* formulas:

State Formulas:

$$(\mathcal{K}, s) \models p_i \text{ iff } \mathcal{K} \models p_i(s)$$
$$(\mathcal{K}, s) \models \mathbf{E}\varphi \text{ iff } (\exists \text{ path } \pi.\pi(0) = s)(\mathcal{K}, \pi) \models \varphi$$
$$(\mathcal{K}, s) \models \mathbf{A}\varphi \text{ iff } (\forall \text{ path } \pi.\pi(0) = s)(\mathcal{K}, \pi) \models \varphi$$

Path Formulas:

$$(\mathcal{K}, \rho) \models \alpha \text{ iff } (\mathcal{K}, \rho(0)) \models \alpha; \text{ for } \alpha \in \mathcal{S}$$
$$(\mathcal{K}, \rho) \models \varphi \wedge \psi \text{ iff } (\mathcal{K}, \rho) \models \varphi \text{ and } (\mathcal{K}, \rho) \models \psi$$
$$(\mathcal{K}, \rho) \models \neg\varphi \text{ iff } (\mathcal{K}, \rho) \not\models \varphi$$
$$(\mathcal{K}, \rho) \models \mathbf{X}\varphi \text{ iff } (\mathcal{K}, \rho|1) \models \varphi$$
$$(\mathcal{K}, \rho) \models \varphi \mathbf{U} \psi \text{ iff } (\exists i)(\mathcal{K}, \rho|i) \models \psi \wedge (\forall j < i)(\mathcal{K}, \rho|j) \models \varphi$$
$$(\mathcal{K}, \rho) \models \varphi \mathbf{B} \psi \text{ iff } (\forall i)\Big((\mathcal{K}, \rho|i) \models \neg\psi \Rightarrow (\exists j < i)(\mathcal{K}, \rho|j) \models \varphi\Big)$$

□

The following are examples of typical formulas and their meanings:

$$\mathbf{EF}p \equiv \text{``We can get to a state where } p \text{ holds''}$$
$$\mathbf{AG}p \equiv \text{``}p \text{ holds at all reachable states''}$$

EGp ≡ "Along some path, p always holds"

EGFp ≡ "Along some path, p holds infinitely often"

Frequently, we are interested only in runs of a system that are in some sense fair. A weak meaning of fairness is that if a processor continually requests a service, then it will eventually get it. A stronger fairness requirement is that if a processor requests a service infinitely often, then it will get it infinitely often. These are expressed in CTL* as follows.

AG(**G** request → **F** get) ≡ weak fairness

A(**GF** request → **GF** get) ≡ strong fairness

CTL is the sublanguage of CTL* in which every occurrence of a modal operator: **X, F, G, U, B** is immediately preceded by a path quantifier: **E** or **A**. CTL is simpler than CTL* but it is still quite expressive.

To characterize the complexity of model checking CTL*, we translate CTL* state formulas into first-order logics. Some CTL* formulas are equivalent to first-order formulas. We give a few examples below. Note that state formulas have no variables, but they are true or false in a given state of a Kripke structure. We think of them as formulas with one implicit free variable. We use "y" for this free variable. From this point of view, we can say that CTL formula p_1 is equivalent to the first-order formula $p_1(y)$ in the sense that,

$$(\mathcal{K}, s) \models p \quad \Leftrightarrow \quad (\mathcal{K}, y/s) \models p(y).$$

In this sense, the following CTL formulas are equivalent to the corresponding first-order formulas:

$$p \equiv \text{"}p \text{ holds now"}$$
$$\equiv p(y)$$

$$\mathbf{EX}p \equiv \text{"}p \text{ holds at some next step"}$$
$$\equiv (\exists y')(R(y, y') \wedge p(y'))$$

On the other hand, consider **EF**p, which means that there exists a state reachable from the present state at which p holds. This is not first-order, but it is naturally expressible using transitive closure,

$$\mathbf{EF}p \equiv \text{"We can get to a state where } p \text{ holds"}$$
$$\equiv (\exists y')(\mathrm{TC}(R(y, y'))(y, y') \wedge p(y')).$$

It is slightly more subtle to express the existence of an infinite path along which p always holds, **EG**p. An infinite path in a finite graph must include a cycle. We express this by saying that there exists a point y' such that (1) there is a path from y to y' along which p holds; and, (2) there is a nontrivial path from y' to itself along which p holds.

To do this it is convenient to have TC^s, the *strict transitive closure* operator,

$$TC^s(\varphi)(y, y') \equiv TC(\varphi)(y, y')) \wedge \Big(y \neq y' \vee$$
$$(\exists y')(\varphi(y, y') \wedge TC(\varphi)(y', y))\Big)$$

$$\mathbf{E}\mathbf{G}p \equiv \text{"Along some infinite path p always holds"}$$
$$\equiv (\exists y')(TC(R(y, y') \wedge p(y))(y, y')$$
$$\wedge \ TC^s(R(y, y') \wedge p(y))(y', y'))$$

As we see in the next theorem, all of CTL can be translated to $FO^2(TC)$, first-order logic with transitive closure using only two domain variables.

Theorem 14.29. *Every formula in CTL may be translated to an equivalent formula in* $FO^2(TC)$.

Proof Given $\varphi \in$ CTL, we describe an equivalent $f(\varphi) \in FO^2(TC)$ by induction on the structure of φ:

$$f(p) = p(y), \text{ for predicate symbol } p$$
$$f(\neg\varphi) = \neg f(\varphi)$$
$$f(\varphi \wedge \psi) = f(\varphi) \wedge f(\psi)$$
$$f(\mathbf{E}(\varphi \mathbf{U}\psi)) = (\exists y')(TC(M_{f(\varphi)})(y, y') \wedge f(\psi)(y'))$$
$$\text{where, } M_\alpha(y, y') \equiv R(y, y') \wedge \alpha(y)$$
$$f(\mathbf{E}(\varphi \mathbf{B}\psi)) = (\exists y')\Big[TC(M_{f(\psi)})(y, y') \wedge$$
$$\Big((f(\varphi)(y') \wedge f(\psi)(y')) \vee TC^s(M_{f(\psi)})(y', y')\Big)\Big]$$

One can prove by induction on the structure of φ that φ and $f(\varphi)$ are equivalent. The interesting cases are the last two. $\mathbf{E}(\varphi \mathbf{U}\psi)$ means that there is an infinite path along which ψ holds at some point, and at all points before this, φ holds. This is expressed by $f(\mathbf{E}(\varphi \mathbf{U}\psi))$.

$\mathbf{E}(\varphi \mathbf{B}\psi)$ means that there is an infinite path along which either ψ always holds, or φ holds at some time before ψ fails to hold. This is captured by the formula $f(\mathbf{E}(\varphi \mathbf{B}\psi))$. □

Theorem 14.29 can be extended to translate all of CTL* to $FO^2(TC)$. However, to do this we sometimes need to add a few boolean variables in addition to the two domain variables.

For example, consider the CTL* formula,

$$\alpha \equiv \mathbf{E}(p \rightarrow q\mathbf{U}r)\mathbf{U}t.$$

To express α in $FO^2(TC)$, we cannot just say that there exists a point y' at which t holds and on the path to y', $(p \rightarrow q\mathbf{U}r)$ holds. The problem is that it might be the case that before y', p and q hold, but r does not. In this case, we must remember the obligation to maintain the truth of $q\mathbf{U}r$ along the path past y'.

The solution is to add a boolean variable b that is true when we must make $q\mathbf{U}r$ true. We can then write the formula $\beta \in \mathrm{FO}^2(\mathrm{TC})$ that is equivalent to α as follows,

$$\beta \equiv (\exists y' b')(\mathrm{TC}(\gamma)(y, 0, y', b') \wedge t(y') \wedge \\ b' \to (\exists y)(\mathrm{TC}(M_q)(y', y) \wedge r(y)))$$

$$\gamma(y, b, y', b') \equiv ((p \vee b) \to (r \vee (q \wedge b'))) \wedge R(y, y')$$

$$M_q(y, y') \equiv R(y, y') \wedge q(y)$$

In this way, one can prove that all of CTL* can be translated to $\mathrm{FO}^2(\mathrm{TC})$.

Fact 14.30. *There is a linear-time computable function g that maps any CTL* formula φ to an equivalent formula $g(\varphi) \in \mathrm{FO}^2(\mathrm{TC})$. While $g(\varphi)$ has only two domain variables, it may have a linear number of boolean variables.*

Exercise 14.31 Translate the following CTL* formulas to $\mathrm{FO}^2(\mathrm{TC})$.

1. $\mathbf{E}(p\mathbf{B}q)$
2. $\mathbf{AG}(p \to \mathbf{X}q)$
3. $\mathbf{AG}(\mathbf{G}p \to \mathbf{F}q)$
4. $\mathbf{A}(\mathbf{GF}p \to \mathbf{GF}q)$

□

14.4 Summary

The three applications discussed in this chapter are closely related. In fact, Model Checking and database query evaluation are essentially the same thing. Given a query φ, our task is to efficiently check whether our huge database, or huge Kripke structure, satisfies φ.

One divergence between these two areas is that the typical database query is of low complexity, usually first-order, or first-order plus counting. Model checking queries, on the other hand, are closely tied to reachability and transitive closure queries.

The sizes of Kripke structures of interest vary greatly, from small ones that can be stored entirely in main memory, to huge ones that can be represented only symbolically. For this reason, it is important to tailor the query language to the size of the structure. In descriptive complexity, we can see the complexity of the query on its face. This offers the advantage that we can predict in advance how long it will take to check and we can use richer queries for small structures and simpler queries for large structures.

The connections between dynamic complexity and databases are obvious. Databases are the dynamic object par excellence: very large, long-lived, and thus worthy of effort for optimizing the dynamic structures that will speed up query evaluation.

One of the key dynamic features of databases are integrity constraints. An integrity constraint is a query that must remain true for all instances of the database. Typical integrity constraints might say that every employee has a unique supervisor, or that no flight is booked to more than 110 percent of its capacity. On each action, the database must check very quickly whether any of its integrity constraints will be violated. If so, it must either disallow the action, or couple it with an appropriate further action to maintain the constraint.

Correctness conditions in model checking are very similar to integrity constraints. They are a collection of simple facts that our system must satisfy. As the system is changed, expanded, and modified, we need to know that these constraints remain valid. How to modify the system while maintaining the constraints is a key challenge to a modern design process.

On a simpler level, from the point of view of efficiency, if our design does not meet some correctness condition, and we modify the design, we would like to be able to check quickly that the slightly modified design now satisfies the condition.

Historical Notes and Suggestions for Further Reading

We could only touch a tiny bit of the vast literature on database theory. We recommend the book [AHV95] by Abiteboul, Hull, and Vianu as a place to start digging deeper.

Theorem 14.13 is due to Papadimitriou [Pap85]. Datalog has the same expressive power as that part of Prolog in which there are no new functions symbols. For this reason, Papadimitriou titled his theorem "P = P."

A good place to start reading about the memory hierarchy, i.e., how complexity depends upon getting our data from across the net and then from disk to main memory to cache, is [ACF94] by Alpern, Carter, Feig, and Selker.

The field of dynamic algorithms is vast and growing. Some of the material from this section, including Theorem 14.22 and Exercise 14.26, was drawn from the paper [PI94] by Patnaik and Immerman. Exercise 14.25 is from [DS95], and Theorem 14.20 is from [DS93], both by Dong and Su. Other useful references on dynamic complexity are [MSV94, R96].

The example of a motor coming on during maintenance at a nuclear waste storage facility, discussed in Section 14.3, was a real example of a bug found by model checking [TPP97].

Much of the discussion of model checking, including Theorem 14.29 and Fact 14.30, are drawn from the paper [IV97] by Immerman and Vardi. Model checking was originally proposed by Clarke and Emerson [CE81]. Two rich sources for more information on model checking are Kurshan's book [Kur94] and McMillan's book [McM93].

15
Conclusions and Future Directions

In these last few pages we give our personal view of what descriptive complexity has achieved to date. We also underline some of the current lines of research that seem especially promising. Finally, we outline some possible applications and future directions that although challenging we consider especially worthwhile.

15.1 Languages That Capture Complexity Classes

There has been tremendous progress in understanding computational complexity from a descriptive point of view. Broadly stated, we can understand the *computational complexity* of a query via its *descriptive complexity*: how rich a language do we need to express the query.

The next few theorems include a few of the highlights of the relationship between computational and descriptive complexity that we have seen:

Theorem 15.1. [Theorem 5.2] $FO[t(n)] = CRAM[t(n)]$.

This means that the quantifier depth — or equivalently, the inductive depth — of an optimal description of a query corresponds exactly to the parallel time needed to compute the query, using polynomially much hardware. Parallel time — one of the most fundamental computational resources — is quantifier depth.

Theorem 15.2. [Theorem 10.16] $DSPACE[n^k] = VAR[k+1]$.

The amount of hardware needed to compute a query is closely tied to the number of variables needed to describe the query. When the amount of hardware is measured purely as space, this result is tight.

The fundamental mysteries of complexity theory are the tradeoffs between parallel time and hardware. A consequence of Theorems 15.1 and 15.2 is that these mysteries can all be understood as the descriptive tradeoff between quantifier depth and number of variables. A corollary is that further work on the tradeoff between quantifier depth and number of variables is extremely worthwhile.

We next mention a few of the ways that we have captured complexity classes as logical languages. These results reflect various insights into the fundamental nature of computation.

Theorem 15.3. [Theorems 9.11 and 9.18]

$$L = FO(DTC) = \text{"Primitive Recursion on Structures"}.$$

Deterministic Logspace is the smallest truly robust complexity class. It is characterized by what we can compute by "following our nose". This can be understood as following a path in which at most one edge leaves any vertex. Equivalently, we can compute a function built by recursion on the input structure itself.

Theorem 15.4. [Corollary 9.22 and Theorem 9.32]

$$NL = FO(TC) = \text{SO-Krom}.$$

Nondeterministic Logspace is characterized by taking the transitive closure of a relation. Tied to this descriptive characterization is the realization that nondeterministic space is closed under complementation. As we have argued, FO(TC) is a very feasible and basic class where much model checking and other formal methods can be carried out. Viewed as a restriction of NP, NL corresponds to SO∃ formulas of the Krom type, meaning that there are at most two literals per clause. These are typified by the 2-SAT problem.

Theorem 15.5. [Theorems 4.10, 9.32, and Corollary 9.8]

$$P = FO(LFP) = FO(IFP) = FO[n^{O(1)}] = \text{SO-Horn}.$$

Polynomial time is a clean wrapper for the set of "truly feasible" queries — those that can be computed exactly in a "reasonable" amount of time for all "reasonably-sized" inputs. P is identical to those queries that can be defined inductively. This is typically formalized via a least fixed-point or equivalently an inflationary fixed point operator. A natural rephrasing of this result is that an NP computation is in PTIME if its clauses are Horn clauses, thus expressing an inductive definition. What is pleasing about Theorem 15.5 is that the most fundamental complexity class is thus characterized by perhaps the most fundamental logical construction.

Theorem 15.6. [Theorem 7.8, Corollaries 7.27, 7.22]

$$NP = SO\exists$$
$$PH = SO = \text{CRAM-HARD}[1, 2^{n^{O(1)}}].$$

Fagin's striking theorem began the subject of descriptive complexity: a very important complexity class is exactly characterized in a simple, descriptive way. The generalization characterizing the polynomial-time hierarchy also shows what a strange complexity class it is: the class of queries computable using exponentially much hardware for only constant time.

Theorem 15.7. [Theorem 10.13, Corollary 10.29]

$$\begin{aligned} \text{PSPACE} &= \text{FO(PFP)} = \text{FO}[2^{n^{O(1)}}] \\ &= \text{SO(TC)} = \text{SO}[n^{O(1)}]. \end{aligned}$$

Finally, the extremely robust complexity class PSPACE corresponds to what we can compute with polynomially much hardware and no time bound whatsoever. The various descriptive characterizations underscore a kind of duality between parallel time and hardware, or equivalently between quantifier-depth and number of variables, [Ho86]. Recall that one second-order variable corresponds to polynomially-many first-order variables.

The above descriptive characterizations reveal that computational complexity is a deep mathematical concept, transcending any particular machine characterization of computation. Learning more about the descriptive resources will help us in our journey toward understanding efficient computation.

15.1.1 Complexity on the Face of a Query

Descriptive complexity allows us to predict accurately the time and space needed to compute a query by simply looking at its syntax, e.g., how many variables and what quantifier-depth it uses.

The potential applications of this feature of descriptive complexity are legion:

- programming
- database query evaluation
- web query evaluation
- model checking

In all the above settings we can decrease the computational task by decreasing the descriptive complexity of queries. Current database query optimization takes advantage of some of these insights. However, there is a great deal more that can and should be done. In particular, some of the known bounds on complexity are not tight. Furthermore, automatic methods to simplify a query need to be developed. As an example, when we write a web query, we should get an estimate of the time to compute it and the number of items likely to be returned. The query evaluation system should be able to suggest and help with ways to refine the query to make it more efficient and better aimed. Furthermore, the query should be computed dynamically rather than from scratch as we modify it to make it more accurate and efficient.

15.1.2 Stepwise Refinement

As a formal method for program development, stepwise refinement is frequently proposed, but rarely implemented in a robust way. The notion of stepwise refinement in descriptive complexity has a clear paradigm, as follows:

1. Express the desired query in a rich, expressive language. To us, that would be second-order logic. In a usable setting, of course, one needs a friendly and convenient front end that automatically compiles as second-order logic.
2. We need to next develop automatic tools that can help refine the query into an equivalent expression in a weaker and thus less complex language. Sometimes this could be done automatically. An important example of this kind of work ins in the area of constraint programming [FV98, KV98]. In this setting, some queries can automatically be translated to polynomial time or NSPACE[$\log n$] queries. In other cases, the queries can be recognized as NP-complete. In this case, as we discuss below, one needs to either choose a related but simpler query or find that the NP-complete optimization problem in question has a plausible approximation algorithm. In this case, the system should propose to the user that an efficient approximation algorithm be used instead. The system should be able to explain the known features of the approximation algorithms and ask the user whether or not to use it.

The main assumption for this paradigm to work is that many computational problems can be separated into a sequence of problems that are easy to specify in some rich language. In this case, a kind of formal development of efficient queries via such stepwise refinement could be valuable and plausible. Of course, there is a great deal of research and system development that is needed to carry this paradigm to fruition.

15.2 Why Is *Finite* Model Theory Appropriate?

One of the fundamental theses of descriptive complexity is that finite logical structures are useful for modeling concepts in computer science. We have found again and again that this assumption is confirmed.

The simple explanation for finite model theory's success as a tool for computer science is that programs, data, and machines are all finite. A typical program manipulates a fixed, finite data structure, using small words to hold the numbers and symbols it manipulates.

As a typical example, We feel that modeling the numbers a computer uses as general, unbounded natural numbers as opposed to k-bit natural numbers may add some superficial elegance. However, we feel that such inaccurate models *mislead* us rather than simplifying our tasks.

We have found that modeling computational problems finitely is usually simpler and often much more accurate than the attempt to smooth things out by imagining that the objects being manipulated are of unbounded or even infinite size.

The finite assumption sometimes requires us to be more careful and detailed. In my experience, this effort almost always pays off giving us much more accurate and ultimately often simpler models.

15.3 Deep Mathematical Problems: P versus NP

Some of the open questions concerning complexity are among the deepest and most significant open questions in mathematics today.

Consider the following boolean query:

$$\text{MATH} = \left\{ \varphi \#^r \mid \text{ZFC} \vdash \varphi, \text{proof length} \leq r \right\}.$$

MATH consists of logical formulas φ, together with a pad of length r, for which there is a proof of φ in ZFC using a total of r symbols. Here ZFC is the set of set theory axioms of Zermelo and Fraenkel plus the Axiom of Choice. ZFC is widely believed to be sufficiently powerful to formalize all current mathematical proofs [Coh66].

Problem MATH is easily seen to be NP-complete. It is in NP because we can nondeterministically guess a proof and then deterministically verify that it is a correct proof of φ. Furthermore, since every satisfiable boolean formula has a short proof in ZFC that it is satisfiable, we have that SAT $\leq_{\text{fop}}$ NP.

It is arguable that problem MATH is a perfect oracle for one of the most difficult tasks that a mathematician tries to do. Namely, she conjectures a theorem and tries to prove it. To do this, she considers some simple cases and weaker partial results and tries to prove them. She makes assumptions and tries to see if she can prove a result from these assumptions. Each lemma and small result has a short proof. Here formulas φ will typically be of the form $A_1 \wedge \cdots \wedge A_k \to \psi$, where $A_1, \ldots, A_k$ are known facts and assumptions.

We believe that MATH encorporates the hardest part of the creative process of doing mathematics. Recall that when Gödel proved his Incompleteness theorem, many people argued that this meant that mathematics could not be mechanized and thus mathematicians could not be replaced by machines.

On the contrary, real mathematicians do not write down arbitrary true statements — and only true statements. On the contrary, they make conjectures and develop short proofs and counterexamples from known, relevant results. The process is reducible to boolean query MATH. When someone finally proves that P $\neq$ NP, this will truly show that mathematicians — as well as other creative individuals — cannot uniformly and feasibly be replaced by machine.

Complexity theoretic lower bounds are deep and important. We believe that it is quite worthwhile to develop knowledge and techniques to understand computational complexity. Much ground work will have to be developed before we reach this goal. We are near the beginning of the process, and we believe that finite model theory can help.

15.4 Toward Proving Lower Bounds

The fundamental open questions in complexity theory all come down to understanding the tradeoffs between the three dimensions of complexity:

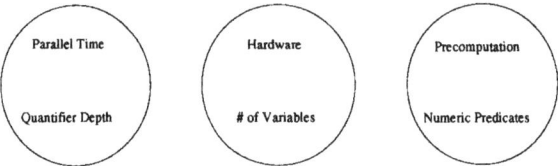

As suggested above, these questions are difficult and deep. We highlight two areas of descriptive complexity that have been lending new insights to this area and deserve further effort.

Ehrenfeucht-Fraïssé Games: As we have seen, many new techniques are being developed to ascertain the exact expressive power of languages using variations of the Ehrenfeucht-Fraïssé game. We are getting somewhat closer to real lower bounds on languages that can truly express computation. One idea among many that we would like to see pursued is the study of the complexity of winning strategies needed by Delilah to win some of these games. Using this tool on itself in this way may lead to valuable new insights.

Reductions: First-order projections are a refined tool for measuring the relationships between queries and thus complexity classes. Consider the following theorem:

Theorem 15.8. *Suppose that A_i is complete via fops for complexity class C_i, $i = 1, 2$, and that $C_1 \subseteq C_2$. Then,*

$$A_2 \leq_{\text{fop}} A_1 \quad \Leftrightarrow \quad C_1 = C_2$$
$$B \leq_{\text{fop}} A_1 \quad \Leftrightarrow \quad B \in C_1 .$$

Theorem 15.8 shows that tools to prove the existence or nonexistence of fops and other weak reductions may help us prove lower bounds and may help us develop algorithms. Programming via reductions — i.e., solving a problem by reducing it to a problem that we already have a good algorithm for — is a valuable way to program. Furthermore, working in both directions: upper bounds by showing that reductions exists and lower bounds by showing that they do not, will help to sharpen these tools.

15.4.1 Role of Ordering

Understanding the role of ordering is fundamental to understanding feasible computation. Indeed, the notion of computation on a finite graph cannot be divorced from the fact that we want to compute an order-independent property of an ordered graph. The search for a language for order-independent P has lead to some of the prettiest work in descriptive complexity. This is one of the many areas that we need to pursue before we have a well-rounded understanding of computation.

One important direction is to extend the language FO(LFP, COUNT) by an appropriate ability to calculate with permutation groups. This would permit the inclusion of the polynomial-time bounded degree and bounded color-class graph canonization algorithms, [Lu82, Ba81]. This would result in some improved graph algorithms as well as an improved understanding of order-independent polynomial time.

15.4.2 Approximation and Approximability

"An NP-completeness proof is typically the first act of the analysis of a computational problem by the methods of the theory of algorithms and complexity, not the last."[1]

Most creative endeavor can be understood as working on NP-complete problems. When we prove that a problem is NP complete, our conclusion should be that we cannot solve it exactly on all reasonably sized instances. The next step should be to look for approximation algorithms or show that these are not available and conclude that we must simplify the stated problem in order to solve it.

We believe that it is feasible to test automatically in many cases whether a query is NP-complete or in P. If it is NP-complete, then we should develop similar ways to test automatically whether it can be efficiently approximated.

There has already been some significant work, started by Papadimitriou and Yannakakis, on determining the approximability of an NP-optimization problem via its syntax [PY] and culminating with the so-called "PCP Theorem".

Fact 15.9. [ALMSS]   NP = PCP[$\log n$, 1].

Fact 15.9 says that every NP query can be accepted by a nondeterministic machine that guesses a proof of membership and hands the proof and the problem instance to a very weak probabilistic verifier. The verifier may flip $O(\log n)$ random coins and, based on this, look at only $O(1)$ bits of the proof in order to decide whether to accept an input or not. This means that "approximate proofs are good enough". A corollary is that a class of NP-optimization problems, including MAX-3SAT, cannot be approximated arbitrarily closely unless P = NP. The class, called SNP, was identified syntactically in [PY].

It is possible to understand Fact 15.9 as a generalization of Fagin's Theorem [MI96]. Much more can and should be discovered concerning the approximability of NP-complete problems via this syntactical approach.

---

[1] C. Papadimitriou, [Pap94], p. 299.

15.5 Applications of Descriptive Complexity

15.5.1 Dynamic Complexity

We argued in Chapter 14 that most modern applications of computers are *dynamic*: a large object is being developed over a long period of time. During this time our system should be helping us. It should be working to maintain information so that it can quickly answer likely questions, when they are asked It should maintain integrity constraints concerning our program, circuit, database, knowledge base, correspondence, lecture notes, etc., warning us when constraints are violated or perhaps even automatically correcting them.

Dynamic complexity is quite different from classical static complexity. It has different computational paradigms, different reductions, different sorts of characteristic problems. A solid theory of upper and lower bounds in dynamic complexity will help us understand this very important dimension of complexity. Many standard algorithmic paradigms of almost the same generality and importance as the index and the counter still need to be developed. This will make programs more useful and more flexible and it will improve our understanding of computation.

15.5.2 Model Checking

Model Checking is a burgeoning field because it has already succeeded in finding many bugs in circuit designs that escaped less formal debugging and testing methods.

In our opinion, the potential for model checking goes beyond debugging circuit designs. As a user designs a circuit, program, web page, lecture, book, course, etc., the system on which it is designed can have a formal model of what is being constructed. The user can ask if the design will have a certain simple property. Automatically, the user can be told, no, and here is a bug, or yes. If the answer is yes, then that can be something that this user or even other users can depend upon and build upon. Furthermore, the system could ask if the property should be checked for dynamically, i.e., the system could issue a warning if a later change in the design caused the desired property to be violated.

In fact, there is huge potential here for rational, computer-aided design and verification. To accomplish this, it is necessary to maintain a formal model and to keep the query language simple.

15.5.3 Abstract State Machines

Recent work by Gurevich and others has introduced a useful method of specifying the operational semantics of computer systems, languages, and protocols. The current state of the system is modeled as a finite logical structure, called an evolving algebra in Gurevich's original treatments, and now called an abstract state machine [Gur91, Gur93].

The advantage of abstract state machines is that the model is simple and accurate, as opposed to overly cumbersome methods such as denotational semantics.

Abstract state machines offer a promising approach to building useful models with which to reason about our systems. More research and development is needed in at least the following areas.

- An easier-to-use specification language is needed.
- Abstract state machines should have model checking built in.
- We need to develop easy-to-use tools for the analysis of the complexity of protocols designed using abstract state machines. Obviously, dynamic complexity is quite relevant here.

15.6 Software Crisis and Opportunity

In 1974 we took a year off between college and graduate school and worked programming for GTE Sylvania. We developed an assembly language program that controlled a hundred phone lines, noting when they went off hook, counting their dial pulses, and sending the numbers dialed to another computer which then ordered my program to make connections. The program ran on a PDP-11 with 4K of memory. That is 4K bytes of memory. Working in assembly language was a pain, but this was a fairly complicated and useful little program.

Today computers are over 1000 times faster than in 1974 and typically have over 10,000 times more memory. Today programming environments are somewhat more convenient. And yet, our ability to produce good, reliable software has not greatly improved.

Yet, some things have changed quite a bit. It is a true watershed that IBM's Deep Blue beat Gary Kasporov, the world chess champion, even if he did not play as well as he might have. Somehow, all that memory and speed has paid off to produce a very nontrivial result. Computers can play chess better than people.

Computers — because of their sheer speed and size — have finally crossed a threshold making them potentially useful in ways that they haven't been up until now. We are thinking of model checking broadly construed: using computers to understand our work and help us with it. In particular it should soon become feasible to use computers to understand our software and help us construct it rationally and maintain it.

Another much talked about change in computers to date is that millions of computers — most of which are currently idle most of the time — are now connected on the web. The potential for smart distributed computation is enormous and difficult to comprehend.

The "software crisis" is currently immense, and in my opinion, spinning out of control. Consider the following three tips of icebergs:

- The Year 2000 problem: A huge amount of object code, created many years ago, is still with us, running many of our machines and data processing systems. Much of this software encoded dates using two digits in order to save memory

space, which used to be expensive. This code is not documented and certainly not formally understood. It is thus a very expensive and error prone job to find all those encoded dates and replace them with four digit dates before some of those dates turn to 00 and appear earlier rather than later than 99. This would be a very funny state of affairs if it weren't so difficult and expensive to fix.
- The U.S. Air Traffic Control System: a huge, complicated and vital conglomeration of software helping to keep our planes aloft and avoiding collisions. The code is very old, and it is running on ancient machines. It needs to be improved, modernized, and ported to new machines. My understanding is that no one really knows how to do this.
- Microsoft: we are often required to use very poor software. It is unreliable and difficult to use.

15.6.1 How can Finite Model Theory Help?

We desperately need ways to conveniently make computers help us build and maintain our software. To do this, we need convenient formal models. Logic, and finite model theory in particular, can be extremely useful in this endeavor.

We need programming environments that are more helpful. They should have a formal understanding of what we are doing. If they did, we could specify the query we want to compute, and our system could go out over the web and find some Java class libraries that already accomplish significant parts of what we need.

Similarly, we need better models, and ones that can be automatically used for the following:

- distributed computing
- security
- privacy
- fault tolerance .

We can and should demand much more of our programs and our programming environments. Logic and formal methods, and in particular descriptive complexity, can provide useful methods in this vast task.

Some of the above goals may seem idealistic. Without question, there is a large amount of worthwhile work to be done using descriptive complexity to improve our understanding of computation.

References

[AV91]   S. Abiteboul and V. Vianu, "Generic Computation And Its Complexity," *32nd IEEE Symposium on FOCS* (1991), 209-219.
[AHV95] S. Abiteboul, R. Hull, and V. Vianu, *Foundations of Databases*, 1995, Addison-Wesley.
[AVV97] S. Abiteboul, M.Y. Vardi, and V. Vianu, "Fixpoint Logics, Relational Machines, and Computational Complexity," *J. Assoc. Comput. Mach.* 44(1) (1997), 30 – 56.
[AAI97] M. Agrawal, E. Allender, R. Impagliazzo, T. Pitassi and S. Rudich, "Reducing the Complexity of Reductions," *ACM Symp. Theory Of Comput.* (1997), 730–738.
[AHU74] A.V. Aho, J.E. Hopcroft and J.D. Ullman, *The Design and Analysis of Computer Algorithms*, Addison- Wesley (1974).
[Ajt83] M. Ajtai, "Σ_1^1 Formulae on Finite Structures," *Annals of Pure and Applied Logic* **24** (1983), 1-48.
[Ajt87] M. Ajtai, "Recursive Construction for 3-Regular Expanders," *28th IEEE Symp. on Foundations of Computer Science* (1987), 295-304.
[Ajt89] M. Ajtai, "First-Order Definability on Finite Structures," *Annals of Pure and Applied Logic* (1989), 211-225.
[AF90]   M. Ajtai and R. Fagin, "Reachability is Harder for Directed than for Undirected Graphs," *J. Symb. Logic*, **55** (1990), 113-150.
[AFS97] M. Ajtai, R. Fagin, and L. Stockmeyer, "The Closure of Monadic NP," IBM Research Report RJ 10092 (1997).
[AG87]   M. Ajtai and Y. Gurevich, "Monotone versus Positive," *J. Assoc. Comput. Mach.* 34 (1987), 1004–1015.
[AKL79] R. Aleliunas, R. Karp, R. Lipton, L. Lovász, and C. Rackoff, "Random Walks, Universal Traversal Sequences, and the Complexity of Maze Problems," *IEEE Found. of Comp. Sci. Symp.* (1979), 218-233.

References

[ABI97]   E. Allender, N. Immerman, J. Balcázar, "A First-Order Isomorphism Theorem," *SIAM J. Comput.* 26(2) (1997), 557-567.

[ACF94]   B. Alpern, L. Carter, E. Feig, and T. Selker, "The Uniform Memory Hierarchy Model of Computation," *Algorithmica, 12(2-3)* (1994), 72–109.

[AG94]   G. Almasi and A. Gottlieb, *Highly Parallel Computing (Second Edition)* 1994, Benjamin-Cummings.

[ASE92]   N. Alon, J. Spencer, and P. ErdHos, *The Probabilistic Method,* 1992, John Wiley and Sons, Inc.

[ADN95]   H. Andréka, I. Düntsch, and I. Németi, "Expressibility of Properties of Relations," *J. Symbolic Logic* 60(3) (1995), 970 - 991.

[AF97]   S. Arora and R. Fagin, "On Winning Strategies in Ehrenfeucht-Fraïssé Games," *Theoret. Comp. Sci.* 174(1-2) (1997), 97-121.

[ALMSS]   S. Arora, C. Lund, R. Motwani, M. Sudan, and M. Szegedy, "Proof Verification and Intractability of Approximation Problems," *IEEE Found. of Comp. Sci. Symp.* (1992), 14 – 23.

[AS92]   S. Arora and S. Safra, "Probabilistic Checking of Proofs: a New Characterization of NP," *J. Assoc. Comput. Mach.* 45(1) (1998), 70-122. A preliminary version appeared in *IEEE Found. of Comp. Sci. Symp.* (1992), 2-13.

[Ba81]   L. Babai, "Moderately Exponential Bound for Graph Isomorphism," in *Proc. Int. Conf. Fundamentals of Computation Theory* (1981), Springer LNCS, 34 – 50.

[BK80]   L. Babai and L. Kučera, Canonical Labelling of Graphs in Linear Average Time," *20th IEEE Symp. on Foundations of Computer Science* (1980), 39-46.

[BL83]   L. Babai and E. Luks, "Canonical Labelling of Graphs," *15th ACM STOC Symp.*, (1983), 171-183.

[BDG]   J. Balcázar, J. Días, and J. Gabarró, *Structural Complexity,* Vols. I and II, EATCS Monographs on Theoretical Computer Science, 1988, Springer-Verlag.

[BI97]   D.M. Barrington and N. Immerman, "Time, Hardware, and Uniformity," in *Complexity Theory Retrospective II,* L. Hemaspaandra and A. Selman, editors, 1997, Springer-Verlag, 1-22.

[BIS88]   D.M. Barrington, N. Immerman, and H. Straubing, "On Uniformity Within NC^1," *Third Annual Structure in Complexity Theory Symp.* (1988), 47-59.

[BBI97]   D.M. Barrington, J. Buss, and N. Immerman, "Capturing Deterministic Space via Number of Variables," in preparation.

[Bar77]   J. Barwise, "On Moschovakis Closure Ordinals," *J. Symb. Logic* 42 (1977), 292-296.

[Bea86]   P. Beame, "Limits on the Power of Concurrent-Write Parallel Machines," *18th ACM STOC* (1986), 169-176.

[Bea96]   P. Beame, "A Switching Lemma Primer," manuscript, http://www.cs.washington.edu/homes/beame/papers.html.

[Ben62]   J. Bennett, "On Spectra" (1962), Ph.D. thesis, Princeton University.

[BGK85]   A. Blass, Y. Gurevich and D. Kozen, "A Zero–One Law for Logic With a Fixed Point Operator," *Information and Computation* 67 (1985), 70-90.

[Bol82]   Béla Bollobás, *Random Graphs,* Academic Press (1982).

[BH92]   R. Boppana and M. Halldórsson, "Approximating Maximum Independent Sets by Excluding Subgraphs," *BIT* 32(2) (1992), 180-196.

[BS90]   R. Boppana and M. Sipser, "The Complexity of Finite Functions," in *Handbook of Theoretical Computer Science, Vol. A* 1990, Jan van Leeuwen, ed., Elsevier, Amsterdam and M.I.T. Press, Cambridge, MA.

References

[B82] E. Börger, "Decision Problems in Predicate Logic," in *Logic Colloquium '82,* G. Lolli, G. Longo and A. Marcia (editors) North-Holland, 1984, 263 – 301.

[BCD88] A. Borodin, S.A. Cook, P.W. Dymond, W.L. Ruzzo, and M. Tompa, "Two Applications of Complementation via Inductive Counting," *Third Annual Structure in Complexity Theory Symp.* (1988), 116-125.

[BCH86] P. Beame, S. Cook, H.J. Hoover, "Log Depth Circuits for Division and Related Problems," *SIAM J. Comput. 15:4* (1986), 994-1003.

[BCP83] A. Borodin, S. Cook, and N. Pippenger, "Parallel Computation for Well-Endowed Rings and Space-Bounded Probabilistic Machines," *Information and Control,* 58 (1983), 113-136.

[Bra96] J. Bradfield, "On the Expressivity of the Modal Mu-Calculus," *Symp. Theoretical Aspects Comp. Sci.* (1996).

[Büc60] R. Büchi, "Weak Second-Order Arithmetic and Finite Automata," *Zeit. Math. Logik. Grund. Math. 6* (1960), 66-92.

[CFI92] J.-Y. Cai, M. Fürer, N. Immerman, "An Optimal Lower Bound on the Number of Variables for Graph Identification," *Combinatorica* **12** (4) (1992) 389-410.

[CH80a] A. Chandra and D. Harel, "Computable Queries for Relational Databases," *JCSS* **21**, No. 2, October, 1980, (156-178).

[CH80b] A. Chandra and D. Harel, "Structure and Complexity of Relational Queries – preliminary version" *21st Symp. on Foundations of Computer Science* (1980), 333-347.

[CH82] A. Chandra and D. Harel, "Structure and Complexity of Relational Queries," *JCSS* **25** (1982), 99-128.

[CKS81] A. Chandra, D. Kozen, and L. Stockmeyer, "Alternation," *JACM,* **28**, No. 1, (1981), 114-133.

[CSV84] A. Chandra, L. Stockmeyer and U. Vishkin, "Constant Depth Reducibility," *SIAM J. of Comp.* **13**, No. 2, 1984, (423-439).

[CE81] E. Clarke and E.A. Emerson, "Design and Synthesis of Synchronization Skeletons Using Branching Time Temporal Logic," in *Proc. Workshop on Logic of Programs*, LNCS 131, 1981, Springer-Verlag, 52-71.

[Coh66] P. Cohen, *Set Theory and the Continuum Hypothesis,* 1966, Benjamin.

[Co88] K. Compton, "0-1 laws in logic and combinatorics," in *NATO Adv. Study Inst. on Algorithms and Order,* I. Rival, editor, 1988, D. Reidel, 353–383.

[Coo71] S. Cook, "The Complexity of Theorem Proving Procedures," *Proc. Third Annual ACM STOC Symp.* (1971), 151-158.

[Coo85] S. Cook, "A Taxonomy of Problems with Fast Parallel Algorithms," *Information and Control* **64** (1985), 2-22.

[Cop94] D. Coppersmith, "A Left Coset Compoosed of n-cycles," Research Report RC19511 IBM (1994).

[Cou90] B. Courcelle, "The Monadic Second-Order Logic of GraphsI: Recognizable Sets of Finite Graphs," *Information and Computation* 85 (1990), 12 - 75.

[Cou97] B. Courcelle, "On the Expression of Graph Properties in Some Fragments of Monadic Second-Order Logic," in *Descriptive Complexity and Finite Models,* N. Immerman and Ph. Kolaitis, eds., 1997, American Mathematical Society, 33 - 62.

[Da84] E. Dahlhaus, "Reduction to NP-Complete Problems by Interpretations," in *Logic and Machines: Decision Problems and Complexity,* Börger, Rödding, and Hasenjaeger eds., Lecture Notes In Computer Science 171, Springer-Verlag (1984), 357-365.

References

[Daw93] A. Dawar, "Feasible Computation Through Model Theory," PhD Dissertation, University of Pennsylvania (1993).

[DGH98] A. Dawar, G. Gottlob, L. Hella, "Capturing Relativized Complexity Classes without Order," to appear in *Mathematical Logic Quarterly*.

[DH95] Anuj Dawar and Lauri Hella, "The Expressive Power of Finitely Many Generalized Quantifiers," *Information and Computation* 123(2) (1995), 172-184.

[DLW96] A. Dawar, S. Lindell, and S. Weinstein, "Elementary Properties of the Finite Ranks," (1996), Technical Report MS-CIS-96-24, University of Pennsylvania, (http://www.cis.upenn.edu/ weinstei/mypapers.html).

[DLW95] A. Dawar, S. Lindell, and S. Weinstein, "Infinitary logic and inductive definability over finite structures," *Information and Computation*, 119 (1995), 160-175.

[DGS86] L. Denenberg, Y. Gurevich and S. Shelah, "Definability by Constant-Depth Polynomial-Size Circuits", *Information and Control* 70 (1986), 216-240.

[deR84] M. de Rougemont, "Uniform Definability on Finite Structures with Successor," *16th ACM STOC Symp.*, (1984), 409-417.

[DL98] W. Diffie and S. Landau, *Privacy on the Line: the Politics of Wiretapping and Encryption*, MIT Press, 1998.

[DS95] G. Dong, J. Su, "Space-Bounded FOIES," *ACM Symp. Principles Database Systems* (1995), 139-150.

[DS93] G. Dong, J. Su, "Incremental and Decremental Evaluation of Transitive Closure by First-Order Queries," *Information and Computation*, 120(1) (1995), 101-106.

[EF95] H.-D. Ebbinghaus, J. Flum, *Finite Model Theory* 1995, Springer 1995.

[EFT94] H.-D. Ebbinghaus, J. Flum, and W. Thomas, *Mathematical Logic*, 2nd edition 1994, Springer-Verlag.

[Ehr61] A. Ehrenfeucht, "An Application of Games to the Completeness Problem for Formalized Theories," Fund. Math. 49 (1961), 129-141.

[End72] H. Enderton, *A Mathematical Introduction to Logic*, Academic Press, 1972.

[ES74] P. Erdös and J. Spencer, *Probabilistic Methods in Combinatorics*, 1974, Academic Press.

[Ete95] K. Etessami, "Counting Quantifiers, Successor Relations, and Logarithmic Space," *IEEE Structure in Complexity Theory Symp.* (1995), 2-11.

[Ete95a] K. Etessami, "Ordering and Descriptive Complexity" Ph.D. thesis, 1995, UMass, Amherst.

[EI95] K. Etessami and N. Immerman, "Reachability and the Power of Local Ordering," *Theoret. Comp. Sci.* 148(2) (1995), 261-279.

[EI95a] K. Etessami and N. Immerman, "Tree Canonization and Transitive Closure," to appear in *Information and Computation*. A preliminary version appeared in *IEEE Symp. Logic In Comput. Sci.* (1995), 331-341.

[Fag73] R. Fagin, "Contributions to the Model Theory of Finite Structures", Ph.D. Thesis (1973), U. C. Berkeley.

[Fag74] R. Fagin, "Generalized First-Order Spectra and Polynomial-Time Recognizable Sets," in *Complexity of Computation*, (ed. R. Karp), *SIAM-AMS Proc.* 7 (1974), 27-41.

[Fag75] R. Fagin, "Monadic generalized spectra," *Zeitschr. f. math. Logik und Grundlagen d. Math.* 21 (1975), 89-96.

[Fag76] R. Fagin, "Probabilities on Finite Models," *J. Symbol. Logic* 41, No. 1 (1976), 50-58.

References

[Fag93] R. Fagin, "Finite-Model Theory – a Personal Perspective," *Theoret. Comp. Sci.* 116 (1993), 3-31.

[Fag97] R. Fagin, "Easier Ways to Win Logical Games," in *Descriptive Complexity and Finite Models,* N. Immerman and Ph. Kolaitis, eds., 1997, American Mathematical Society, 1 - 32.

[FSV95] R. Fagin, L. Stockmeyer, and M. Vardi, "On monadic NP vs. monadic co-NP," *Information and Computation* 120(1) (1995), 78-92.

[FV98] T. Feder and M. Vardi, "The Computational Structure of Monotone Monadic SNP and Constraint Satisfaction: A Study through Datalog and Group Theory," (1998).

[Fel50] W. Feller, *An Introduction to Probability Theory and Its Applications,* Vol. 1, 1950, John Wiley, New York.

[FRW84] F. Fich, Prabhakar Ragde, and, Avi Wigderson (1984), "Relations Between Concurrent-Write Models of Parallel Computation," *Third ACM Symp. on Principles of Distributed Computing,* 179-189.

[FW78] S. Fortune and J. Wyllie, "Parallelism in Random Access Machines," *ACM Symp. Theory Of Comput.* (1978), 114-118.

[Fra54] R. Fraïssé, "Sur les Classifications des Systems de Relations," Publ. Sci. Univ. Alger I (1954).

[Fur87] M. Fürer, "A Counterexample In Graph Isomorphism Testing – Extended Abstract," manuscript (October, 1987).

[FSS84] M. Furst, J.B. Saxe, and M. Sipser, "Parity, Circuits, and the Polynomial-Time Hierarchy," *Math. Systems Theory,* 17 (1984), 13-27.

[Ga81] D. Gabbay, "Expressive Functional Completeness in Tense Logic," in: *Aspects of Philosophical Logic,* 1981, ed. Monnich, D. Reidel, Dordrecht, 91-117.

[Ga81] H. Gaifman, "On Local and Non-Local Properties," *Proc. Herbrand Logic Colloq.* (1981), 105-135.

[GH97] F. Gire and H. Hoang, "An Extension of Fixpoint Logic with a Symmetry-Based Choice Construct," to appear in *Information and Computation* .

[Göd30] Gödel, K., "Die Vollständigkeit der Axiome des Logischen Funktionenkalküls," *Monatshefte für Mathematik und Physik 37* (1930), 349 - 360, (English translation in [vH]).

[Go82] L. Goldschlager, "A Universal Interconnection Pattern for Parallel Computers," JACM, October 1982.

[Go77] L. Goldschlager, "The Monotone and Planar Circuit Value Problems are Log Space Complete for P," *SIGACT News* 9(2) (1977).

[Grä9ll] E. Grädel, f"Capturing Complexity Classes by Fragments of Second Order Logic," *Theoret. Comp. Sci.* 101 (1992), 35-57.

[Grä92] E. Grädel, "On Transitive Closure Logic," *Computer Science Logic* (1992), LNCS, Springer, 149–163.

[GM95] E. Grädel and G. McColm, "On the Power of Deterministic Transitive Closures," *Information and Computation* 119 (1995), 129-135.

[GM96] E. Grädel and G. McColm, "Hierarchies in Transitive Closure Logic, Stratified Datalog and Infinitary Logic," *Annals of Pure and Applied Logic* 77 (1996), 166–199.

[GO93] E. Grädel and M. Otto, "Inductive Definability with Counting on Finite Structures," *Computer Science Logic* 1993, LNCS 702, Springer, 231–247.

[Gra84] E. Grandjean, "The Spectra of First-Order Sentences and Computational Complexity," SIAM J. of Comp. **13**, No. 2 (1984), 356-373.

References

[Gra85] E. Grandjean, "Universal quantifiers and time complexity of Random Access Machines," *Math. Syst. Th.* (1985), 171-187.

[Gra89] E. Grandjean, "First-order spectra with one variable," to appear in *J. Comput. Syst. Sci.*

[Gr95] M. Grohe, "Complete Problems for Fixed-Point Logics," *J. Symbolic Logic* 60 (1995), 517-527.

[Gro96] M. Grohe, "Equivalence in Finite-Variable Logics is Complete for Polynomial Time," *IEEE Found. of Comp. Sci. Symp.* (1996).

[Gro96a] M. Grohe, "Arity Hierarchies," *Annals of Pure and Applied Logic* 82 (1996), 103-163.

[Gro97] M. Grohe, "Large Finite Structures With Few L^k-Types," *IEEE Symp. Logic In Comput. Sci.* (1997).

[Gro97a] M. Grohe, "Canonization for L^k-Invariants is Hard," *Annual Conference of the European Association for Computer Science Logic* (1997), M. Nielsen and W. Thomas, eds., 185-200.

[Gro97b] M. Grohe, "Fixed-Point Logics on Planar Graphs," manuscript (1997).

[Gur83] Y. Gurevich, "Algebras of feasible functions," *IEEE Found. of Comp. Sci. Symp.* (1983), 210-214.

[Gur84] Y. Gurevich, "Toward Logic Tailored for Computational Complexity," *Computation and Proof Theory* (M.M. Ricther et. al., eds.). Springer-Verlag Lecture Notes in Math. 1104 (1984), 175-216.

[Gur88] Y. Gurevich, "Logic and the Challenge of Computer Science," in *Current Trends in Theoretical Computer Science,* ed. E. Börger, Computer Science Press (1988), 1-57.

[Gur91] Y. Gurevich, " Evolving Algebras: A Tutorial Introduction," *Bulletin of EATCS,* 43 (1991), 264-284.

[Gur93] Y. Gurevich, "Evolving Algebras 1993: Lipari Guide", *Specification and Validation Methods,* ed. E. Börger, Oxford University Press, 1995, 9–36.

[GS85] Y. Gurevich and S. Shelah, "Fixed-Point Extensions of First-Order Logic," *Annals of Pure and Applied Logic* 32 (1986), 265–280.

[GS96] Y. Gurevich and S. Shelah, "On Finite Rigid Structures," *J. Symbolic Logic* 61(2) (1996), 549 - 562.

[HP93] P. Hajek and P. Pudlak, *Metamathematics of First-Order Arithmetic,* 1993, Springer, Berlin.

[Ha65] W. Hanf, Model-Theoretic Methods in the Study of Elementary Logic," in J. Addison, L. Henkin, and A. Tarski, eds., *The Theory of Models,* 1965, North Holland, 105-135.

[HIM78] J. Hartmanis, N. Immerman, and S. Mahaney, "One-Way Log Tape Reductions," *IEEE Found. of Comp. Sci. Symp.* (1978), 65-72.

[Has86] J. Hastad, "Almost Optimal Lower Bounds for Small Depth Circuits," *18th ACM STOC Symp.,* (1986), 6-20.

[He96] L. Hella, "Logical Hierarchies in PTIME," *Information and Computation* 129(1) (1996), 1-19.

[HKL97] L. Hella, Ph. Kolaitis, and K. Luosto, "How to Define a Linear Order on Finite Models,"*Annals of Pure and Applied Logic* 87 (1997), 241-267.

[HKL96] L. Hella, Ph. Kolaitis, and K. Luosto, "Almost Everywhere Equivalence of Logics in Finite Model Theory," *Bulletin of Symbolic Logic* 2(4) (1996), 422 - 443.

[HLN97]   L. Hella, L. Libkin, and J. Nurmonen, "Notions of locality and their logical characterizations over finite models," manuscript.
[Hil85]   D. Hillis, *The Connection Machine* 1985, MIT Press.
[Hon82]   J.-W. Hong, "On Some Deterministic Space Complexity Problems," *SIAM J. Comput.* **11** (1982), 591-601.
[Ho86]   J.-W. Hong *Computation: Computability, Similarity, and Duality*, 1986, John Wiley & Sons.
[HT72]   J. Hopcroft and R. Tarjan, "Isomorphism of Planar Graphs," in *Complexity of Computer Computations*, R. Miller and J.W. Thatcher, eds., (1972), Plenum Press, 131-152.
[HU79]   J. Hopcroft and J. Ullman, *Introduction to Automata Theory, Languages, and Computation*, Addison-Wesley (1979).
[I79]   N. Immerman, "Length of Predicate Calculus Formulas as a New Complexity Measure," *20th IEEE FOCS Symp.* (1979), 337-347. Revised version: "Number of Quantifiers is Better than Number of Tape Cells," *JCSS* 22(3), June 1981, 65-72.
[I80]   N. Immerman, "Upper and Lower Bounds for First Order Expressibility,"*21st IEEE FOCS Symp.* (1980), 74-82. Revised version: *JCSS* 25(1) (1982), 76-98.
[I82]   N. Immerman, "Relational Queries Computable in Polynomial Time," *14th ACM STOC Symp.* (1982), 147-152. Revised version: *Information and Control*, 68 (1986), 86-104.
[I83]   N. Immerman, "Languages Which Capture Complexity Classes," *15th ACM STOC Symp.* (1983), 347-354. Revised version: "Languages That Capture Complexity Classes," *SIAM J. Comput.* 16(4) (1987), 760-778.
[I87]   N. Immerman, "Expressibility as a Complexity Measure: Results and Directions," *Second Structure in Complexity Theory Conf.* (1987), 194-202.
[I88]   N. Immerman, "Nondeterministic Space is Closed Under Complementation," *SIAM J. Comput.* 17(5) (1988), 935-938. Also appeared in *Third Structure in Complexity Theory Conf.* (1988), 112-115.
[I89]   N. Immerman, "Descriptive and Computational Complexity,"in *Computational Complexity Theory*, ed. J. Hartmanis, Lecture Notes for AMS Short Course on Computational Complexity Theory, *Proc. Symp. in Applied Math.* 38, American Mathematical Society (1989), 75-91.
[I89a]   N. Immerman, "Expressibility and Parallel Complexity," *SIAM J. of Comput.* 18 (1989), 625-638.
[I91]   N. Immerman, "DSPACE[n^k] = VAR[$k + 1$]," *Sixth IEEE Structure in Complexity Theory Symp.* (July, 1991), 334-340.
[IKL95]   N. Immerman, Ph. Kolaitis, and J. Lynch, "A Tutorial on Finite Model Theory," DIMACS, August, 1995.
[IK87]   N. Immerman and D. Kozen, "Definablitity with Bounded Number of Bound Variables," *Second LICS Symp.* (1987), 236-244.
[IL95]   N. Immerman, S. Landau, "The Complexity of Iterated Multiplication," *Information and Computation* 116(1) (1995), 103-116.
[IL90]   N. Immerman and E. Lander, "Describing Graphs: A First-Order Approach to Graph Canonization," in *Complexity Theory Retrospective*, Alan Selman, ed., Springer-Verlag (1990), 59-81.
[IPS96]   N. Immerman, S. Patnaik and D. Stemple, "The Expressiveness of a Family of Finite Set Languages," *Theoretical Computer Science* 155(1) (1996), 111-

140. A preliminary version appeared in *Tenth ACM Symposium on Principles of Database Systems* (1991), 37-52.
[IV97] N. Immerman and M. Vardi, "Model Checking and Transitive Closure Logic," *Proc. 9th Int'l Conf. on Computer-Aided Verification* (1997), Lecture Notes in Computer Science, Springer-Verlag, 291 - 302.
[JL77] N. Jones and W. Laaser, "Complete Problems for Deterministic Polynomial Time," *Theoret. Comp. Sci.* 3 (1977), 105-117.
[JLL76] N. Jones, E. Lien and W. Laaser, "New Problems Complete for Nondeterministic Logspace," *Math. Systems Theory* 10 (1976), 1-17.
[JS74] N. Jones and A. Selman, "Turing Machines and the Spectra of First-Order Formulas," *J. Symbolic Logic* **39** (1974), 139-150.
[Ka79] R. Karp, "Probabilistic Analysis of a Canonical Numbering Algorithm for Graphs," *Relations between combinatorics and other parts of mathematics*, Proceedings of Symposia in Pure Mathematics 34, 1979, American Mathematical Society, 365 - 378.
[KL82] R. Karp and R. Lipton, "Turing Machines That Take Advice," *Ensiegn. Math.* 28 (1982), 192-209.
[KV95] Ph. Kolaitis and J. Väänänen, "Generalized Quantifiers and Pebble Games on Finite structures, *Annals of Pure and Applied Logic*, 74(1) (1995), 23–75. ¿
[KV98] Ph. Kolaitis and M. Vardi, "Conjunctive-Query Containment and Constraint Satisfaction," *ACM Symp. Principles Database Systems* (1998).
[KV92a] Ph. Kolaitis and M. Vardi, "0-1 Laws for Fragments of Second-Order Logic: an Overview," in Y. Moschovakis, editor, *Logic From Computer Science* 1992, Springer-Verlag, 265–286.
[Ku87] L. Kučer, "Canonical Labeling of Regular Graphs in Linear Average Time," *28th IEEE FOCS Symp.* (1987), 271-279.
[Kur64] S. Kuroda, "Classes of Languages and Linear-Bounded Automata," *Information and Control* **7** (1964), 207-233.
[Kur94] R. Kurshan, *Computer-Aided Verification of Coordinating Processes*, 1994, Princeton University Press, Princeton, NJ.
[L75] R. Ladner, "The Circuit Value Problem is log space complete for P," *SIGACT News*, 7(1) (1975), 18 – 20.
[LR96] R. Lassaigne and M. de Rougemont, *Logique et Complexité*, 1996, Hermes.
[LJK87] K.J. Lange, B. Jenner, and B. Kirsig, "The Logarithmic Hierarchy Collapses: $A\Sigma_2^L = A\Pi_2^L$," *14th ICALP* (1987).
[Lei87] D. Leivant, "Characterization of Complexity Classes in Higher-Order Logic," *Second Structure in Complexity Theory Conf.* (1987), 203–217.
[Lei89] D. Leivant, "Descriptive Characterizations of Computational Complexity," *J. Comput. Sys. Sci.* 39 (1989), 51-83.
[LP81] H. Lewis and C. Papadimitriou, *Elements of the Theory of Computation* 1982, Prentice-Hall.
[LP82] H. Lewis and C. Papadimitriou, "Symmetric Space Bounded Computation," *Theoret. Comput. Sci.* **19** (1982),161-187.
[LV93] M. Li and P. Vitányi, *An Introduction to Kolmogorov Complexity and its Applications*, 1993, Springer-Verlag, New York.
[L92] S. Lindell, "A purely logical characterization of circuit uniformity," *7th Structure in Complexity Theory Conf.* (1992), 185-192.
[L] S. Lindell, "How to define exponentiation from addition and multiplation in first-order logic on finite structures," manuscript.

References

[LG77]   L. Lovász and P. Gács, "Some Remarks on Generalized Spe ctra," *Zeitchr. f. math, Logik und Grundlagen d. Math,* 23 (1977), 547-554.

[Lu82]   E. , "Isomorphism of Graphs of Bounded Valence Can be Tested in Polynomial Time," *J. Comput. Sys. Sci.* 25 (1982), pp. 42-65.

[Lyn82] J. Lynch, "Complexity Classes and Theories of Finite Models," *Math. Sys. Theory* 15 (1982), 127-144.

[MP94]   J. Makowsky, Y. Pnueli, "Arity versus alternation in Second-Order Logic," in *Logical Foundations of Computer Science,* A. Nerode, Y. Matiyasevich eds., Springer LNCS 813 1994, 240 - 252.

[McM93] K. McMillan, *Symbolic Model Checking,* 1993, Kluwer.

[MP71]   R. McNaughton and S. Papert, *Counter-Free Automata,* 1971, MIT Press, Cambridge, MA.

[Man89] U. Manber, *Introduction to Algorithms: A Creative Approach,* Addison-Wesley, (1989).

[MT97]   O. Matz and W. Thomas, "The Monadic Quantifier Alternation Hierarchy Over Graphs is Infinite," *IEEE Symp. Logic In Comput. Sci.* (1997), 236-244.

[MI94]   J.A. Medina and N. Immerman, "A Syntactic Characterization of NP-Completeness," *IEEE Symp. Logic In Comput. Sci.* (1994), 241-250.

[MI96]   J.A. Medina and N. Immerman, "A Generalization of Fagin's Theorem," *IEEE Symp. Logic In Comput. Sci.* (1996), 2 - 12.

[MSV94] P. Miltersen, S. Subramanian, J. Vitter, and R. Tamassia, "Complexity Models for Incremental Computation," *Theoret. Comp. Sci.* (130:1) (1994), 203-236.

[Mos74] Y. Moschovakis, *Elementary Induction on Abstract Structures,* North Holland (1974).

[Mos80] Y. Moschovakis, *Descriptive set theory,* 1980, North-Holland Pub. Co., Amsterdam, 637 p.

[NT95]   N. Nisan and A. Ta-Shma, "Symmetric Logspace is Closed Under Complement," *Chicago J. Theoret. Comp. Sci.* (1995).

[Ott96] M. Otto, "The Expressive Power of Fixed-Point Logic with Counting," *J. Symbolic Logic* 61(1) (1996), 147 - 176

[Ott97] M. Otto, *Bounded Variable Logics and Counting: A Study in Finite Models,* 1997, Lecture Notes in Logic, vol. 9, Springer-Verlag.

[Pap94] C. Papadimitriou, *Computational Complexity* 1994, Addison-Wesley.

[Pap85] C. Papadimitriou, "A Note on the Expressive Power of Prolog," *EATCS Bulletin* 26 (1985), 21-23.

[PY]   C. Papadimitriou and M. Yannakakis, "Optimization, Approximation, and Complexity Classes," *J. Comput. Sys. Sci.* , 43 (1991), 425-440.

[PI94]   S. Patnaik and N. Immerman, "Dyn-FO: A Parallel, Dynamic Complexity Class," *J. Comput. Sys. Sci.* 55(2) (1997), 199-209. A preliminary version appeared in *ACM Symp. Principles Database Systems* (1994), 210-221.

[Poi82] B. Poizat, "Deux ou trois chose que je sais de Ln" *JSL* 47 (1982), 641-658.

[R96]   G. Ramalingam *Bounded Incremental Computation,* 1996, Springer LNCS 1089.

[Raz87] A. Razborov, "Lower Bounds on the Size of Bounded Depth Networks Over a Complete Basis With Logical Addition," *Matematischi Zametki* 41 (1987), 598-607 (in Russian). English translation in *Mathematical Notes of the Academy of Sciences of the USSR* 41, 333-338.

[Rei87] J. Reif, "On Threshold Circuits and Polynomial Computation," *Second Annual Structure in Complexity Theory Symp.* (1987), 118-123.

References

[RS72] D. Rödding and H. Schwichtenberg, "Bemerkungen zum Spektralproblem," *Zeitschrift f̈ math. Logik und Grundlagen der Mathematik* 18 (1972), 1-12.

[Ruz81] L. Ruzzo, "On Uniform Circuit Complexity," *J. Comp. Sys. Sci.*, 21, No. 2 (1981), 365-383.

[Sav70] W. Savitch, "Relationships Between Nondeterministic and Deterministic Tape Complexities," *J. Comput. System Sci.* 4 (1970), 177-192.

[Sav73] W. Savitch, "Maze Recognizing Automata and Nondeterministic Tape Complexity," *J. Comput. Sys. Sci.* 7 (1973), 389-403.

[Sch97] N. Schweikardt, "The Monadic Quantifier Alternation Hierarchy over Grids and Pictures," *Annual Conference of the European Association for Computer Science Logic* (1997), M. Nielsen and W. Thomas, eds., 383-397.

[Sch94] T. Schwentick, "Graph Connectivity and Monadic NP," *IEEE Found. of Comp. Sci. Symp.* (1994), 614-622.

[Sch97] T. Schwentick, "Padding and the Expressive Power of Existential Second-Order Logics," *Annual Conference of the European Association for Computer Science Logic* (1997), M. Nielsen and W. Thomas, eds., 399-412.

[SB98] T. Schwentick and K. Barthelmann, "Local Normal Forms for First-Order Logic with Applications to Games and Automata," to appear in *Symp. Theoretical Aspects Comp. Sci.* (1998).

[See95] D. Seese, "FO-Problems and Linear Time Computability," Tech Report, Institut f̈ Informatik und Formale Beschreibungsverfahren, Universität Karlsruhe, Germany (1995).

[Sip83] M. Sipser, "Borel Sets and Circuit Complexity," *15th Symp. on Theory of Computation* (1983), 61-69.

[Smo87] R. Smolensky, "Algebraic Methods in the Theory of Lower Bounds for Boolean Circuit Complexity," *19th ACM STOC* (1987), 77-82.

[Spe93] J. Spencer, "Zero-One Laws With Variable Probability," *J. Symbolic Logic* 58 (1993), 1–14.

[Ste91] I. Stewart, "Comparing the Expressibility of Languages Formed Using NP-Complete Operators", *J. Logic and Computation* 1(3) (1991), 305-330.

[Sto77] L. Stockmeyer, "The Polynomial-Time Hierarchy," *Theoretical Comp. Sci.* 3 (1977), 1-22.

[SV84] L. Stockmeyer and U. Vishkin, "Simulation of Parallel Random Access Machines by Circuits," *SIAM J. of Comp.* 13, No. 2 (1984), 409-422.

[Str94] H. Straubing, *Finite Automata, Formal Logic, and Circuit Complexity*, 1994, Birkhäuser.

[Sze88] R. Szelepcsényi, "The Method of Forced Enumeration for Nondeterministic Automata," *Acta Informatica* 26 (1988), 279-284.

[Tar36] A. Tarksi, "Der Wahrheitsbegriff in den Formalisierten Sprachen," *Studia Philosophica 1* (1936).

[Tar55] A. Tarksi, "A Lattice-Theoretical Fixpoint Theorem and its Applications," *Pacific. J. Math.*, 55 (1955), 285-309.

[Tho86] S. Thomas, "Theories With Finitely Many Models," *J. Symbolic Logic,* 51, No. 2 (1986), 374-376.

[Tra50] B. Trahtenbrot, "The Impossibility of an Algorithm for the Decision Problem for Finite Domains," *Doklady Academii Nauk SSSR*, n.s., vol 70 (1950), 569-572 (in Russian).

[Tur84] G. Turán, "On the Definability of Properites of Finite Graphs," *Discrete Math.* 49 (1984), 291-302.

References

[TPP97] A. Turk, S. Probst, and G. Powers, "Verification of a Chemical Process Leak Test Procedure," in *Computer Aided Verification, 9th International Conf.*, O. Grumberg, ed. 1997, Springer, 84-94.
[Tys97] J. Tyszkiewicz, "The Kolmogorov Expression Complexity of Logics," *Information and Computation* 135(2) (1997), 113-136.
[Val82] L. Valiant, "Reducibility By Algebraic Projections," *L'Enseignement mathématique*, **28**, 3-4 (1982), 253-68.
[vD94] D. van Dalen, *Logic and Structure, Third Edition*, 1994, Springer-Verlag.
[vH] J. van Heijenoort, *From Frege to Gödel: A Source Book in Mathematical Logic, 1879 - 1931* 1967, Harvard University Press.
[Var82] M. Vardi, "Complexity of Relational Query Languages," *14th Symposium on Theory of Computation* (1982), 137-146.
[Wei76] B. Weisfeiler, ed., *On Construction and Identification of Graphs*, Lecture Notes in Mathematics 558, Springer, 1976.
[Wra76] C. Wrathall, "Complete Sets and the Polynomial Hierarchy," *Theoret. Comp. Sci.* **3** (1976).
[Yao85] A. Yao ,"Separating the Polynomial-Time Hierarchy by Oracles," *26th IEEE Symp. on Foundations of Comp. Sci.* (1985), 1-10.

Index

Underlined page numbers indicate where that index entry is defined.

[[ ]] ($w[[i]]$, bit i of w), 172
$\Leftrightarrow$, $\underline{7}$
$\equiv$, $\underline{7}$
$\leftrightarrow$, $\underline{8}$
2-SAT, $\underline{30}$
3-COLOR, $\underline{113}$
3-SAT, $\underline{31}$, 119

0, 12
1, 12

Abbreviation, 8
Abiteboul, S., 168, 197, 201, 240
Abiteboul-Vianu Theorem, 197
abstract state machines, 248
AC, $\underline{80}$, 82
accept
  by circuit, $\underline{79}$
  by Turing machine, $\underline{24}$
advice, $\underline{176}$
aggregate operator, 224
Agrawal, M., 179
Ajtai, M., 66, 127, 129, 137, 218
Ajtai-Fagin game, $\underline{127}$
Ajtai-Fagin Methodology Theorem, 127
Aleliunas, R., 55
Allender, E., 179

Alpern, B., 240
alternating space, *see* ASPACE
alternating time, *see* ATIME
arithmetic, 13, 48
arithmetic hierarchy, $\underline{43}$
Arity Hierarchy Theorem, 219
Arora, S., 129, 137, 247
ASPACE, $\underline{35}$
ASPACE-TIME, $\underline{40}$
ATC (alternating transitive closure), 148, $\underline{175}$
ATIME, $\underline{35}$
ATIME-ALT, $\underline{40}$

Babai, L., 184, 189, 192, 201, 247
Balcázar, J.L., 43, 179
Barrington, D.M., 21, 155, 168, 179
Barthelmann, K., 111
Barwise, J., 111
Bennett, J., 21
bijection game, $\underline{189}$, 219
bin, $\underline{24}$
BIT, $\underline{13}$, 14, 16, 69
Blass, A., 111
Bollobas, B., 137
bool, $\underline{14}$, 64
boolean circuit, $\underline{37}$

boolean variable, 174
  in first-order formula, 14
Boppana, R., 124
Borodin, A., 88
bound variable, 7
bounded
  expressive power, 109
Bradfield, J., 155
Büchi, R., 22
Buss, J., 168

Cai, J., 219
canonization, 183
Carter, L., 240
CC_k, 184
$C(c, m, \tau)$, 126
Chandra, A., 88, 139
characterize, 182
circuit, 77–88
$\mathcal{C}^k$, 187
CLIQUE, 28, 31, 96, 114
CNF, 32
Cohen, P., 245
COLOR, 66, 113
color valence, 184
color-class size, 184
colored graph, 182
Compactness Theorem, 22, 104
complete
  problem, 30, 31
  query language, 19
  sentence, 99
Completeness Theorem, 21
concurrent, 68
configuration, 35, 39
  of a game, 92, 182
CONNECTED, 128
constraint programming, 244
constructible, 26
Cook's theorem, 119
Cook, S., 43, 88, 119
Coppersmith, D., 137
counting game ($\mathcal{G}_C^k$), 187
counting quantifier, 186
Courcelle, B., 137
CRAM, 68, 68–71, 82
CVP, 30, 37, 54, 66
CYCLE, 30

DAG, 37
Dahlhaus, E., 123
database, 6, 221
  genealogical, 6, 19, 111
Datalog, 225, 224–226
Dawar, A., 21, 199, 201
De Morgan law, 78
de Rougemont, M., 137
decision tree, 204
Deep Blue, 249
degree, 102
Delilah, 91
depth
  inductive, 61–63
descriptive complexity
  origin, 115
deterministic space, *see* DSPACE
deterministic transitive closure, *see* DTC
DNF, 204
dom, 92
Dong, G., 240
DSPACE, 26
DTC (deterministic transitive closure), 143
DTIME, 26
dynamic, 229, 248
DynFO, 229

Ebbinghaus, H.-D., 154, 155
$E_C^k(\mathcal{A})$, 198
Ehrenfeucht, A., 91–112
Ehrenfeucht-Fraïssé game, 91, 91–112
$E^k(\mathcal{A})$, 196
$\equiv_d$, 98
Etessami, K., 219
ETIME, 34
exclusive or ($\oplus$), 11
existential
  formula, 12
  ($\exists x.\alpha$), 10
exponential time, *see* EXPTIME
EXPTIME, 26
extension axioms, 107

Fagin's Theorem, 115, 181, 247
Fagin, R., 1, 111, 115, 123, 127, 129, 137, 181, 242, 247
Feder, T., 244

Feig, E., 240
finite automaton, 80
Flum, J., 154, 155
FO, 19
FO(COUNT), 186
FO(DTC), 143
FO(LFP), 59, 139–143
FO(M), 84
FO(NUMBER), 186
FO(PFP), 161
FO(TC), 143
FO-VAR$[t(n), v]$, 163
FO^k(COUNT), 195
fop, *see* projection, first-order
($\forall x.\alpha$), 10
forest, 132, 191
formula
  complete, 126
FO-VAR$[t(n), v(n)]$, 170
Fraïssé, R., 91–112
free variable, 7, 9
Fürer, M., 219
Furst, M., 218

$\mathcal{G}_M^K$, 91
Gödel, K., 21, 245
Gács, P., 123
Gabbay, D., 104
Gaifman graph, 100
Gaifman, H., 100, 111
GEOGRAPHY, 31, 158
Gire, F., 202
$\mathcal{G}_C^k$, 187
GO, 31
Goldschlager, L., 55
Grädel's Theorem, 117
Grädel, E., 117, 151, 155, 201, 219
Grandjean, E., 119, 123
Grohe, M., 201, 202, 219
Gurevich, Y., 66, 111, 142, 148, 154, 201, 248

Hajek, P., 21
Halldórsson, M., 124
Hanf's Theorem, 100, 111, 189, 201
Harel, D., 139
Hartmanis, J., 55
HC (Hamiltonian Circuit), 114
Hella, L., 198, 201, 219

HEX, 31
Hoang, H., 202
Hong, J-W., 168, 243
Hopcroft, J., 189
Horn, 225
Horn formula, 117, 151
Hull, R., 240

ID, 35
IFP (inflationary fixed point operator), 142
Impagliazzo, R., 179
IND, 62, 82
induction, 57–66
  simultaneous, 59
inflationary fixed point, *see* IFP
inherently sequential, 213
init(τ, n), 227
Inj(f), 114
integrity constraint, 240
interpretation, 7
invariant, 196, 198
isomorphism, 17, 93
ITER, 160

Jones, N., 123

Kanellakis, P., 129
Karp, R., 55, 198
Kasporov, G., 249
Kleene, S., 154
Knaster, B., 58, 66
Knaster-Tarski Theorem, 58
Kolaitis, Ph., 111, 180, 198, 244
Kolmogorov Complexity, 180
Kozen, D., 104, 111, 201
K_r, 96
Kripke structure, 234
Krom formula, 151
Kuroda, S., 151
Kurshan, R., 240
Kučera, L., 192, 201

L (logspace), 26, 148
Ladner, R., 55
Landau, S., 179
Lander, E., 201
least fixed point, *see* LFP
Lehman, A., 219

Leivant, D., 155
Lewis, H., 55
LFP, 59
LH (logarithmic-time hierarchy), 85
Lindell, S., 21, 88, 201
linear time, 103
Lipton, R., 55
literal, 32
$\mathcal{L}^k$, 94
$\mathcal{L}^k$-type, 195
$\mathcal{L}_m$, 94
logarithmic hierarchy, see LH
logspace, see L
Lovász, L., 55, 123
$\mathcal{L}(\text{wo BIT})(\tau)$, 13
$\mathcal{L}(\text{wo}\leq)(\tau)$, 13
Luks, E., 247
Lund, C., 247
Luosto, K., 198
Lynch, J., 118, 123

M (majority quantifier), 84
Mahaney, S., 55
MAJ (majority query), 31, 81
majority quantifier, see M
Makowsky, J., 180
MATH, 245
Matz, O., 137
max, 12
McColm, G., 155, 219
McMillan, K., 240
McNaughton, R., 22
MCVP, 37, 54
Medina, J.A., 123, 179, 247
memory hierarchy, 226, 240
methodology, 99, 111, 126, 127
Methodology Theorem, 99
Miltersen, P., 240
model checking, 234–239, 248
 symbolic, 234
monadic, 6
monotone, 58, 58
Moschovakis, Y., 66, 141, 154
Motwani, R., 247
Mu-Calculus, 155

N, 25
$\hat{n}$, 25
NC, 41, 80, 82

NETWORK-FLOW, 30
NL, 26
nondeterministic polynomial time, see NP
nondeterministic space, see NSPACE
nonuniform, 176
NP, 26
NSPACE, 26
NTIME, 26
number, 186

Ω, 207, 215
oracle, 27
order-independent, 181
ordering ($\leq$), 12, 25, 93, 181–202
Ordering Proviso, 13
Otto Theorem, 199
Otto, M., 199, 202

P, 26
Papadimitriou, C., 55, 123, 240, 247
Papert, S., 22
PARITY, 29, 63, 228
parity quantifier ($\oplus$), 198
partial fixed point, see PFP
Patnaik, S., 240
PCP Theorem, 247
PFP, 161
PH, 22, 41
Pippenger, N., 41, 43, 88
Pitassi, T., 179
planar graph, 201
PLUS, 9, 14
Pnueli, Y., 180
polynomial space, see PSPACE
polynomial time, see P
polynomial-time hierarchy, see PH
positive, 58, 58
precedence, 8
prenex form, 12
preservation theorem, 12
primitive recursive, 149
problem, see query, boolean
projection, 171
 first-order, 172, 246
 quantifier-free, 55, 172
PSPACE, 26
Pudlak, P., 21

$Q(C)$, 26
$Q(FO)$, 19
qfp, *see* projection, quantifier-free
QSAT, 31, 38
quantifier, 7, 8
 generalized, 178
 restricted, 10, 11, 47
quantifier rank, 94, 94, 95
query, 17, 26
 boolean, 26
 first-order
  dual of $(\widehat{I})$, 47
 order-independent, 18
 safe, 223

Rackoff, C., 55
RAM (Random Access Machine), 37, 54, 217, 228–230
Ramalingam, G., 240
Razborov, A., 218
REACH, 30
REACH(acyclic), 230
REACH$_d$, 30
REACH$_{d\ell}$, 159, 171
reduction
 first-order ($\leq_{fo}$), 28, 28–34, 45–55
  closure under, 49
 first-order, Turing, 178
 logspace ($\leq_{log}$), 28
 many-one, 28
 polynomial-time ($\leq_p$), 28
 projection, *see* projection
 Turing, 27
regular language, 80
relation
 input, 13
 numeric, 13
requests
 basic, *see* $\mathcal{R}_{n,\tau}$
restriction, 11, 203
rigid, 201
$\mathcal{R}_{n,\tau}$, 227
Rudich, S., 179
Ruzzo, L., 88

Samson, 91
SAT, 31, 32, 114, 119
Savitch's Theorem, 27, 38, 39
Savitch, W., 27, 38, 39, 55

Saxe, J., 218
Schweikardt, N., 137
Schwentick, T., 111, 137
Seese, D., 103, 111
Selker, T., 240
Selman, A., 123
semantics of first-order logic, 7
separator, 217
Shelah, S., 142, 154, 201
$\sim_m^k$, 93
Sipser, M., 218
Smolensky, R., 218
S_n, 134
SO, 113
SO(TC), 166
SO∃, 113, 115
software crisis, 249
Space Hierarchy Theorem, 27
Spector, C., 154
spectrum (of a first-order formula), 122
SQL, 222
stable coloring, 189, 194
Stage Comparison Theorem, 141
state explosion problem, 234
static, 226
Stewart, I., 180
Stockmeyer, L., 88, 111, 137
Straubing, H., 21, 22
structure, 5
 initial, *see* **init**(τ, n)
Su, J., 240
Subramanian, S., 240
substitution, 7, 9, 64
substructure, 11
 induced, 93, 96
SUC, 13
Sudan, M., 247
Suslin, M., 154
synchronous, 68
Szegedy, M., 247
Szelepcsényi, R., 155

Tamassia, R., 240
Tarjan, R., 189
Tarski, A., 21, 58, 66
τ_{ag} (vocabulary of alternating graphs), 53, 175
τ_c (vocabulary of circuits), 37, 79
τ_{cg} (vocabulary of colored graphs), 182

τ_g (vocabulary of graphs), 5
τ_{gk}, 28, 114
τ_s (vocabulary of strings), 5
τ_{thc} (vocabulary of threshold circuits), 79
TC (transitive closure), 143
TC^s (strict transitive closure), 237
ThC, 82
ThC, 80
Thomas, W., 137
threshold circuit, 79
TIMES, 9, 14
Tompa, M., 88
Trahtenbrot, B., 22
transition system, 234
transitive closure, *see* TC
truth, 7
traveling salesperson problem (TSP), 115
Tyszkiewicz, J., 112, 198, 202

Ultracomputer, 84
uniform
  circuit, 79
  first-order, 79
universal
  formula, 12

Väänänen, J., 111, 180
Valiant, L., 171
var, 192
Vardi, M., 66, 111, 137, 168, 240, 244
variable
  boolean, 163
vc, 192
vertex refinement, 189
Vianu, V., 168, 197, 201, 240
view, 222
Vishkin, U., 88
Vitter, J., 240
vocabulary, 5
  relational, 6

Weinstein, S., 21, 201
Weisfeiler, B., 219

Yannakakis, M., 247
Yao, A., 218

Zero-One Law, 108, 198
ZFC, 245

MIX
Papier aus verantwortungsvollen Quellen
Paper from responsible sources
FSC® C105338

If you have any concerns about our products,
you can contact us on
ProductSafety@springernature.com

In case Publisher is established outside the EU,
the EU authorized representative is:
**Springer Nature Customer Service Center GmbH
Europaplatz 3, 69115 Heidelberg, Germany**

Printed by Libri Plureos GmbH
in Hamburg, Germany